Biology and Aquaculture of Tilapia

Editors

José Fernando López-Olmeda
Department of Physiology, Faculty of Biology
University of Murcia, Murcia, Spain

Francisco Javier Sánchez-Vázquez
University of Murcia, Murcia, Spain

Rodrigo Fortes-Silva
Department of Animal Science, University of Viçosa (UFV)
Viçosa, Brazil

CRC Press
Taylor & Francis Group
Boca Raton London New York

CRC Press is an imprint of the
Taylor & Francis Group, an **informa** business
A SCIENCE PUBLISHERS BOOK

Cover credit: Main cover picture: José Antonio Oliver
Small cover pictures, from left to right: Gonzalo de Alba, Rosimery Frisso, José Antonio Oliver and Bruno O. de Mattos.

First edition published 2021
by CRC Press
6000 Broken Sound Parkway NW, Suite 300, Boca Raton, FL 33487-2742

and by CRC Press
2 Park Square, Milton Park, Abingdon, Oxon, OX14 4RN

Library of Congress Cataloging-in-Publication Data
Names: López-Olmeda, José Fernando, 1980- editor.
Title: Biology and aquaculture of tilapia / editors, José Fernando López-Olmeda, Department of Physiology, Faculty of Biology, University of Murcia, Murcia, Spain, F. Javier Sanchez Vazquez, Murcia University, Murcia, Spain, Rodrigo Fortes da Silva, Department of Animal Science, University of Viçosa (UFV), Viçosa, Brazil.
Description: First edition. | Boca Raton : CRC Press, 2021. | Includes bibliographical references and index.
Identifiers: LCCN 2021009939 | ISBN 9780367420635 (hardcover)
Subjects: LCSH: Tilapia.
Classification: LCC QL638.C55 B58 2021 | DDC 597/.74--dc23
LC record available at https://lccn.loc.gov/2021009939

ISBN: 978-0-367-42063-5 (hbk)
ISBN: 978-1-032-12533-6 (pbk)
ISBN: 978-1-003-00413-4 (ebk)

Typeset in Times New Roman
by Innovative Processors

Preface

Tilapia is the common name of a group of fishes from the Cichlid family. The term derives from the word *thlapi,* the Tswana (a Bantu language) word for fish. Tilapia comprises around a hundred of species, divided into four main genera: *Tilapia, Oreochromis, Coptodon* and *Sarotherodon.* The tilapias are endemic to Africa and to the Jordan River valley.

Tilapia is probably the first farmed fish in human history. A bas-relief found in the Egyptian tomb of Nakht, from 2,000 BC (thus 4,000 years ago), depicts two small fish that can be identified as tilapia. Indeed, the ancient Egyptians created a hieroglyph for tilapia, which reveals the importance that this species had for them. Tilapia is also very important for Christian mythology. For instance, it is believed that this species was the fish caught by Peter, the apostle, and the fish used by Jesus in the story that tells how he fed 5,000 people with five loaves and two fish.

Modern tilapia culture begins around 100 years ago, in the 1920s of the past century. Since then, tilapia has shown a great potential for aquaculture thanks to a variety of advantages like fast growth, very low feed conversion ratios, tolerance of high stocking densities, adaptability to different culture systems, tolerance to different habitats, high rate of reproduction and an omnivorous diet that allows the use of feeds based on vegetables or cereals. All of these advantages have boosted tilapia aquaculture, growing exponentially in the last three decades. Nowadays, according to the last report on "The state of world fisheries and aquaculture", published by FAO in 2020, one tilapia species, the Nile tilapia (*Oreochromis niloticus*), is the third most cultured fish worldwide, with a production of 4.5 million tonnes in 2018. Taking all tilapia species together, their global production in 2018 was 5.55 million tonnes, close to the first spot in the list of major species produced in aquaculture held by the grass carp (5.7 million tonnes).

The objective of the present book is to provide a comprehensive revision of the biology of tilapia and link this knowledge to its application to tilapia aquaculture. The book content is mainly focused on the Nile tilapia, due to its prominent prevalence in the aquaculture of tilapiine fish. Nevertheless, it also includes information from other species that are used in both aquaculture and research such as the Mozambique tilapia (*Oreochromis mossambicus*), the blue tilapia (*Oreochromis aureus*), and other tilapia species and hybrids.

The book has 15 chapters that cover a wide variety of aspects of the biology of tilapia and its culture. Chapters 1 and 2 are devoted to tilapia genetics and the design of selective breeding programs. Chapters 3, 4 and 5 are focused on the digestive physiology and nutrition of tilapia, including recommendations for feeding practices and feed management. Chapter 6 reviews the osmoregulatory mechanisms in tilapia, providing also a comparative analysis between the Nile and Mozambique tilapia. Chapter 7 is devoted to the most common pathologies that affect tilapia, and the use of vaccines and immunostimulants. Chapters 8 and 9 are focused on the reproduction of tilapia, covering the gonadal development, the reproductive cycles in mature fish and their physiological control. Chapters 10 and 11 review tilapia embryonic and larval development (chapter 10), and larviculture (chapter 11), thus providing a complete view of this process and its management under culture conditions. Chapter 12 is devoted to biological rhythms in tilapia, which are present in many of the biological processes described in the previous chapters. Finally, the last three chapters are focused on new technologies of tilapia aquaculture and production. Chapters 13 and 14 provide a description of two new culture technologies that are rapidly growing in importance: aquaponics (chapter 13) and bioflocs (chapter 14). They also include examples of their application to tilapia aquaculture. Chapter 15 reviews the methods of sacrifice and new procedures of tilapia processing.

The present book counts on the collaboration of experts from both academia and research institutes, who provide their expertise to make a complete and up to date revision of the biology and aquaculture of tilapia. We hope that it will become a great handbook for tilapia farmers, researchers, students or curious readers that want to delve into the knowledge of this amazing species.

José Fernando López-Olmeda
Francisco Javier Sánchez-Vázquez
Rodrigo Fortes-Silva

Contents

Tilapia Genetic Resources: Conservation and Use for Aquaculture

Alexandre Wagner Silva Hilsdorf[1]* **and Eric M. Hallerman**[2]*

[1] Integrated Center of Biotechnology, University of Mogi das Cruzes, Mogi das Cruzes, P.O. Box 411, 08701-970, SP, Brazil
[2] Department of Fish and Wildlife Conservation, Virginia Polytechnic Institute and State University, Blacksburg, VA, 23061-0321, USA

1. Introduction

Tilapia is the generic term for a group of cichlid fishes, drawn from the Latinization of *thlapi*, the Tswana (a Bantu language) word for "fish." The taxonomy of tilapias has been an issue of much debate. Initially, the genus *Tilapia* covered all the different species. Taxonomists subsequently divided this group into four genera based mainly on breeding and brooding behaviors, as well as morphological and feeding characters. According to Trewavas' (1983) classification system, these four genera are *Tilapia*: substrate spawners; *Sarotherodon*: paternal or biparental mouthbrooders; *Oreochromis*: maternal mouthbrooders and *Danakilia*: mouthbrooders. The three most important tilapia genera for aquaculture are *Coptodon* (*Tilapia*), *Sarotherodon*, and *Oreochromis*, which include 112 recognized species and subspecies of cichlid fishes, 41 of them listed as being under some level of threat (Hallerman and Hilsdorf 2014). The tilapias are endemic to the African continent (excluding Madagascar), and to the Jordan River valley and coastal rivers. The genus *Tilapia* occurs throughout West and Central Africa (e.g. *T. rendalli, T. zillii*). The genus *Sarotherodon* is distributed in West Africa, eastwards towards the Nile River and the more northerly Great Rift lakes (e.g. *S. galilaeus*). The genus *Oreochromis* occurs mostly in Central and Eastern Africa (e.g. *O. mossambicus, O. aureus, O. niloticus, O. machrochir,* and *O. shiranus*) (Philippart and Ruwet 1982).

Tilapiine fishes have a long history as cultivated organisms; a bas-relief from an Egyptian tomb, dating 2000 B.C. (Figure 1), shows a pair of small fish that can be identified as tilapia, probably *Oreochromis niloticus,* a species still abundant in the Nile River valley (Hickling 1963). The modern history of tilapia cultivation seems to have begun around 1924, when the first experimental attempts

*Corresponding authors: wagner@umc.br and ehallerm@vt.edu

to cultivate them were made (Balarin and Hatton 1979). At present, tilapias are among the most cultivated aquaculture species, farmed in over 120 countries worldwide (Barroso et al. 2019). The FAO (2019c) global aquaculture production database lists 12 tilapia species (Figure 2) as important to aquaculture production: (i) *Oreochromis aureus* (Steindachner 1864) – blue tilapia; (ii) *Oreochromis andersonii*

Fig. 1. (A) The circled hieroglyph in the tomb of the ancient Egyptian official Nakht depicts a tilapia-like fish above the head of the central figure (Source: Wikipedia). (B) A painting in the tomb of the scribe Nebamun depicts a pond with tilapias, circa 1400 BC (British Museum collection).

Fig. 2. The most important tilapia species farmed worldwide (FAO 2019c).

(Castelnau 1861) – three-spotted tilapia; (iii) *Oreochromis niloticus* (Linnaeus 1757) – Nile tilapia; (iv) *Oreochromis macrochir* (Boulenger 1912) – longfin tilapia; (v) *Oreochromis mossambicus* (Peters 1852) – Mozambique tilapia; (vi) *Oreochromis shiranus* (Boulenger 1897) – *shiranus tilapia*; (vii) *Oreochromis spirulus* (Günther 1894) – sabaki tilapia; (viii) *Oreochromis tanganicae* (Günther 1894) – Tanganyika tilapia; (ix) *Sarotherodon galilaeus* (Linnaeus 1758) – mango tilapia; (x) *Sarotherodon melanotheron* (Rüppell 1852) – blackchin tilapia; (xi) *Coptodon rendalli* (Boulenger 1897) (= *Tilapia rendalli*) redbelly tilapia; and (xii) *Coptodon zillii* (Gervais 1848) (= *Tilapia zillii)* – redbreast tilapia.*Oreochromis urolepis* (Norman 1922) (= *Oreochromis urolepis hornorum*) – wami, despite not being listed in the FAO global aquaculture production database (FAO 2019c),is also an important species due to its use in inter-specific hybridization to improve aquaculture productivity (Wohlfarth 1994, Lovshin et al. 1990). Although *O. niloticus* is by far most famous for aquaculture in many countries, other tilapia species represent the mainstay of tilapia farming in many local communities in Africa, at all levels of production from subsistence production to highly intensive farming.

2. Tilapia genetic resources for aquaculture

Tilapias have become an essential source of valuable animal protein over recent decades, with culture methods progressing from extensive production and practically no industrial production to global production and distribution as a white-fish commodity. Among all farmed tilapia species, Nile tilapia stands out as the main species produced worldwide. That is because of its higher growth rate, resistance to diseases, acceptance of a wide range of natural foods and artificial feeds, lack of intermuscular bones, ability to withstand low oxygen tension, and above all its highly acceptable fillet (Fitzsimmons 2000, Young and Muir 2002, Prabu et al. 2019). Despite controversies regarding global introductions of non-native tilapias into aquatic ecosystems (Vicente and Fonseca-Alves 2013, Costa-Pierce 2003, Zengeya et al. 2013), Nile tilapia is farmed on practically all continents and plays a central role in aquaculture production in many countries. It stands third in finfish production worldwide behind grass carp (*Ctenopharyngodo nidellus)* and silvercarp (*Hypophthalmichthys molitrix*) (FAO 2019a). The significance of Nile tilapia for food security and economic development is growing in many developing countries (Ansah et al. 2014). This is demonstrated by the steady growth of Nile tilapia production from 1950 to the present (Figure 3).

The large and growing importance of tilapias to aquaculture draws attention to the genetic variation that underlies its adaptation to culture conditions and response to selective breeding for valued traits. This genetic variation has come to be termed fish genetic resources, or FiGR. The importance of FiGR to global food security was acknowledged during the 1980s in two symposia organized to discuss finfish genetic resources conservation, the Fish Gene Pools Symposium held in Stockholm (Ryman 1981) and the FAO/UNEP Symposium on Conservation of the Genetic Resources of Fish held in Rome (FAO/UNEP 1981). An Expert Consultation on Utilization and Conservation of Aquatic Genetic Resources was held in Grottaferrata, Italy in 1992 (FAO/FIRI 1993). Another important meeting held in Italy in 1998 addressed the

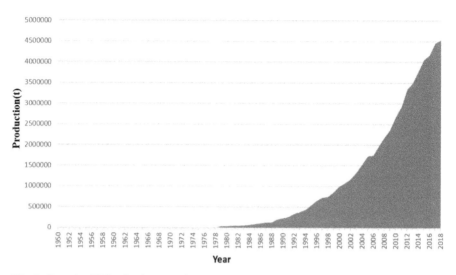

Fig. 3. Growth of Nile tilapia aquaculture production between 1950 and 2016 (FAO 2019c).

international, regional, and national policies for the conservation and sustainable use of aquatic genetic resources (Pullin et al. 1999). FAO has organized an international working group on aquatic genetic resources and technology that meets to advance FAO and member countries' activities regarding aquatic genetic resources; its third meeting was held in Rome in 2019. Global awareness about the importance of FiGR to aquaculture and fisheries has been attested to by many other contributions addressing concerns and challenges facing the conservation of diverse FiGR and its sustainable use for food production (Penman et al.2005, Olesen et al. 2007, Solar 2009, Lind et al. 2012a, Hilsdorf and Halleman 2017, Pilling et al., 2020). Against this background, aquatic genetic resources (AqGR) for food and agriculture include "DNA, genes, chromosomes, tissues, gametes, embryos, and other early life-history stages, individuals, strains, stocks, and communities of organisms of actual or potential value for food and agriculture." This definition was stated in the seminal publication by FAO (2019d), "The State of the World's Aquatic Genetic Resources for Food and Agriculture", the first document to compile the efforts of 92 countries to gather information of the current status of AqGR as an essential step towards developing and implementing policies for their present and future conservation, sustainable use and development.

Awareness of the importance of tilapia genetic resources to food production and to understanding of the natural distributions of economically important tilapias date back to the 1980s with the works of Philippart and Rewet (1982) and Trewavas (1983). In 1987, a historic meeting took place in Bangkok, Thailand, where for the first time the importance of tilapia genetic resources was acknowledged in the Workshop on Tilapia Genetic Resources for Aquaculture organized by the International Center for Living Aquatic Resources Management (ICLARM) (Pullin 1988). This workshop brought together prominent scientists, including Drs. Rosemary Lowe-McConnell, a distinguished English ichthyologist, ecologist, and limnologist known for research on tilapia and aquaculture; Roger S.V. Pullin, lead scientist at ICLARM who played an essential role in the conservation of genetic resources for aquaculture, and Ethelwynn Trewavas, ichthyologist at the British Museum of Natural History, known for her seminal work on tilapia taxonomy, who although absent had sent material for discussion. This meeting foresaw the importance of tilapia genetic resources' conservation to the future contribution of tilapia to global food security. Another workshop convened by ICLARM and Ghanaian and German institutions in 1997 drew attention to Ghanaian and sub-Saharan fish genetic resources and raised and discussed common interests in biodiversity and genetic resources conservation and sustainable use in tilapia and other species (Pullin et al. 1997). Since then, other publications have discussed the conservation and sustainable use of aquatic genetic resources with particular attention to tilapia species, such as those presented as proceedings of the meeting 'Genetics and Aquaculture in Africa' convened at the Centre de Recherches Océanologiquesd'Abidjan (Agnèse 1998). Seeking a basis for international policy, the 2006 Workshop on Status and Trends in Aquatic Genetic Resources (Bartley et al. 2007) concluded that information on FiGR should be global, authoritative, free and objective. Among work needed is assessment of the status of FiGR in capture fisheries and aquaculture; improved technical capacities for scientists, technical persons, governments and industry; genetically improved farmed types of aquatic species; appropriate policy instruments on use and conservation of FiGR; and prioritization of species, geographic areas and production systems for expenditure of limited resources for conservation and use of FiGR. The article "Use and Exchange of Genetic Resources of Nile Tilapia (*Oreochromis niloticus*)" by Eknath and Hulata (2009) was published in a special issue of Reviews in Aquaculture on use and exchange of genetic resources of cultured aquatic organisms. More recently, at the Tenth International Symposium on Tilapia in Aquaculture, Hallerman and Hilsdorf (2014) reviewed knowledge regarding conservation of different tilapia species, addressing the molecular genetics and adaptive differentiation of populations in the light of the evolutionary significant unit (ESU) and management unit (MU) concepts drawn from conservation genetics.

Tilapias have been introduced outside Africa since the 1930s when the accidental introduction of *O. moss ambicus* was noticed in the Serang River in Java and then in other countries (Atz 1954, Riedel 1965). Subsequently, many tilapia species have been introduced into nearly all tropical and subtropical countries of the world. In Asia, introductions of *O. niloticus* have been reported in many different countries since 1974 (see Philippart and Ruwet 1982). Following these introductions, locally farmed strains of Nile tilapia were established throughout Asia; these strains show different levels of genetic differentiation as a result of selection, adaptation to local farming conditions, and genetic drift. The Chitralada strain, shipped from Japan in 1965 and kept isolated at the Chitralada Palace in Thailand was one of the first well-known Nile tilapia strains used in aquaculture (Chotiyarnwong 1971). During the 1990s, the GIFT Project combined the genetic variation in eight local farmed stocks into the base population subsequently subjected to selection to yield an increase in growth rate of 64% over nine generations (Eknath et al. 1993, Ponzoni et al. 2011). The subsequent distribution of the GIFT strain for commercial farming in Asia revolutionized tilapia aquaculture in Asia and contributed to increased global tilapia production (Eknath et al. 1993, 1998, 2007, Dey 2000, Ponzoni et al. 2011). Production of this strain resulted in substantial positive socio-economic impact (ADB 2005, see Figure 3). The experience brought by the GIFT project led to other tilapia selective breeding programs with outstanding results, such as the program developed in Abbassa, Egypt named the Genetically Improved Abbassa Nile Tilapia (GIANT) (Ibrahim et al. 2019). Many other genetically improved tilapia strains have been developed for aquaculture production, including the Genetically Enhanced Tilapia-Excellent (GET-EXCEL), Brackishwater Enhanced Saline Tilapia (BEST), Genetically Male Tilapia (GMT), Chitralada, YY-male, Cold-tolerant tilapia (COLD), and Florida red strains (Ordoñez et al. 2017).

Due to the risk of contamination of locally adapted native genetic stocks, the WorldFish Center and other development partners responsible for the GIFT strain adopted a policy that did not allow the dissemination of GIFT to African countries where the original parental stocks were collected (Gupta and Acosta 2004). The strain was placed under evaluation at a government research station and its production was subsequently detected at other farms, and its presence was detected in the wild in Ghana (Anane-Taabeah et al. 2019). Tilapias have been found as feral populations on nearly all continents of the world, including Australia, the southern USA, and the Neotropical region of Central and South America. In South America, particularly in Brazil, tilapia introductions date back to the 1930s when individuals of *Coptodon rendalli* (=*Tilapia rendalli*) were brought from the Democratic Republic of the Congo to be introduced into reservoirs of the Brazilian Traction Light and Power Company for weed control in the state of São Paulo (Azevedo 1955). In 1971, the Brazilian government shipped juveniles of *O. niloticus* and *O. urolepishornorum* from Bouaké Station in the Ivory Coast to the aquaculture research station at Pentecostes, Ceará (DNOCS – National Department of Drought Alleviation) to be stocked into northeastern reservoirs for food production (Pullin 1988, Saint-Paul 2017). More recently, genetically improved lines of the Chitralada and GIFT strains were introduced into the Brazilian aquaculture industry in 1996 and 2005, respectively (Barroso et al. 2015).

3. Research on genetic variation of tilapiine species

Within their natural distributions, tilapia populations show genetic differentiation that reflects their natural history of dispersal and adaptation to regional and local conditions. For example, study of the population genetics of *O. niloticus* range-wide (Agnese et al. 1997) showed that natural populations clustered into three groups: west African populations (in the Senegal, Niger, Volta, and Chad drainages), Ethiopian Rift populations (Lakes Ziway, Awasa, Koka, and Sodore hot springs in the Awash River), and Nile drainage (Nile, Lakes Tana and Edward) and Kenyan Rift populations (lakes Turkana and Baringo and River Suguta). The authors hypothesized that *O. niloticus* originated in the Nile and then independently colonized East and West Africa. Bezault et al. (2011) screened nine microsatellite DNA loci across 350 samples from ten natural populations representing four subspecies (*O. n. niloticus*,

O. n. vulcani, O. n. cancellatus and *O. n. filoa*), finding high genetic differentiation among populations across the Ethiopian, Nilotic and Sudano-Sahelian regions and ichthyofaunal provinces, suggesting the predominant effect of paleo-geographic events at the macrogeographic scale in defining putative evolutionary significant units. Intermediate levels of divergence were found between populations in rivers and lakes within the regions, presumably reflecting relatively recent interruptions of gene flow between hydrographic basins, suggesting different management units. Using allozyme markers to investigate population genetic structure within West Africa, Rognon et al. (1996) observed modest levels of genetic differentiation among seven wild populations. Hallerman and Hilsdorf (2014) reviewed the state of knowledge on genetic resources of all tilapias.

Culture of tilapias often occurs within ecosystems containing native, locally adapted populations. Many, if not most, tilapia production systems do not effectively confine their stock within the farm. For example, Anane-Taabeah et al. (2019) found that the non-native cultured tilapias have escaped into the wild and interbred with local populations. Their results show that aquaculture can be a vector in the spread of invasive non-native species and strains as well as underscore the importance of genetic baseline studies to guide conservation planning for wild populations. Continuing impacts on native tilapia populations may jeopardize the long-term conservation of genetic resources of critical tilapia species, particularly those of importance for aquaculture, as shown by publications cited above. The long-term capacity of a given species to persist and adapt to changing environments relies on the extant genetic diversity harbored by the species and how this diversity is distributed among the populations (Lande and Shannon 1996, Meffe and Carroll 1997). In particular, locations that contain isolated populations with distinct and high levels of genetic diversity can be a source of local adaptive genetic variation that can act as "gene banks." The genetic diversity is fundamental even for species with a long history of domestication. Diverse breeds have been established by humans, according to the Domestic Animal Diversity Information System (DAD-IS) currently representing 8,800 breeds of livestock and poultry across the world (FAO 2019b). However, farmed genetic resources have been lost as a result of local breed extinctions and genetic erosion. An estimated 25% of the local farm animal breeds have been reported extinct since 1900, and 27% are at risk of imminent extinction (FAO 2019b). However, farmed fishes are not typically maintained as breeds with closed populations and breeding records, as is commonly practiced for livestock. The use of strains in aquaculture species is typical, although no consensus exists. Gunnes and Gjedrem (1978) defined a strain as "a discrete breeding population from a river, river system, or a fjord leading to a river." Ponzoni et al. (2013) extended the definition to include "any discrete breeding population from a hatchery may also be termed a strain for evaluation purposes." Therefore, strain selection for a selective breeding program relies on local community knowledge, environmental characteristics of the region where the strain is found, production performance, and levels of intra and inter-strain genetic diversity.

Since the 1960s, development of DNA markers and statistical metrics to quantify population genetic differentiation have been added to such classical methods as morphology- and meristics-based assessments and variation in allozymes (Hallerman and Hilsdorf 2014). A general search in Web of Science and Google Scholar resulted in 74 published studies regarding tilapia genetic resources evaluation, e.g. genetic variability, species geographic delimitation, hybrid occurrence, and population genetic diversity of tilapias (Table 1). What we can observe across these studies is an evolution of characterization markers from morphological, meristic and proteins up to DNA-based markers as mitochondrial DNA RFLP (restriction fragment length polymorphism), RAPD (random amplification of polymorphic DNA), microsatellites and SNPs (single nucleotide polymorphisms). Most population genetic surveys of tilapias to date have used classical mitochondrial sequence and nuclear microsatellite markers. Future work might also use SNP markers to cover the entire genome. The genomic approach potentially would allow demonstration of genetic variation of importance to aquaculture performance or ecological adaptation. While several studies (Van Bers et al. 2012, Xia et al. 2014, Delomas et al. 2019, Peñaloza et al. 2020, Bartie et al. 2020) have established SNP markers for *O. niloticus*, considerable work is still needed for characterization of genetic resources of tilapias.

Human-induced hybridization due to transfers of tilapia species beyond their natural distributions was already detected during the 1960s using morphometric assessments (Table 1). The compiled DNA-based studies shown in Table 1 addressed diverse levels of genetic variation among wild relatives, feral, and farmed type populations or stocks (Çiftci and Okumuş 2002). Regarding wild populations, the different genetic markers and methodology of analysis used over the years showed levels of genetic differentiation among the population in various tilapia species across the African continent. In the case of *O. niloticus*, the parameters of population structure (F_{ST}, D_{EST}, G_{ST}, among others) show moderate to high genetic differentiation (0.05 to 0.15 and 0.15 to 0.25, respectively) (Wright 1978). These outcomes suggest the putative presence of genetically distinctive populations across the natural occurrence of Nile tilapia, in which integrity must be kept for the continuous use of this species for food production. The loss of a particular genetic strain or population implies depletion of adaptive and sometimes unique genetic resources, perhaps useful for future genetic improvement and farming of Nile tilapia.

Molecular tools for genetic resources evaluation have been reviewed elsewhere (Toro et al. 1999, Talle et al. 2005, FAO 2011, Lenstra et al. 2012, Yang et al. 2013, Socol et al. 2015, Hilsdorf and Hallerman 2017). Advances have been achieved, and new and more precise tools are available for genetic assessment of tilapia genetic resources, such as SNPs. A growing body of research has sought to characterize variation within the tilapia genome that underlies key performance traits, such as growth rate, salinity tolerance, and sex ratio, i.e. to detect Mendelian segregation of quantitative trait loci (QTLs). Beyond the advance in scientific knowledge, the intent of these studies is to provide the basis for genetic marker-assisted selection (MAS). Experimental designs for QTL detection and MAS in tilapia have been discussed by Poompuang and Hallerman (1997), Liu and Cordes (2004), and Cnaani and Hulata (2008). Genetic linkage studies have detected QTLs relating to fish size (Cnaani et al. 2003), cold tolerance (Cnaani et al. 2003, Moen et al. 2004), sex determination (Shirak et al. 2006, Khan 2011), hypoxia tolerance (Li et al. 2017), salinity tolerance (Gu et al. 2018a, b, Jiang et al. 2019), and omega-3 fatty acid content (Lin et al. 2018) in Nile tilapia and its hybrids. The incorporation of marker-QTL linkages into MAS

Table 1. Development and evolution of the population-/species molecular markers for tilapia genetic resources characterization

Species	Locality	Primary and secondary farmed types	Marker	Population origin	Goals/Outcomes	References
Tilapia nigra *Tilapia leucosticta*	Kenya	No strain defined	Morphometric/ Merisitcs	Wild relatives	Hybridization occurrence	Elder and Garrod (1961)
Tilapia nigra *Tilapia mossambica*	Kenya	No strain defined	Morphometric/ Merisitcs	Wild relatives	Hybridization occurrence	Whitehead (1962)
Tilapia spp.	Lake Victoria	No strain defined	Species assessment	Wild relatives	Introduction and hybridization occurrence	Welcomme (1966)
Tilapia mossambica *Tilapia hornorum* *Tilapia zillii* *Tilapia melanopleura*	Malaysia/ Canada	No strain defined	Allozyme/ isozyme	Farmed	Marker development/ species differentiation	Chen and Tsuyuk (1970)
Tilapia nilotica *Tilapia zilli* *Tilapia galilaea* *Tilapia aurea*	Egypt	No strain defined	Proteins (blood serum)	Wild relatives/ Feral	Species differentiation	Badawi (1971)
Tilapia aurea *Tilapia mossambica* *Tilapia niloticus* *Tilapia vulcani*	Israel	No strain defined	Allozyme/ isozyme	Farmed	Species differentiation	Herzberg (1978)
Tilapia zilli *Sarotherodon aureus* *Sarotherodon galilaeus*	Israel	No strain defined	Allozyme /isozyme	Feral and wild	Species differentiation	Komfield (1979)
Tilapia zillii	England	No strain defined	Allozyme/ isozyme	Feral	Stock identification	Cruz et al. (1982)

(Contd.)

Table 1. (*Contd.*)

Species	Locality	Primary and secondary farmed types	Marker	Population origin	Goals/Outcomes	References
Sarotherodon aureus *Sarotherodon niloticus*	Israel	No strain defined	Allozyme/isozyme	Wild relatives	Species and hybrids differentiation	Avtalion (1982)
Sarotherodon andersoni *Sarotherodon aureus* *Sarotherodon galilaeus* *Sarotherodon jipe* *Sarotherodon macrochir* *Sarotherodon mossambicus* *Sarotherodon niloticus* *Sarotherodon spilurus* *Tilapia zilli*	Botswana Egypt, Kenya Botswana Mozambique	No strain defined	Allozyme/isozyme	Wild relatives	Marker development Mean heterozygosity (across species) = 0.058	McAndrew and Majumdar (1983)
Oreochromis niloticus *Tilapia zilli*	Japan	No strain defined	Allozyme/isozyme	Farmed	Genetic variation *O. niloticus* He = 0.082 Na = 1.34 *T. zilli* He = 0.022 Na = 1.057	Basiao and Taniguchi (1984)
Oreochromis niloticus *Oreochromis mossambicus*	Sri Lanka	No strain defined	Allozyme/isozyme	Feral	Hybridization occurrence	De Silva and Ranasinghe (1989)

Species	Country	Strain	Method	Type	Purpose/Results	Reference
Oreochromis niloticus *Oreochromis spilurus*	Egypt Kenya Ethiopia	No strain defined	RFLP Mitochondrial	Wild relatives	Species differentiation	Seyoum and Kornfield (1992)
Oreochromis niloticus *Oreochromis mossambicus* *Oreochromis aureus*	Egypt Kenya	Vulcani, Kenya Baringo, Kenya Manzala, Egypt Baobab, Kenya	RAPD	Wild relatives/ Farmed	Species differentiation	Bardakci and Skibinski (1994)
Oreochromis niloticus	Philippines	Swansea strain (*O. n niloticus*) Turkana strain (*O. n vulcani*) Baringo strain Hybrid ancestry strains Red strains	RAPD VNTR	Farmed	Strain differentiation RAPD/GD = 0.2178 VNTR/GD = 0.5745	Naish et al. (1995)
Oreochromis niloticus	Philippines	African and Asian strains	Allozyme/ isozyme	Wild relatives/ Farmed	Population structure and genetic diversity $D = 0.017$	Macaranas et al. (1995)
Oreochromis niloticus *Tilapia zillii*	Egypt Kenya Niger Ivory Coast Mali	Bouake strain	Allozyme/ isozyme	Wild relatives/ Farmed	Population structure and hybridization *O. niloticus* $D = 0.214$ *T. zilli* $D = 0.634$	Rognon et al. (1996)
Oreochromis niloticus *Oreochromis esculentus* *Oreochromis leucostictus* *Tilapia zillii* *Sarotherodon galilaeus*	Lake Victoria and satellite lakes	No strain defined	RAPD	Wild relatives	Population structure and genetic variation *O. niloticus*: $F_{ST} = 0.69$ *O. leucostictus* $F_{ST} = 0.69$ *O. esculentus* $F_{ST} = 0.80$	Mwanja et al. (1996)

(Contd.)

Table 1. (*Contd.*)

Species	Locality	Primary and secondary farmed types	Marker	Population origin	Goals/Outcomes	References
Oreochromis niloticus *Oreochromis aureus*	Senegal Niger Ghana Egypt Ethiopia Kenya	No strain defined	Allozyme/ isozyme RFLP mtDNA (D-loop)	Wild relatives	Populations and subspecies genetic characterization	Agnèse et al. (1997)
Oreochromis niloticus	Egypt Kenya Mali Senegal Burkina Faso	*O. n. vulcani* *O. n. eduardianus* *O. n. cancellatus* *O. n. baringoensis* *O. n. sugutae*	mtDNA ND5/6 region RFLP	Wild relatives	Population structure K = 13-16%	Rognon and Guyomard (1997)
Oreochromis niloticus *Oreochromis mossambicus*	Sri Lanka	No strain defined	Allozyme/ isozyme	Feral	Hybridization occurrence	De Silva (1997)
Oreochromis mossambicus *Oreochromis mortimeri* *Oreochromis macrochir*	Zimbabwe	No strain defined	Allozyme/ isozyme	Wild relatives	Hybridization occurrence	Gregg et al. (1998)
Oreochromis niloticus *Oreochromis esculentus*	Kenya	No strain defined	Allozymes/ Microsatellites	Wild relatives	Genetic variation and hybridization	Agnèse et al. (1999)
Oreochromis niloticus	Lake Victoria region	No strain defined	Microsatellites	Wild relatives	Population structure and genetic diversity $F_{ST} = 0.081$	Fuerst et al. (2000)

Species	Location	Strain	Marker	Status	Purpose/Findings	Reference
Oreochromis niloticus *Oreochromis hornorum* *Tilapia rendalli* Hybrids	Brazil	No strain defined	RAPD	Farmed	Species and hybrids differentiation	Lima et al. (2000)
Oreochromis mossambicus *Oreochromis niloticus*	Mozambique	No strain defined	Allozyme/isozyme	Wild relatives	Hybridization occurrence	Moralee et al. (2000)
Oreochromis mossambicus	Southern Africa	No strain defined	Microsatellites	Wild relatives, Farmed and Feral	Population structure and genetic diversity $F_{ST} = 027$	Hall (2001)
Oreochromis niloticus *Oreochromis aureus*	Egypt Kenya Niger Mali Burkina Faso Senegal Ivory Coast	No strain defined	Allozyme/isozyme mtDNA (cytochrome b D-loop)	Wild relatives	Population structure and hybridization occurrence *O. niloticus* $D = 0.037$; $K = 0{,}062$ *O. aureus* $D = 0.023$; $K = 0{,}0462$	Rognon et al. (2003)
Tilapia zillii *Oreochromis niloticus* *Oreochromis aureus* *Sarotherodon galilaeus*	Egypt	No strain defined	18S ribosomal nuclear DNA	Wild relatives	Species differentiation	El-Serafy et al. (2003)
Oreochromis niloticus	Ghana	No strain defined	RAPD mtDNA (D-loop)	Wild relatives	Population structure and genetic diversity	Falk and Abban (2004)
Oreochromis niloticus *Oreochromis aureus*	Mexico	Stirling, red and hybrids	Meristic/ Allozyme/isozyme	Farmed/Feral	Genetic differentiation among strains $\theta_S = 0.632$	Barriga-Sosa et al. (2004)

(Contd.)

Table 1. (*Contd.*)

Species	Locality	Primary and secondary farmed types	Marker	Population origin	Goals/Outcomes	References
Oreochromis niloticus	Egypt	No strain defined	RAPD	Wild relatives	Population structure and genetic diversity RAPD/GD = 0.1906	Hassanien et al. (2004)
Oreochromis niloticus	Netherlands	GÖTT, AIT, IDRC and GIFT	Microsatellite	Farmed	Genetic diversity among strains $F_{ST} = 0.178$	Rutten et al. (2004)
Oreochromis niloticus and Red hybrids	Philippines	Domesticated: NIFI and Israel Improved: GIFT, GMT, FAC-selected and SEAFDEC-selected Red hybrid: BFS, FACred, NIFIred, HL, and PF	Microsatellite/ mtDNA-RFLP (D-loop)	Farmed	Genetic diversity among strains O. niloticus $\Phi_{ST} = 0.0038$ Red hybrids $\Phi_{ST} = 0.0063$	Romana-Eguia et al. (2004)
Oreochromis niloticus	Brazil	Bouaké e Chitralada	RAPD	Farmed	Genetic differentiation among strains ShI = 0,222	Povh et al. (2005)
Oreochromis niloticus	Egypt	No strain defined	Microsatellite	Wild relatives	Population structure and genetic diversity $F_{ST} = 0.035$	Hassanien and Gilbey (2005)
Oreochromis niloticus *Oreochromis aureus* *Sarotherodon galilaeus* *Tilapia zillii*	Egypt	No strain defined	Allozyme/ isozyme	Wild relatives	Species differentiation	El-Serafy et al. (2006)

Species	Location	Strain	Marker	Sample	Purpose/Results	Reference
Oreochromis niloticus *Oreochromis. leucostictus*	Kenya	No strain defined	Microsatellite/D-loop	Wild relatives	Hybridization occurrence	Nyingi and Agnése (2007)
Oreochromis niloticus	Brazil	Tai Chitralada Red Stirling	Microsatellite	Farmed	Genetic differentiation among strains $F_{ST} = 0.131$	Moreira et al. (2007)
Oreochromis mossambicus *Oreochromis niloticus*	Southern Africa and northern Nile River	No strain defined	Microsatellite/mtDNA (D-loop)	Wild relatives/Farmed	Population structure and hybridization occurrence *O. mossambicus* $F_{ST} = 0.1684$ *O. niloticus* $F_{ST} = 0.1573$	D'Amato et al. (2007)
Oreochromis niloticus	Lake Victoria Lake Kyoga Lakes Edward George	No strain defined	RAPD	Wild relatives	Population structure and genetic diversity RAPD/GD = 0.5415	Mwanja et al. (2008)
Oreochromis niloticus	Egypt	No strain defined	RAPD	Wild relatives	Population structure and genetic diversity $G_{ST} = 0.23$	Rashed et al. (2008)
Oreochromis niloticus	Kenya	*O. n. vulcani* *O. n. eduardianus* *O. n. baringoensis* *O. sugutae* *O. n. niloticus*	Microsatellite mtDNA (D-loop)	Wild relatives	Population structure and genetic diversity Microsatellite $F_{ST} = 0.256$ mtDNA $\pi = 0.013$	Nyingi et al. (2009)
Oreochromis niloticus *Oreochromis aureus*	Egypt	No strain defined	Allozyme/isozyme	Wild relatives	Hybridization occurrence	Bakhoum et al. (2009)

(Contd.)

Table 1. (*Contd.*)

Species	Locality	Primary and secondary farmed types	Marker	Population origin	Goals/Outcomes	References
Oreochromis aureus *Tilapia rendalli* *O. mossambicus*	Mexico	No strain defined	Morphometric/ Allozyme/ isozymes mtDNA-RFLP (16S rRNA - cytb)	Feral	Species identification and population structure *O. aureus* Microsatellite $F_{ST} = 0.410$ mtDNA $\Phi_{ST} = 0.4343$	Espinosa-Lemus et al. (2009)
Oreochromis sp.	*Malaysia*	No strain defined	RAPD	Farmed/Feral	Farmed and feral genetic differences RAPD/GD = 0.12-0.17	Othman et al. (2010)
Oreochromis niloticus *Oreochromis esculentus* *Oreochromis. leucostictus* *Oreochromis variabilis* *Tilapia zillii*	Lake Victoria Region	No strain defined	Microsatellite	Wild relatives	Population structure and genetic diversity *O. niloticus* $F_{ST} = 0.150$ *O.esculentus* $F_{ST} = 0.216$ *O. leucostictus* $F_{ST} = 0.156$ *O. variabilis* $F_{ST} = 0.133$ *T. zillii* $F_{ST} = 0.215$	Mwanja et al. (2010)

Species	Location	Strain	Marker/Method	Type	Purpose/Results	Reference
Oreochromis spp.	Colombia	Red hybrids	Microsatellite	Farmed	Genetic differentiation among strains $\Phi_{ST} = 0.176$	Briñez et al. (2011)
Oreochromis niloticus *Oreochromis esculentus*	Kenya	No strain defined	Microsatellite mtDNA (D-loop)	Wild relatives/ Feral	Population structure and genetic diversity Microsatelite *O. niloticus* $F_{ST} = 0.033$ *O. esculentus* $F_{ST} = 0.057$ mtDNA *O. niloticus* $\pi = 0.077$ $\Phi_{ST} = 0.195$ *O. esculentus* $\pi = 0.003$ $\Phi_{ST} = 0.244$	Angienda et al. (2011)
Oreochromis niloticus	Sudano-Sahelian and Nilo–Sudanian river basins	*O. n. niloticus,* *O. n. vulcani,* *O. n. cancellatus* *O. n. filoa*	Microsatellite	Wild relatives	Population structure and genetic diversity $R_{ST} = 0.297$	Bezault et al. (2011)
Oreochromis aureus *Oreochromis mossambicus* *Oreochromis niloticus* *Oreochromis urolepis* *Sarotherodon melanotheron* *Tilapia redalli*	Hawai	No strain defined	mtDNA D-loop and COI	Farmed/Feral	Species identification	Wu and Yang (2012)

(Contd.)

Table 1. (*Contd.*)

Species	Locality	Primary and secondary farmed types	Marker	Population origin	Goals/Outcomes	References
Oreochromis niloticus	Brazil	Panamá-UNISUL RED (O. ssp) GIFT	Microsatellite	Farmed	Genetic diversity among strains $F_{ST} = 0.13$	Petersen et al. (2012)
Oreochromis niloticus *Oreochromis aureus* *Tilapia zillii* *Sarotherodon galilaeus*	Egypt	No strain defined	Inter-simple sequence repeat (ISSR)	Farmed	Genetic diversity within and among species O. niloticus GS = 0.969 O. aureus GS = 0.972 T. zillii GS = 0.878 S. galilaeus GS = 0.961	Saad et al. (2012)
Oreochromis niloticus	Thailand	Chitralada GIFT	Microsatellite	Farmed and feral	Population structure and genetic diversity $K_{structure} = 3$	Sukmanomon et al. (2012)
Oreochromis niloticus	Kenya	No strain defined	Microsatellite mtDNA (D-loop)	Wild relatives	Population structure and genetic diversity Microsatellite $F_{ST} = 0.164$ mtDNA: $\pi = 0.024$	Ndiwa et al. (2014)
Oreochromis niloticus *Oreochromis andersonii* *Oreochromis macrochir*	Zambia	No strain defined	Microsatellite	Wild relatives	Hybridization occurrence	Deines et al. (2014)

Species	Country	Strain	Marker	Category	Purpose/Results	Reference
Oreochromis mossambicus	Mozambique	No strain defined	Microsatellite	Wild relatives	Population structure and genetic diversity $D_{EST} = 0.0327$	Simbine et al. (2014)
Sarotherodon melanotheron Tilapia guineensis	Nigeria	No strain defined	Morphometric/ Merisites	Wild relatives	Species identification	Kuton and Adeniyi (2014)
Tilapia zillii	Nigeria	No strain defined	RAPD	Wild relatives	Population structure and genetic diversity $D = 0.763$	Oladimeji et al. (2015)
Oreochromis niloticus	Sri Lanka	GIFT	Microsatellite	Farmed	Genetic diversity between strain lines	De Silva (2015)
Oreochromis niloticus	Brazil	Thai Chitralada Red-Stirling GIFT UFLA (Boiaké)	Microsatellite	Farmed	Genetic differentiation among strains $D_{EST} = 0.373$	Dias et al. (2016)
Sarotherodon melanotheron Sarotherodon galilaeus Oreochromis niloticus Tilapia zillii	Nigeria	No strain defined	mtDNA (D-loop)	Farmed	Species genetic diversity *S. melanotheron* $\pi = 0.21$ *S. galilaeus,* $\pi = 0.01$ *O. niloticus* $\pi = 0.04$ *T. zilli* $\pi = 0.05$	Agbebi et al. (2016)
Tilapia guineensis	Nigeria	No strain defined	Microsatellite	Wild relatives	Population structure and genetic diversity $D = 0.118$	Ukenye et al. (2016)
Oreochromis mossambicus Oreochromis niloticus Oreochromis aureus	Philippines	No strain defined	mtDNA COI	Farmed/Feral	Species identification	Maranan et al. (2016)

(Contd.)

Table 1. (*Contd.*)

Species	Locality	Primary and secondary farmed types	Marker	Population origin	Goals/Outcomes	References
Oreochromis niloticus *Oreochromis aureus* *Tilapia zillii*	Egypt	No strain defined	Allozyme/ isozymes	Wild relatives	Population structure and genetic diversity	El-Fadly et al. (2016)
Oreochromis niloticus	Brazil	Chitralada, GST, and GIFT	Microsatellite	Farmed	Genetic differentiation among strains $F_{ST} = 0.2915$	Baggio et al. (2016)
Coptodon zillii	Egypt	No strain defined	mtDNA (COI, D-loop, CYTB)	Wild relatives	Population structure and genetic diversity $\Phi_{ST} = 0.4231$ $\pi = 0.0014$	Soliman et al. (2017)
Oreochromis niloticus	Egypt	No strain defined	mtDNA COI	Wild relatives	Population variation in different salinities environments	Mohamed-Geba et al. (2017)
Sarotherodon melanotheron	Benin	No defined genetic varieties or strains	Microsatellite	Wild relatives	Population structure and genetic diversity $G'_{ST} = 0.286$	Amoussou et al. (2018)
Oreochromis urolepis *Oreochromis niloticus* *Oreochromis leucostictus*	Tanzania	No strain defined	Microsatellite/ Morphometric	Wild and Feral	Hybridization occurrence	Shechonge et al. (2018)
Oreochromis niloticus	Uganda	No strain defined	Microsatellite	Wild relatives	Population structure and genetic diversity	Tibihika et al. (2019)

Species	Location	Strain	Marker	Wild/Farmed	Purpose/Results	Reference
Oreochromis niloticus	Tanzania	No strain defined	Microsatellite/ Morphometric	Wild relatives	Population structure and genetic diversity $F_{ST} = 0.249$	Shechonge et al. (2019)
Oreochromis niloticus *Oreochromis aureus* *Oreochromis mossambicus* *Oreochromis u. hornorum*	Scotland Africa Scotland Singapore Israel	Stirling strain Wild relatives populations	SNP	Wild relatives/ Farmed	Marker development Species identification	Syaifudin et al. (2019)
Oreochromis spp.	Africa	No strain defined	Nuclear genes markers/ mtDNA (ND2)	Wild relatives	Phylogenetic and molecular species determination	Ford et al. (2019)
O. niloticus	West Africa countries	No strain defined	SNP	Wild relatives	Population structure and genetic diversity Global $F_{ST} = 0.144$	Lind et al. 2019
O. niloticus	Malaysia	GIFT	SNP	Farmed	number of informative SNP markers= 40,930	Peñaloza et al. (2020)
O. niloticus	Egypt	Wild relatives Abbassa Strain	SNP	wild relatives/ Farmed	Evidence for signatures of selection, and genetic diversity related to domestication in the AS compared to wild Egyptian Nile River $F_{ST} = -0.008$-0.058 no genes of major effect were presently detected	Nayfa et al. 2020

(Contd.)

Table 1. (*Contd.*)

Species	Locality	Primary and secondary farmed types	Marker	Population origin	Goals/Outcomes	References
Coptodon zillii *O. niloticus* *O. variabilis* *Sarotherodon galilaeus* *O. esculentus* *O. jipe*	East and West African countries	No strain defined	Microsatellites	wild relatives/ Farmed	Testing the species boundaries of all Tilapiine species interacting with Nile tilapia. Thirty-eight SSR previously developed for *O. niloticus* were tested for cross-species amplification resulting in detectable working markers	Kariuki et al. 2021
O. niloticus	Benin	Wild relatives CEFRA strain	SNP	wild relatives/ Farmed	Population structure and genetic diversity. $F_{ST} = 0.018\text{-}0.143$	Fagbémi. et al 2021

*All taxonomy nomenclatures are according to their publication year. Some of them do not correspond to the current taxonomy classification.
RAPD = Random Amplified Polymorphic DNA; **VNTR** = Variable Number of Tandem Repeats; **Na** = Alleles number per locus; **D or** G_{ST} = Nei's genetic distance (Nei 1987); **K** = Nucleotide divergence; θ_S = Genetic differentiation among the analyzed samples (Weir and Cockerham 1984); **Sh1** = Shannon index; F_{ST} = Wright's population differentiation index (*AMOVA*); Φ_{ST} = Inter population differentiation index by the Analysis of Molecular Variance (*AMOVA*); D_{EST} = Jost's population differentiation index; R_{ST} = Slatkin's population differentiation index; G_{ST} = Hedrick's population differentiation index; **RAPD/GD** = RAPD genetic differentiation (Lynch and Milligan 1994); **VNTR/GD** = Minisatellites genetic differentiation (Jin and Chakraborty 1993; $K_{structure}$ = number of population assessed using Structure software; **GS** = Genetic similarity; **COI** = Cytochrome c oxidase I; **16S rRNA** = 16S ribosomal RNA; **Cytb** = Cytochrome b; **ND2, 5 and 6** = NADH-ubiquinone dehydrogenase 5 and 6. . **Primary and secondary farmed types:** refers to a farmed aquatic organism below the level of species that could be a strain, variety, hybrid, triploid, monosex group, or other genetically altered form or wild type; **Wild relatives:** they are all farmed species can still be found in nature; **Feral populations:** individuals produced in captive that return to the wild. (standardized terminology of AqGR used by FAO, see Mair and Lucente, 2020)

for genetic improvement of aquaculture species was described in detail by Sonesson et al. (2007); for tilapias, MAS is yet before us. Whole-genome resequencing has shown signatures of selection in cultured Nile, Mozambique, and red tilapias (Xia et al. 2015).

Besides, new opportunities are offered to evaluate the current status of tilapia genetic resources chiefly within their native distributions. Genomic tools also can be applied to detect the segregation of genes affecting ecological adaptation. Considerable work has been done with model systems (e.g., three spine stickleback Gasterosteus aculeatus, Hohenlohe et al. 2010), with systems of interest to management (e.g., lake trout Salvelinus nemaycush, Goetz et al. 2010, Chinook salmon Oncorhynchus tshawytscha, Larson et al. 2014), or conservation (e.g., Pacific lamprey Entosphenus tridentatus Hess et al. 2013). Allendorf et al. (2010) provide a useful review of principles. The application of genomic approaches to detecting adaptation-related genes in natural tilapia populations is yet to be done but may provide insights useful for the conservation of critical fish genetic resources.

There is a growing consensus that the management of genetic resources is based on a three-step linear model, progressing from conservation to evaluation to utilization (Berthaud 1997). The genetic characterization of wild relatives and farmed types tilapia genetic resources is but a part of monitoring genetic variation for conservation and sustainable use of this globally important food source (Lind et al. 2012b).

Acknowledgments

Alexandre W.S. Hilsdorf gratefully acknowledges the Foundation for Supporting Teaching and Research (FAEP), São Paulo Research Foundation (FAPESP), and University of Mogi das Cruzes (UMC) for supporting this work. Funding for Eric Hallerman's participation in this work was provided in part by the Virginia Agricultural Experiment Station and the Hatch Program of the National Institute of Food and Agriculture, U.S. Department of Agriculture. AWSH is recipient of National Council for Scientific and Technological Development (CNPq) productivity scholarship (304662/2017-8).

References cited

ADB. 2005. An Impact Evaluation of the Development of Genetically Improved Farmed Tilapia and Their Dissemination in Selected Countries. Asian Development Bank, Manila, Philippines. Available at: https://www.adb.org/sites/default/files/evaluation-document/35050/files/ies-tilapia-dissemination-0.pdf.

Agbebi, O.T., C.J. Echefu, I.O. Adeosun, A.H. Ajibade, E.A. Adegbite, A.O. Adebambo, et al. 2016. Mitochondrial diversity and time of divergence of commonly cultured cichlids in Nigeria. Br. Biotechnol. J. 13: 1–7.

Allendorf, F.W., P.A. Hohenlohe and G. Luikart. 2010. Genomics and the future of conservation genetics. Nature Rev. Genet. 11: 697–709.

Anane-Taabeah, G., E.A. Frimpong and E. Hallerman. 2019. Aquaculture-mediated invasion of the Genetically Improved Farmed Tilapia (GIFT) into the lower Volta Basin of Ghana. Diversity 11: 188.

Angienda, P.O., H.J. Lee, K.R. Elmer, R. Abila, E.N. Waindi and A. Meyer. 2011. Genetic structure and gene flow in an endangered native tilapia fish (*Oreochromis esculentus*) compared to invasive Nile tilapia (*Oreochromis niloticus*) in Yala Swamp, East Africa. Conserv. Genet. 12: 243–255.

Ansah, Y.B., E.A. Frimpong and E.M. Hallerman. 2014. Genetically-improved tilapia strains in Africa: Potential benefits and negative impacts. Sustainability 6: 3697–3721.

Agnèse, J.F., B. Adépo-Gourène, E.K. Abban and Y. Fermon. 1997. Genetic differentiation among natural populations of the Nile tilapia *Oreochromis niloticus* (Teleostei, Cichlidae). Heredity 79: 88–96.

Agnèse, J.F. 1998. Genetics and Aquaculture in Africa [ed.]. Editions de l'Orstom, Paris.

Agnése, J.F., B. Adépo-Gourène, J. Owino, L. Pouyaud and R. Aman. 1999. Genetic characterization of a pure relict population of *Oreochromis esculentus*, an endangered tilapia. J. Fish Biol. 54: 1119–1123.

Amoussou, T.O., I.Y.A. Karim, G.K. Dayo, I.I. Toko, M. Séré, A. Chikou, et al. 2018. Genetic characterization of Benin's wild populations of *Sarotherodon melanotheron melanotheron* Rüppell, 1852. Mol. Biol. Rep. 45: 1981–1994.

Atz, J.M. 1954. The peregrinating tilapia. Anim. Kingdom. 57: 148–155.

Avtalion, R.R. 1982. Genetic markers in *Sarotherodom* and their use for sex and species identification. pp. 269–277. *In*: Pullin, R.S.V. and R.H. Lowe-McConnell [eds.]. The Biology and Culture of Tilapias. ICLRM Conference Proceedings 7, International Center for Living Aquatic Resources Management, Manila, Philippines.

Azevedo, P. 1955. Aclimação da tilápia no Brasil. Chácaras e Quintais 92: 190–192.

Bakhoum, S.A., M.A. Sayed-Ahmed and E.A. Ragheb. 2009. Genetic evidence for natural hybridization between Nile tilapia (*Oreochromis niloticus*; Linnaeus, 1757) and blue tilapia (*Oreochromis aureus*; Steindachner, 1864) in Lake Edku, Egypt. Glob. Vet. 3: 91–97.

Badawi, H.K. 1971. Electrophoretic studies of serum proteins of four *Tilapia* species (Pisces). Mar. Biol. 8: 96–98.

Baggio, R.A., R. Orélis-Ribeiro and W.A. Boeger. 2016. Identifying Nile tilapia strains and their hybrids farmed in Brazil using microsatellite markers. Pesqu. Agropecu. Bras. 51: 1744–1750.

Balarin, J.D. and J.P. Hatton. 1979. Tilapia: A Guide to Their Biology and Culture in Africa. Unit of Aquatic Pathobiology, University of Stirling, Stirling, UK.

Bardakci, F. and D.O.F. Skibinski. 1994. Application of the RAPD technique in tilapia fish: Species and subspecies identification. Heredity 73: 117–123.

Barriga-Sosa, I.D.L.A., M.D.L. Jiménez-Badillo, A.L. Ibáñez and J.L. Arredondo-Figueroa. 2004. Variability of tilapias (*Oreochromis* spp.) introduced in Mexico: Morphometric, meristic and genetic characters. J. Appl. Ichthyol. 20: 7–14.

Barroso, R.M., R.A. Tenório, M.X. Pedroza-Filho, D.C. Webber, L.S. Belchior, E.F. Tahim, et al. 2015. Gerenciamento genético da tilápia nos cultivos comerciais. EMBRAPA Pesca e Aquicultura-Documentos (INFOTECA-E). Available at: https://www.infoteca.cnptia.embrapa.br/infoteca/bitstream/doc/1036709/1/CNPASA2015doc23.pdf.

Barroso, R.M., A.E.P. Muñoz and J. Cai. 2019. Social and economic performance of tilapia farming in Brazil. FAO Fisheries and Aquaculture Circular No. 1181. FAO, Rome.

Bartie, K.L., K. Taslima, M. Bekaert, S. Wehner, M. Syaifudin, J.B. Taggart, et al. 2020. Species composition in the Molobicus hybrid tilapia strain. Aquaculture, p. 735433.

Bartley, D.M., B.J. Harvey and R.S.V. Pullin (eds.). 2007. Workshop on Status and Trends in Aquatic Genetic Resources: A basis for international policy. Victoria, British Columbia, Canada, 8–10 May 2006. FAO Fisheries Proceedings. No. 5. FAO, Rome. 179 pp.

Basiao, Z.U. and N. Taniguchi. 1984. An investigation of enzyme and other protein polymorphisms in Japanese stocks of the tilapias *Oreochromis niloticus* and *Tilapia zillii*. Aquaculture 38: 335–345.

Berthaud, J. 1997. Strategies for conservation of genetic resources in relation with their utilization. Euphytica 96: 1–12.

Bezault, E., P. Balaresque, A. Toguyeni, Y. Fermon, H. Araki, J.F. Baroiller, et al. 2011. Spatial and temporal variation in population structure of wild Nile tilapia (*Oreochromis niloticus*) across Africa. BMC Genetics 12: 102.

Briñez, R.R., X.O. Caraballo and M.V. Salazar. 2011. Genetic diversity of six populations of red hybrid tilapia, using microsatellite genetic markers. Rev. MVZ Cordoba 16: 2491–2498.

Chen, F.Y. and H. Tauyuki. 1970. Zone electrophoretic studies on the plasma proteins of *Tilapia mossambica* and *T. hornorum* and their F_1 hybrids, *T. zillii*, and *T. melanopleura*. J. Fish. Res. Board Can. 27: 2167–2177.

Chotiyarnwong, A. 1971. Studies on *Tilapia nilotica* Linnaeus, *Tilapia mossambicus* Peters and their hybrids. Kasetsart University, Bangkok, Thailand.

Çiftci, Y. and İ. Okumuş. 2002. Fish population genetics and applications of molecular markers to fisheries and aquaculture: I. Basic principles of fish population genetics. Turk. J. Fish. Aquat. Sci. 2: 145–155.

Cnaani, A., E.M. Hallerman, M. Ron, J.I. Weller, M. Indelman, Y. Kashi, et al. 2003. Detection of a chromosomal region with two quantitative trait loci, affecting cold tolerance and fish size, in an F_2 tilapia hybrid. Aquaculture 223: 117–128.

Cnaani, A. and G. Hulata. 2008. Tilapias. pp. 101–116. *In*: Genome Mapping and Genomics in Fishes and Aquatic Animals [ed.]. Springer, Berlin.

Costa-Pierce, B.A. 2003. Rapid evolution of an established feral tilapia (*Oreochromis* spp.): The need to incorporate invasion science into regulatory structures. Biol. Invasions 5: 71–84.

Cruz, T.A., J.P. Thorpe and R.S.V. Pullin. 1982. Enzyme electrophoresis in *Tilapia zillii*: A pattern for determining biochemical genetic markers for use in tilapia stock identification. Aquaculture 29: 311–329.

D'Amato, M.E., M.M. Esterhuyse, B.C.W. Van der Waal, D. Brink and F.A.M. Volckaert. 2007. Hybridization and phylogeography of the Mozambique tilapia *Oreochromis mossambicus* in southern Africa evidenced by mitochondrial and microsatellite DNA genotyping. Conserv. Genet. 8: 475–488.

Deines, A.M., I. Bbole, C. Katongo, J.L. Federand and D.M. Lodge. 2014. Hybridization between native *Oreochromis* species and introduced Nile tilapia *O. niloticus* in the Kafue River, Zambia. Afr. J. Aquat. Sci. 39: 23–34.

Delomas, T.A., B. Gomelsky, N. Vu, M.R. Campbell and N.D. Novelo. 2019. Single-nucleotide polymorphism discovery and genetic variation in YY-male and mixed-sex strains of Nile tilapia available in the United States. N. Amer. J. Aquacult. 81: 183–188.

De Silva, C.D. and J. Ranasinghe. 1989. Biochemical evidence of hybrid gene introgression in some reservoir populations of tilapia in southern Sri Lanka. Aquac. Res. 20: 269–277.

De Silva, C.D. 1997. Genetic variation in tilapia populations in man-made reservoirs in Sri Lanka. Aquac. Int. 5: 339–349.

De Silva, M.P.K.S.K. 2015. Genetic diversity of genetically improved farmed tilapia (GIFT) broodstocks in Sri Lanka. Int. J. Sci. Res. Innov. Tech. 2: 66–76.

Dey, M.M. 2000. The impact of genetically improved farmed Nile tilapia in Asia. Aquac. Econ. Manag. 4: 107–124.

Dias, M.A.D., R.T.F. Freitas, S.E. Arranz, G.V. Villanova and A.W.S. Hilsdorf. 2016. Evaluation of the genetic diversity of microsatellite markers among four strains of *Oreochromis niloticus*. Anim. Genet. 47: 345–353.

Eknath, A.E., H.B. Bentsen, R.W. Ponzoni, M. Rye, N.H. Nguyen, J. Thodesen, et al. 2007. Genetic improvement of farmed tilapias: Composition and genetic parameters of a synthetic base population of *Oreochromis niloticus* for selective breeding. Aquaculture 273: 1–14.

Eknath, A.E., M.M. Dey, M. Rye, B. Gjerde, T.A. Abella, R. Sevilleja, et al. 1998. Selective breeding of Nile tilapia for Asia. *In*: Proceedings of the 6th World Congress of Genetics Applied to Livestock Production, Armidale, Australia, 11–16 January 1998; ICLARM Contribution No. 1397; International Center for Living Aquatic Resources Management: Manila, The Philippines, 1998.

Eknath, A.E. and G. Hulata. 2009. Use and exchange of genetic resources of Nile tilapia (*Oreochromis niloticus*). Rev. Aquacult. 1: 197–213.

Eknath, A.E., M.M. Tayamen, M.S. Palada-de Vera, J.C. Danting, R.A. Reyes, E.E. Dionisio, et al. 1993. Genetic improvement of farmed tilapias: The growth performance of eight strains of *Oreochromis niloticus* tested in different farm environments. Aquaculture 111: 171–188.

Elder, H.Y. and D.J. Garrod. 1961. A natural hybrid of *Tilapia nigra* and *Tilapia leucosticte* from Lake Naivasha, Kenya Colony. Nature 191: 722–724.

El-Fadly, G., M.I. Rehan-Khatab and A. Kalboush. 2016. Muscle protein and liver esterases banding patterns as biochemical markers to determine genetic diversity in Egyptian populations of tilapia species. Egyptian J. Genet. Cytol. 45: 187–203.

El-Serafy, S.S., M.H. Awwad, N.A.H. Abdel-Hamide and M.S. Azab. 2003. Restriction fragment length polymorphisms (RFLPs) of the small-subunit ribosomal DNA as a tool for identification of Tilapia spp. Egypt J. Aquat. Biol. Fish. 4: 465–482.

El-Serafy, S.S., N.A.H. Abdel-Hamide, M.H. Awwad and M.S. Azab. 2006. Comparative study on electrophoretic protein pattern characterization of *Tilapia* species in the River Nile, Egypt. Egypt J. Aqual. Biol. Fish. 10: 147–178.

Espinosa-Lemus, V., J.L. Arredondo-Figueroa and I.A. Angeles Barriga-Sosa. 2009. Morphometric and genetic characterization of tilapia (Cichlidae: Tilapiini) stocks for effective fisheries management in two Mexican reservoirs. Hidrobiológica 19: 95–107.

Fagbémi, M.N.A., L.M. Pigneur, A. André, N. Smitz, V. Gennotte, J.R. Michaux, et al. 2021. Genetic structure of wild and farmed Nile tilapia (*Oreochromis niloticus*) populations in Benin based on genome wide SNP technology. Aquaculture 535: 736432.

Falk, T.M. and E.K. Abban. 2004. Genetic diversity of the Nile tilapia *Oreochromis niloticus* (Teleostei, Cichlidae) from the Volta system in Ghana. pp. 13–15. *In*: Abban, E.K., P. Dugan, C.M.V. Casal, T.M. Falk [eds.]. Biodiversity, Management and Utilization of West African Fishes, Malaysia: WorldFish Center.

FAO. 2011. Molecular genetic characterization of animal genetic resources. FAO Animal Production and Health Guidelines. No. 9. FAO, Rome.

FAO. 2019a. FAO Yearbook: Fishery and Aquaculture Statistics 2017/FAO Annuaire: Statistiques des Pêches et de L'aquaculture 2017/FAO Anuario: Estadísticas de Pesca y Acuicultura 2017. FAO, Rome.

FAO. 2019b. Domestic Animal Diversity Information System (DAD-IS). Food and Agriculture Organization of the United Nations. Available at: http://www.fao.org/dad-is/en/.

FAO. 2019c. Fisheries Global Information System. Global Capture Production 1950-2016. Food and Agriculture Organization of the United Nations, Rome.

FAO. 2019d. The State of the World's Aquatic Genetic Resources for Food and Agriculture. FAO Commission on Genetic Resources for Food and Agriculture Assessments. FAO, Rome. Available at: http://www.fao.org/cgrfa/topics/aquatic/en/.

FAO/FIRI. 1993. Report of the Expert Consultation on Utilization and Conservation of Aquatic Genetic Resources. Grottaferrata, Italy, 9-13 November 1992. FAO Fish. Rep. 491, 58 p.

FAO/UNEP. 1981. Conservation of the genetic resources of fish: Problems and recommendations. Report of the Expert Consultation on the Genetic Resources of Fish. Rome, 9–13 June 1980. FAO Fish. Tech. Pap. (217): 43 p. FAO, Rome.

Fitzsimmons, K. 2000. Tilapia: The most important aquaculture species of the 21st Century. pp. 3–8. *In*: Tilapia Aquaculture in the 21st Century. Proceedings from the Fifth International Symposium on Tilapia in Aquaculture (Vol. 1), Rio de Janeiro, Brazil.

Ford, A.G., T.R. Bullen, L. Pang, M.J. Genner, R. Bills, T. Flouri, et al. 2019. Molecular phylogeny of *Oreochromis* (Cichlidae: Oreochromini) reveals mito-nuclear discordance and multiple colonisation of adverse aquatic environments. Mol. Phylogenetics Evol. 136: 215–226.

Fuerst, P.A., W.W. Mwanja and L. Kaufman. 2000. The genetic history of the introduced Nile tilapia of Lake Victoria (Uganda–E. Africa): The population structure of *Oreochromis niloticus* (Pisces: Cichlidae) revealed by DNA microsatellite markers. pp. 30–41. *In:* Tilapia Aquaculture in the 21st Century. Proceedings from the Fifth International Symposium on Tilapia in Aquaculture (Vol. 1), Rio de Janeiro, Brazil.

Goetz, F., D. Rosauer, S. Sitar, G. Goetz, C. Simchick, S. Roberts, et al. 2010. A genetic basis for the phenotypic differentiation between siscowet and lean lake trout (*Salvelinus namaycush*). Molec. Ecol. 19: 176–196.

Gregg, R.E., J.H. Howard and F. Snhonhiwa, 1998. Introgressive hybridization of tilapias in Zimbabwe. J. Fish Biol. 52: 1–10.

Gu, X.H., D.L. Jiang, Y. Huang, B.J. Li, C.H. Chen, H.R. Lin, et al. 2018a. Identifying a major QTL associated with salinity tolerance in Nile tilapia using QTL-seq. Marine Biotechnol. 20: 98–107.

Gu, X.H., B.J. Li, H.R. Lin and J.H. Xia. 2018b. Unraveling the associations of the tilapia DNA polymerase delta subunit 3 (POLD3) gene with saline tolerance traits. Aquaculture 485: 53–58.

Gunnes, K. and T. Gjedrem. 1978. Selection experiments with salmon. IV. Growth of Atlantic salmon during two years in the sea. Aquaculture 15: 19–33.

Gupta, M.V. and B.O. Acosta. 2004. From drawing board to dining table: The success story of the GIFT project. NAGA WorldFish Center Q. 27: 4–14.

Hall, E.G. 2001. An analysis of population structure using microsatellite DNA in twelve southern African populations of the Mozambique tilapia, *Oreochromis mossambicus* (Peters). M.Sc. Thesis, University of Stellenbosch, Stellenbosch, South Africa.

Hallerman, E. and A.W.S. Hilsdorf. 2014. Conservation genetics of tilapias: Seeking to define appropriate units for management. Isr. J. Aquacult.-Bamidgeh 66: 2–19.

Hassanien, H.A., M. Elnady, A. Obeida and H. Itriby. 2004. Genetic diversity of Nile tilapia populations revealed by randomly amplified polymorphic DNA (RAPD). Aquacult. Res. 35: 587–593.

Hassanien, H.A. and J. Gilbey. 2005. Genetic diversity and differentiation of Nile tilapia (*Oreochromis niloticus*) revealed by DNA microsatellites. Aquacult. Res. 36: 1450–1457.

Herzberg, A. 1978. Electrophoretic esterase patterns of the surface mucus for the identification of tilapia species. Aquaculture 13: 1–83.

Hess, J.E., N.R. Campbell, D.A. Close, M.F. Docker and S.R. Narum. 2013. Population genomics of Pacific lamprey: Adaptive variation in a highly dispersive species. Molec. Ecol. 22: 2898–2916.

Hickling, C.F. 1963. The cultivation of tilapia. Scientific American 208: 143–152.

Hilsdorf, A.W.S. and E.M. Hallerman. 2017. Genetic Resources of Neotropical Fishes. Cham, Switzerland [ed.]. Springer International Publishing.

Hohenlohe, P.A., S. Bassham, P.D. Etter, N. Stiffler, E.A. Johnson and W.A. Cresko. 2010.

Population genomics of parallel adaptation in three spine stickleback using sequenced RAD tags. PLoS Genet. 6: e1000862.

Ibrahim, N.A., A. Mohamed Nasr-Allah and H. Charo-Karisa. 2019. Assessment of the impact of dissemination of genetically improved Abbassa Nile tilapia strain (GIANT-G9) versus commercial strains in some Egyptian governorates. Aquacult. Res. 50: 2951–2959.

Jiang, D.L., X.H. Gu, B.J. Li, Z.X. Zhu, H. Qin, N.Z. Meng, et al. 2019. Identifying a long QTL cluster across chrLG18 associated with salt tolerance in tilapia using GWAS and QTL-seq. Mar. Biotechnol. 21: 250–261.

Jin, L. and R. Chakraborty. 1993. A bias-corrected estimate of heterozygosity for single-probe multilocus DNA fingerprints. Mol. Biol. Evol. 10: 1112–1114.

Kariuki, J., P.D. Tibihika, M. Curto, E. Alemayehu, G. Winkler and H. Meimberg. 2021. Application of microsatellite genotyping by amplicon sequencing for delimitation of African tilapiine species relevant for aquaculture. Aquaculture 537: 736501.

Khan, M.G. 2011. Marker-assisted selection in enhancing genetically male Nile tilapia (*Oreochromis niloticus* L.) production. Ph.D. dissertation, University of Stirling, Stirling, Scotland. https://dspace.stir.ac.uk/handle/1893/2980#.Xclr_UnsZaQ.

Kornfield, I.L., U. Ritte, C. Richler and J. Wahrman. 1979. Biochemical and cytological differentiation among cichlid fishes of the Sea of Galilee. Evolution 33: 1–14.

Kuton, M.P. and B.T. Adeniyi. 2014. Morphological variations of *Tilapia guineensis* (Bleeker 1862) and *Sarotherodon melanotheron* (Ruppell 1852) (Pisces: Cichlidea) from Badagry and Lagos lagoon, South-West, Nigeria. J. Fish. Livest. Prod. 2: 2332–2608.

Lande, R. and S. Shannon. 1996. The role of genetic variation in adaptation and population persistence in a changing environment. Evolution 50: 434–437.

Larson, W.A., L.W. Seeb, M.V. Everett, R.K. Waples, W.D. Templin and J.E. Seeb. 2014. Genotyping by sequencing resolves shallow population structure to inform conservation of Chinook salmon (*Oncorhynchus tshawytscha*). Evol. Appl. 7: 355–369.

Lenstra, J.A., L.F. Groeneveld, H. Eding, J. Kantanen, J.L. Williams, P. Taberlet, et al. 2012. Molecular tools and analytical approaches for the characterization of farm animal genetic diversity. Anim. Genet. 43: 483–502.

Li, H.L., X.H. Gu, B.J. Li, C.H. Chen, H.R. Lin and J.H. Xia. 2017. Genome-wide QTL analysis identified significant associations between hypoxia tolerance and mutations in the GPR132 and ABCG4 genes in Nile tilapia. Mar. Biotechnol. 19: 441–453.

Lima, F.M., F.H.F. Costa, A.H. Sampaio, S. Saker-Sampaio, B.S. Cavada, I.R.C.B. Rocha, et al. 2000. Genetic variability using molecular markers (RAPD) in species and hybrids of tilápias (Pisces, Cichlidade). pp. 41–47. *In:* Tilapia Aquaculture in the 21st Century. Proceedings from the Fifth International Symposium on Tilapia in Aquaculture (Vol. 1), Rio de Janeiro, Brazil.

Lin, G., L. Wang, T.S. Ngoh, L. Ji, L. Orbán and G.H. Yue. 2018. Mapping QTL for omega-3 content in hybrid saline tilapia. Mar. Biotechnol. 20: 10–19.

Lind, C.E., R.E. Brummett and R.W. Ponzoni. 2012. Exploitation and conservation of fish genetic resources in Africa: Issues and priorities for aquaculture development and research. Rev. Aquacult. 4: 125–141.

Lind, C.E., R.W. Ponzoni, N.H. Nguyen and H.L. Khaw. 2012. Selective breeding in fish and conservation of genetic resources for aquaculture. Reprod. Domest. Anim. 47: 255–263.

Lind, C.E., S.K. Agyakwah, F.Y. Attipoe, C. Nugent, R.P.M.A. Crooijmans and A. Toguyeni. 2019. Genetic diversity of Nile tilapia (*Oreochromis niloticus*) throughout West Africa. Scientific Reports 9: 16767.

Liu, Z.J. and J.F. Cordes. 2004. DNA marker technologies and their applications in aquaculture genetics. Aquaculture 238: 1–37.

Lovshin, L.L., A.B. Da Silva, A. Carneiro-Sobrinho and F.R. Melo. 1990. Effects of *Oreochromis niloticus* females on the growth and yield of male hybrids (*O. niloticus* female × *O. hornorum* male) cultured in earthen ponds. Aquaculture 88: 55–60.

Lynch, M. and B.G. Milligan. 1994. Analysis of population structure with RAPD markers. Mol. Ecol. 3: 91–99.

Macaranas, J.M., L.Q. Agustin, M.C.A. Ablan, M.J.R. Pante, A.A. Eknath and R.S. Pullin. 1995. Genetic improvement of farmed tilapias: Biochemical characterization of strain differences in Nile tilapia. Aquacult. Int. 3: 43–54.

Mair, G. and D. Lucente. 2020. What are "farmed types" in aquaculture and why do they matter? FAO Aquaculture Newsletter 61: 40–42. url: http://www.fa o.org/3/ca8302en/CA8302EN.pdf.

Maranan, J.B.D., Z.U. Basiao and J.P. Quilang. 2016. DNA barcoding of feral tilapias in Philippine lakes. Mitochondrial DNA A 27: 4302–4313.

McAndrew, B.J. and K.C. Majumdar. 1983. Tilapia stock identification using electrophoretic markers. Aquaculture 30: 249–261.

Meffe, G.K. and C.R. Carroll. 1997. Genetics: Conservation of diversity within species, pp. 161–202. *In*: G.K. Meffe and C.R. Carroll [eds.]. Principles of Conservation Biology. Sinauer Associates, Sunderland, MA, USA.

Moen, T., J.J. Agresti, A. Cnaani, H. Moses, T.R. Famula, G. Hulata, et al. 2004. A genome scan of a four-way tilapia cross supports the existence of a quantitative trait locus for cold tolerance on linkage group 23. Aquacult. Res. 35: 893–904.

Mohammed-Geba, K., S.El-S.H. El-Nab, E. Awad and A.I. Nofal. 2017. DNA barcoding identifies a unique haplotype of Nile tilapia *Oreochromis niloticus* thriving in Egyptian freshwater and brackish water lakes. Int. J. Ecotox. Ecobiol. 2: 172–177.

Moralee, R.D., F.H. Van der Bank and B.C.W. Van der Waal. 2000. Biochemical genetic markers to identify hybrids between the endemic *Oreochromis mossambicus* and the alien species, *O. niloticus* (Pisces: Cichlidae). Water Sa 26: 263–268.

Moreira, A.A., A.W.S. Hilsdorf, J.V. Silva and V.R. Souza. 2007. Variabilidade genética de duas variedades de *Tilápia nilótica* por meio de marcadores microsatélites. Pesq. Agropec. Bras. 42: 521–526.

Mwanja, W., G. Booton, L. Kaufman, M. Chandler, P. Fuerst, E. Donaldson, et al. 1996. Population and stock characterization of Lake Victoria tilapine fishes based on RAPD markers. pp. 115–124. *In*: Donaldson, E.M. and D.D. Mackinlay [eds.]. Aquaculture Biotechnology Symposium: Proceedings of the International Congress on the Biology of Fishes. American Fisheries Society, Bethesda, MD, USA.

Mwanja, W., G.C. Booton, L. Kaufman and P.A. Fuerst. 2008. A profile of the introduced *Oreochromis niloticus* (Pisces: Teleostei) populations in Lake Victoria Region in relation to its putative origin of Lakes Edward and Albert (Uganda-E. Africa) based on random amplified polymorphic DNA analysis. Afr. J. Biotechnol. 7: 1769–1773.

Mwanja, W.W., L. Kaufman and P.A. Fuerst. 2010. Comparison of the genetic and ecological diversity of the native to the introduced tilapiines (Pisces: Cichlidae), and their population structures in the Lake Victoria region, East Africa. Aquat. Ecosyst. Health Manage. 13: 442–450.

Naish, K.-A., M. Warren, F. Bardakci, D.O.F. Skibinski, G.R. Carvalho and G.C. Mair. 1995. Multilocus DNA fingerprinting and RAPD reveal similar genetic relationships between strains of *Oreochromis niloticus* (Pisces: Cichlidae). Mol. Ecol. 4: 271–274.

Nayfa, M.G., D.B. Jones, J.A. Benzie, D.R. Jerry and K.R. Zenger. 2020. Comparing genomic signatures of selection between the Abbassa Strain and eight wild populations of Nile tilapia (*Oreochromis niloticus*) in Egypt. Front. Genet. 11: 567969.

Ndiwa, T.C., D.W. Nyingi and J.-F. Agnese. 2014. An important natural genetic resource of *Oreochromis niloticus* (Linnaeus, 1758) threatened by aquaculture activities in Loboi Drainage, Kenya. PLoS ONE 9: e106972.

Nei, M. 1987. Molecular Evolutionary Genetics. Columbia University Press, New York, NY.

Nyingi, D., L. De-Vos, R. Aman and J.-F. Agnèse. 2009. Genetic characterization of an unknown and endangered native population of the Nile tilapia *Oreochromis niloticus* (Linnaeus, 1758) (Cichlidae; Teleostei) in the Loboi Swamp (Kenya). Aquaculture 297: 57–63.

Nyingi, D.W. and J-F. Agnèse. 2007. Recent introgressive hybridization revealed by exclusive mtDNA transfer from *Oreochromis leucostictus* (Trewavas, 1933) to *Oreochromis niloticus* (Linnaeus, 1758) in Lake Baringo, Kenya. J. Fish BioLake 70: 148–154.

Oladimeji, T.E., M.O. Awodiran and O.O. Komolafe. 2015. Genetic differentiation studies among natural populations of *Tilapia zillii*. Not. Sci. Biol. 7: 423–429.

Olesen, I., G.K. Rosendal, M.W. Tvedt, M. Bryde and H.B. Bentsen. 2007. Access to and protection of aquaculture genetic resources: Structures and strategies in Norwegian aquaculture. Aquaculture 272: S47–S61.

Ordoñez, J.F.F., M.F.H. Ventolero and M.D. Santos. 2017. Maternal mismatches in farmed tilapia strains (*Oreochromis* spp.) in the Philippines as revealed by mitochondrial *COI* gene. Mitochondrial DNA Part A 28: 526–535.

Othman, R., M.S.M. Zan and M. Remely. 2010. Genetic variation among river and hatchery populations of *Oreochromis* sp. using random amplified polymorphic DNA (RAPD) fingerprinting. pp. 63–65. *In*: Proceedings of the International Academy of Biology (IAB) and Islamic Academy of Sciences (IAS) International Symposium, Shah Alam, Malaysia.

Peñaloza, C., D. Robledo, A. Barría, T.Q. Trịnh, M. Mahmuddin, P. Wiener, et al. 2020. Development and validation of an open access snp array for Nile tilapia (*Oreochromis niloticus*). G3-Genes Genom. Genet. 10: 2777-2785.

Penman, D.J., M.V. Gupta and M.M. Dey. 2005. Carp genetic resources for aquaculture in Asia. WorldFish Center Technical Report 65 [ed.]. WorldFish Center, Penang, Malaysia.

Petersen, R.L., J.E. Garcia, G. Mello, A.M. Liedke, T.C. Sincero and E.C. Grisard. 2012. Análise da diversidade genética de tilápias cultivadas no estado de Santa Catarina (Brasil) utilizando marcadores microssatélites. Bol. Inst. Pesca 38: 313–321.

Philippart, J.C. and J.C. Ruwet. 1982. Ecology and distribution of tilapias. pp. 15–59. *In*: Pullin and Lowe-McConel [eds.]. The Biology and Culture of Tilapias, ICLARM Conference Proceedings 7, Center for Living Aquatic Resources Management, Manila, Philippines.

Pilling, D., J. B´elanger, S. Diulgheroff, J. Koskela, G. Leroy, G. Mair and I. Hoffmann. 2020. Global status of genetic resources for food and agriculture: challenges and research needs. Genet. Resour. 1: 4-16.

Ponzoni, R.W., J.W. James, N.H. Nguyen, W. Mekkawy and H.L. Khaw. 2013. Strain comparisons in aquaculture species: A manual 2013–12. WorldFish; CGIAR Research Program Livestock and Fish, Penang, Malaysia.

Ponzoni, R.W., N.H. Nguyen, H.L. Khaw, A. Hamzah, K.R.A. Bakar and H.Y. Yee. 2011. Genetic improvement of Nile tilapia (*Oreochromis niloticus*) with special reference to the work conducted by the WorldFish Center with the GIFT strain. Aquaculture 3: 27–41.

Poompuang, S. and E.M. Hallerman. 1997. Toward detection of quantitative trait loci and marker-assisted selection in fish. Rev. Fish. Sci. 5: 253–277.

Povh, J.A., H.L.M. Moreira, R.P. Ribeiro, A.J. Prioli, L. Vargas, D.V. Blanck, et al. 2005. Estimativa da variabilidade genética em linhagens de tilápia do Nilo (*Oreochromis niloticus*) com a técnica de RAPD. Acta Sci. Anim. Sci. 27: 1–10.

Prabu, E., C.B.T. Rajagopalsamy, B. Ahilan, I. Jegan Michael Andro Jeevagan and M. Renuhadevi. 2019. Tilapia, an excellent candidate species for world aquaculture: A review. Annu. Res. Rev. Biol. 31: 1–14.

Pullin, R.S. 1988. Tilapia genetic resources for aquaculture. pp. 16–108. *In*: ICLARM Conference Proceedings. International Center for Living Aquatic Resources Management, Manila, Philippines.

Pullin, R.S., C.M.V. Casal, E.K. Abban and T.M. Falk. 1997. The characterization of Ghanaian tilapia genetic resources for use in fisheries and aquaculture [ed.]. ICLARM Conference Proceedings 52, International Center for Living Aquatic Resources Management, Manila, Philppines.

Pullin, R.S.V., D.M. Battley and J. Kooiman. 1999. Towards policies for conservation and sustainable use of aquatic genetic resources [ed.]. ICLARM Conf. Proc. 59, FAO, Bellagio, Italy.

Rashed, M.A., Y.M. Saad, M.M. Ibraim and A.A. El-Seoudy. 2008. Genetic structure of natural Egyptian *Oreochromis niloticus* evaluated using dominant DNA markers. Glob. Vet. 2: 87–91.

Riedel, D. 1965. Some remarks on the fecundity of Tilapia (*T. mossambicus* Peters) and its introduction into Middle Central America (Nicaragua) together with a first contribution towards the limnology of Nicaragua. Hydrobiologia 25: 357–388.

Rognon, X., M. Andriamanga, B. McAndrew and R. Guyomard. 1996. Allozyme variation in natural and cultured populations in two tilapia species: *Oreochromis niloticus* and *Tilapia zillii*. Heredity 76: 640–650.

Rognon, X. and R. Guyomard. 1997. Mitochondrial DNA differentiation among East and West African Nile tilapia populations. J. Fish Biol. 51: 204–207.

Rognon, X. and R. Guyomard. 2003. Large extent of mitochondrial DNA transfer from *Oreochromis aureus* to *O. niloticus* in West Africa. Mol. Ecol. 12: 435–445.

Romana-Eguia, M.R.R., M. Ikeda, Z.U. Basiao and N. Taniguchi. 2004. Genetic diversity in farmed Asian Nile and red hybrid tilapia stocks evaluated from microsatellite and mitochondrial DNA analysis. Aquaculture 236: 131–150.

Rutten, M.J.M., H. Komen, R.M. Deerenberg, M. Siwek and H. Bovenhuis. 2004. Genetic characterization of four strains of Nile tilapia (*Oreochromis niloticus* L.) using microsatellite markers. Anim. Genet. 35: 93–97.

Ryman, N. 1981. Fish Gene Pools (Ecological Bulletins, 34). Stockholm [ed.]. Editorial Service, FRN.

Saad, Y.M., M.A. Rashed, A.H. Atta and N.E. Ahmed. 2012. Genetic diversity among some tilapia species based on ISSR markers. Life Sci. J. 9: 4841–4846.

Saint-Paul, U. 2017. Native fish species boosting Brazilian's aquaculture development. Acta Fish. 5: 1–9.

Seyoum, S. and I. Kornfield. 1992. Identification of the subspecies of *Oreochromis niloticus* (Pisces: Cichlidae) using restriction endonuclease analysis of mitochondrial DNA. Aquaculture 102: 29–42.

Shechonge, A., B.P. Ngatunga, R. Tamatamah, S.J. Bradbeer, J. Harrington, A.G. Ford, et al. 2018. Losing Cichlid fish biodiversity: Genetic and morphological homogenization of tilapia following colonization by introduced species. Conserv. Genet. 19: 1199–1209.

Shechonge, A., B.P. Ngatunga, R. Tamatamah, S.J. Bradbeer, E. Sweke, A. Smith, et al. 2019. Population genetic evidence for a unique resource of Nile tilapia in Lake Tanganyika, East Africa. Environ. Biol. Fishes 102: 1107–1117.

Shirak, A., E. Seroussi, A. Cnaani, A.E. Howe, R. Domokhovsky, N. Zilberman, et al. 2006. *Amh* and *Dmrta2* genes map to tilapia (*Oreochromis* spp.) linkage group 23 within quantitative trait locus regions for sex determination. Genetics 174: 1573–1581.

Simbine, L., J.V. Silva and A.W.S. Hilsdorf. 2014. The genetic diversity of wild *Oreochromis mossambicus* populations from the Mozambique southern watersheds as evaluated by microsatellites. J. Appl. Ichthyol. 30: 272–280.

Socol, C.T., I. Lelior, I. Mihalca and F.L. Criste. 2015. Molecular and population genetics tools for animal resources conservation: A brief overview. Scientific Papers Animal Science and Biotechnologies 48: 95–102.

Solar, I.I. 2009. Use and exchange of salmonid genetic resources relevant for food and aquaculture. Rev. Aquacult. 1: 174–196.

Soliman, T., W. Aly, R.M. Fahim, M.L. Berumen, H. Jenke-Kodama and G. Bernardi. 2017. Comparative population genetic structure of redbelly tilapia (*Coptodon zillii* (Gervais, 1848)) from three different aquatic habitats in Egypt. Ecol. Evol. 7: 11092–11099.

Sonesson, A.K., 2007. Possibilities for marker-assisted selection in aquaculture breeding schemes. pp. 309–328. In: Guimaraes, E.P.; Ruane, J., Scherf, B.D., Sonnino, A., and Dargie, J.D. [eds.]. Marker-Assisted Selection: Current Status and Future Application in Crops, Livestock, Forestry and Fish. Food and Agriculture Organization of the United Nations (FAO), Rome.

Sukmanomon, S., W. Senanan, A.R. Kapuscinski and U. Na-Nakorn. 2012. Genetic diversity of feral populations of Nile tilapia (*Oreochromis niloticus*) in Thailand and evidence of genetic introgression. Kasetsart J. (Nat. Sci.) 46: 200–216.

Syaifudin, M., M. Bekaert, J.B. Taggart, K.L. Bartie, S. Wehner, C. Palaiokostas, et al. 2019. Species-specific marker discovery in tilapia. Sci. Rep. 9: 1–11.

Talle, S.B., W.S. Chenyabuga, E. Fimland, O. Syrstad, T. Meuwissen and H. Klungland. 2005. Use of DNA technologies for the conservation of animal genetic resources: A review. Acta Agric. Scand. A Anim. Sci. 55: 1–8.

Tibihika, P.D., M. Curto, E. Dornstauder-Schrammel, S. Winter, E. Alemayehu, H. Waidbacher, et al. 2019. Application of microsatellite genotyping by sequencing (SSR-GBS) to measure genetic diversity of the East African *Oreochromis niloticus*. Conserv. Genet. 20: 357–372.

Toro, M.A., L. Silio, J. Rodriganez, C. Rodriguez and J. Fernandez. 1999. Optimal use of genetic markers in conservation programmes. Genet. Sel. Evol. 31: 255–261.

Trewavas, E. 1983. Tilapine Fishes of the Genera *Sarotherodon, Oreochromis* and *Danakilia*. London, British Museum (Natural History).

Ukenye, E.A., I.A. Taiwo, O.R. Oguntade, T.O. Oketoki and A.B. Usman. 2016. Molecular characterization and genetic diversity assessment of *Tilapia guineensis* from some coastal rivers in Nigeria. Afr. J. Biotechnol. 15: 20–28.

Van Bers, N.E.M., R.P.M.A. Crooijmans, B.W. Dibbits, J. Komen and M.A.M. Groenen. 2012. SNP marker detection and genotyping in tilapia. Mol. Ecol. Resour. 12: 932–941.

Vicente, I.S.T. and C.E. Fonseca-Alves. 2013. Impact of introduced Nile tilapia (*Oreochromis niloticus*) on non-native aquatic ecosystems. Pak. J. Biol. Sci. 16: 121–126.

Weir, B.S. and C.C. Cockerham. 1984. Estimating *F*-statistics for the analysis of population structure. Evolution 38: 1358–1370.

Welcomme, R.L. 1966. Recent changes in the stocks of tilapia in Lake Victoria. Nature 212: 52–54.

Whitehead, P.J.P. 1962. The relationship between *Tilapia nigra* (Gunther) and *Tilapia mossambica* (Peters) in the Eastern Rivers of Kenya. Proc. Zool. Soc. Lond. 138: 605–637.

Wohlfarth, G.W. 1994. The unexploited potential of tilapia hybrids in aquaculture. Aquacult. Res. 25: 781–788.

Wright, S. 1978. Evolution and the Genetics of Population IV: Variability Within and Among Natural Populations [ed.]. The University of Chicago Press, Chicago.

Wu, L. and Yang, J. 2012. Identifications of captive and wild tilapia species existing in Hawaii by mitochondrial DNA control region sequence. PLoS ONE 7: e51731.

Xia, J.H., Z.Y. Wan, Z.L. Ng, L. Wang, G.H. Fu, G. Liu, et al. 2014. Genome-wide discovery and in-silico mapping of gene-associated SNPs in Nile tilapia. Aquaculture 432: 67–73.

Xia, J.H., Z. Bai, Z. Meng, Y. Zhang, L. Wan, F. Liu, et al. 2015. Signatures of selection in tilapia revealed by whole-genome resequencing. Sci. Rep. 5: 14168.

Yang, W., X. Kang, Q. Yang, Y. Lin and M. Fang. 2013. Review on the development of genotyping methods for assessing farm animal diversity. J. Anim. Sci. Biotech. 4: 2.

Young, J.A. and J.F. Muir. 2002. Tilapia: Both fish and fowl? Mar. Resour. Econ. 17: 163–173.

Zengeya, T.A., M.P. Robertson, A.J. Booth and C.T. Chimimba. 2013. A qualitative ecological risk assessment of the invasive Nile tilapia, *Oreochromis niloticus* in a sub-tropical African river system (Limpopo River, South Africa). Aquat. Cons.: Mar. Freshwat. Ecosys. 23: 51–64.

Selective Breeding of Farmed Tilapia

Rafael Vilhena Reis Neto[1]*, Rilke Tadeu Fonseca de Freitas[2], José Manuel Yáñez[3], Carlos Antônio Lopes de Oliveira[4], Eduardo Maldonado Turra[5] and Fabio Luiz Buranelo Toral[5]

[1] UNESP Aquaculture Center, São Paulo State University, Via de Acesso Prof. Paulo Donato Castellane, s/n - Jaboticabal/SP Brazil - CEP 14884-900
[2] Department of Animal Science, Federal University of Lavras, Lavras, MG Brazil
[3] Facultad de Ciencias Veterinarias y Pecuarias, Universidad de Chile, Santiago, Chile
[4] Department of Animal Science, State University of Maringá, Maringá, PR Brazil
[5] Department of Animal Science, Federal University of Minas Gerais, Belo Horizonte, MG Brazil

1. Introduction

Genetic improvement aims to increase the productivity of a population by selecting the best individuals from one generation to be the parents of the next generation. A key question for this process is: Who are the 'best individuals'? Initially, we may think that they are the ones with the best performance: the fastest growth, best body traits, greatest food efficiency, disease resistance, etc. However, all of these traits, from a genetic standpoint, are phenotypes, that is, they can be observed or measured. Each phenotype can be described as the result of the sum of genetic factors and environmental factors (Phenotype = Genotype + Environment or $P = G + E$). Thus, the best performance (phenotype) of an animal can be due to environmental factors that are not inheritable. Given this possibility, the animal with the best performance would not transmit this superiority to its offspring.

Furthermore, there are also non-inheritable genetic factors. Considering the equation $P = G + E$, it is still possible to separate genetic factors into additives, dominance, and epistasis ($G = A + D + I$). Only additive genetic factors are passed from one generation to the next by selection. Thus, answering the question we initially proposed, the 'best individuals' are those that have the highest additive genetic values, or as it is said in animal breeding, the highest breeding value (BV). Statistical designs and models as well as molecular information are used in breeding programs to estimate the BV of animals in different environments as accurately as possible. The next sections will discuss these topics in detail.

*Corresponding author: rafael.vilhena@unesp.br

2. Design of breeding programs in farmed Tilapia

Breeding programs aim to promote consistent and lasting genetic gains. Therefore, the correct evaluation and selection of animals, a mating system that optimizes genetic gains while maintaining inbreeding at acceptable levels, and an effective dissemination of genetically superior animals to the productive system are aspects that should be considered in the design of a breeding program in tilapia. Different systems, where the tilapia is farmed, and the high fecundity rate of this species are additional factors that should be considered in the implementation of breeding programs (Ponzoni et al. 2010).

Figure 1 shows a detailed design of a fish breeding program proposed by Gjedrem and Baranski (2009). The authors proposed genetic evaluation and selection of animals, mating selection, hatchery and initial growth (using separate hatching trays and hapas for each family in order to maintain the pedigree), individual identification (e.g. passive integrated transponder [PIT]), and sending the identified animals to different evaluation sites (nucleus evaluation station and/or field and challenge stations). They also proposed the dissemination of genetic gain from a nucleus specimen to commercial farmers through the use of multipliers.

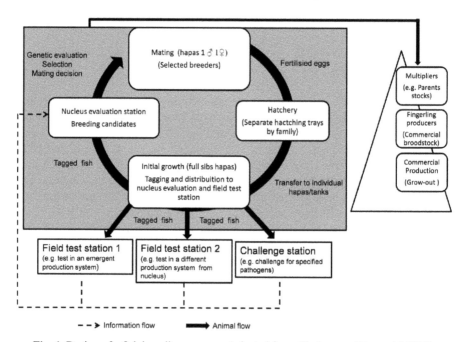

Fig. 1. Design of a fish breeding program (adapted from Gjedrem and Baranski 2009).

2.1. Genetic evaluation: Brief notes about data analysis

According to Mrode and Thompson (2005), the phenotypes of economic traits are determined by environmental, genetic, and residual effects. The appropriate identification of environmental and genetic effects and the use of relationship information—whether ascending, descending, or collateral—will allow a more accurate BV estimation.

Several statistical methodologies have been proposed to estimate the BV of economic traits. According to Lind et al. (2012), combined selection, using an individual's and its relative's information to estimate the BV, has been used in different fish breeding programs. The same authors highlighted that combined selection can be used to estimate the BV for traits that require slaughter and/or to evaluate disease challenge or tolerance to some environmental conditions.

Currently, the mixed model equations (MME) proposed by Henderson (1984) are most often used to estimate the genetic merit in a combined selection in tilapia breeding programs (Thodesen et al. 2011, Oliveira et al. 2016, Garcia et al. 2017). This methodology allows the simultaneous estimation of systematic environmental and genetic effects (BVs), using information from the individual and its relatives, and the variance components estimated previously from the population under evaluation (Henderson 1984, Mrode and Thompson 2005, Gjedrem and Baranski 2009).

A simple animal model takes the equation $y = X\beta + Za + e$, where y is the vector of observations, β is the vector of fixed effects, a is the vector of random animal effects, e is the vector of random residual effects, X is a matrix that relates records to fixed effects, and Z is a matrix that relates the records to random animal effects. Thus, the MME for animal model described above are:

$$\begin{bmatrix} \beta \\ a \end{bmatrix} = \begin{bmatrix} X'R^{-1}X & X'R^{-1}Z \\ Z'R^{-1}X & Z'R^{-1}Z + G^{-1} \end{bmatrix}^{-1} \begin{bmatrix} X'R^{-1}y \\ Z'R^{-1}y \end{bmatrix},$$

where $G = A\sigma_a^2$, $R = I\sigma_e^2$, A is numerator relationship matrix, σ_a^2 is additive genetic variance and σ_e^2 is residual variance, and I is an identity matrix order $n \times n$. The expectations of random effects are zero and cov $(a, e) = 0$, $V = ZGZ' + R$, where V is phenotypic variance and covariance matrix. The statistical properties are described by Henderson (1984).

Due to the peculiarities of animal management, an additional effect is usually added in the statistical model, namely, the common family effect (due to separate rearing of full-sib families from spawning to tagging; Figure 1). Thus, the animal model is described by $y = X\beta + Za + Wc + e$, where y is the vector for an economic trait (e.g. body weight), β is the vector of fixed effects (e.g. sex, age, or cage/tank), a is the vector of random animal effects, c is the vector of random common family effects, e is the vector of random residual effects, X is a matrix that relates records to fixed effects, Z is a matrix that relates the records to random animal effects, and W is a matrix that relates the records to random common family effects.

Several computer programs have been proposed to estimate variance components and solve the mixed model equations. This effort led to the estimated breeding value (EBV) for economic traits (Lind et al. 2011, Masuda 2018). The use of these programs is widespread in tilapia genetic breeding programs worldwide. Thus, this methodology can be utilized to estimate the EBV and promote the selection and mating of selected animals.

2.2. Mating decision: Some considerations

There are two points that should be considered with regards to the mating decision: mating to produce the next generation of the selected nucleus and mating for dissemination to commercial farms/industry.

For the selected nucleus, the use of information from relatives, combined with the MME methodology, results in co-selection of relatives. This application can increase the inbreeding levels for the next generation, especially if one performs truncation selection. Several strategies have been developed to promote matings that produce maximum genetic gain while restricting/controlling inbreeding, including optimum contribution (OC) and mate selection (Yoshida et al. 2017). In the OC method, the intensity of use of the selected animals is optimized considering the progeny's expected genetic merit and the co-ancestry between the parents. For the MS methodology, the selection and mating decisions are made simultaneously so that the inbreeding coefficient of future progeny can define the selection and mating of animals (Yoshida et al. 2016). According to Yoshida et al. (2017, 2020), the OC and MS methods maximize the genetic gains in Nile tilapia and Coho Salmon under a controlled rate of inbreeding. The authors suggest that these methods can be an alternative to selection based exclusively on genetic merit.

2.3. Dissemination of genetic gains

The efficient distribution of genetic material from the selected nucleus to commercial producers is critical to maximize the benefits of breeding programs. According to Gjedrem and Baranski (2009), genetic dissemination occurs by direct transfer of the nucleus or through multipliers. Direct transfer is difficult for species such as tilapia because the amount of fingerlings they produce is not sufficient to meet the demands of producers. Thus, the use of multipliers is necessary to disseminate in volume and quality the demands of the industry (Gjedrem and Baranski 2009). An additional advantage of multipliers is the development of genetic lines to meet specific requirements. Thus, one or more trait(s) can be used to select animals and meet market demands. Additionally, due to the use of small numbers of breeders, this method can produce more standardized offspring.

Another important point is the use of a field test station. The environmental conditions of the nucleus evaluation station are sometimes different from emergent production systems or important farming commercial conditions. Thus, the data collected from these different situations can allow selection of robust animals or help in the development of specific genetic lines.

3. Selective breeding in biofloc systems

The development of freshwater fish farming has been boosted by the expansion of cage-based tilapia production in various existing reservoirs as well as production of it and other species in cages and ponds. Based on these systems, this activity might become untenable due to the spent water volume, the environmental impact caused by the generated effluents, and other environmental issues. The increase in cage production has been limited by conflicts—mainly environmental—related to its legalization. The development of fish farming in ponds requires the need to occupy large areas. This fact leads to great investments in land and dependence on large volumes of water per kilogram of fish produced. Given these constraints, it is necessary to improve aquaculture production systems to use both land and water more efficiently.

Since the mid-1990s, the Veterinary School of the Federal University of Minas Gerais (EV-UFMG), Brazil, and other research institutions around the world have made efforts to study and develop water-reuse fish production systems, such as recirculating aquaculture systems (RAS). The results from this research should allow the production systems to achieve the growing needs for economic and environmental sustainability. However, because they are intensive production systems, with a diverse demand for equipment to maintain proper rearing conditions and electricity for operation, the risks due to potential malfunctions are great and require a short-term response. Thus, efforts have been directed to less complex options regarding structures, equipment, and management. Compared to RAS, biofloc technology (BFT) is advantageous because it has a minimum water expenditure, uses lower protein diets and in a smaller quantity, and requires a much smaller investment in equipment to maintain water quality. Furthermore, BFT is considered to be an environmentally sustainable system (see Chapter 14 of the present book for further details on biofloc technology).

In addition to these favorable aspects, there is evidence that fish raised in BFT have better reproductive performance, greater resistance to infections, and less susceptibility to stress (Ahmad et al. 2017, Avnimelech 2007, 2009). These advantageous aspects justify the efforts to study and improve it, especially for tilapia production, of which there is substantially less information when compared to shrimp production in BFT. NGTAqua, the nucleus of nutrition, genetics, and technology studies for tilapiculture, which is part of the Aquaculture Laboratory from the Federal University of Minas Gerais (LAQUA-UFMG), has been making great efforts to generate information in order to support the sustainable development of this production chain. These endeavors are particularly important given the expectation that BFT can reduce production costs and minimize environmental impacts for tilapia production.

In 2009, a Nile tilapia genetic improvement program was started by NGTAqua. The base genetic material came from two commercial sources that produced fingerlings from the Chitralada line, one from Minas Gerais and the other from Paraná (both Brazilian states). Since the beginning at our facilities, the study and selection of this genetic material has been conducted by the formation of families, with the identification of individuals by microchips. The first research was conducted exclusively in a RAS. Indeed, NGTAqua has more than 100 production units, with volumes from 35 to 4,000 L, which utilize different recirculating water systems for the various studies (reproduction, nutrition, breeding, etc.). These initial studies revealed that the evaluations should examine factors other than body weight, namely, meat yields and measures of fish body conformation. In 2012, a new investment in structures gave NGTAqua the ability to expand its studies on production systems. In the dam at the experimental farm of EV-UFMG, a set of cages was installed for the development of studies in this type of system. At LAQUA-UFMG, four agricultural greenhouses were established in order to fully exploit the possibilities of developing aquaculture activities in BFT. These greenhouses contain 54 units with a 4.000 L volume, 24 units with a 1.000 L volume, and 20 units with 150 L volume to support these studies.

One of the main objectives of our initial proposal was to try to answer whether a selection program would be needed for each type of production system, or if only one improved lineage would be sufficient to achieve the needs and particularities of each system. In the NGTAqua program, the first three generations were evaluated in the three production systems. A total of 308 families were produced in these three generations, and over 9.000 tilapia were evaluated. In each generation, fish from all families were distributed in the three production systems and evaluated for body weight and carcass weight and yield. In each generation, the individuals were identified and weighed at approximately 56 days of age (counted from the yolk sac absorption) and distributed in several tanks of the three production systems. They were again weighed at various ages, from around 150 to 350 days. Additionally, fin samples were collected from most of the fish (over 17.000 are stored for the six generations already evaluated) for future work involving the use of genetic markers.

Tilapia have been challenged to a high stocking density, particularly in the BFT system. No less than 40 individuals per cubic meter have been allocated in each tank. Thus, the tilapia have been raised under a more stressful condition than the most widely practiced commercial stocking densities. By ensuring that animals in the breeding program are selected under harsh conditions, it has been possible to produce a population better adapted to several growth conditions in the biofloc system itself.

Our results suggest that it is possible to select tilapias and have consistent genetic gains for body weight traits from these production systems. Furthermore, up to the age of 168 days (24 weeks), there is a high genetic correlation (> 0.8; i.e. there is little genotype-environment interaction) between body weights in the three production systems (Turra et al. 2016). Thus, no specific breeding program is required to achieve an optimal body weight at this age for each evaluated production system. However, for older ages, the results are apparently different (Fernandes et al. 2019). As shown in Table 1, the genetic correlation for the body weight trait at 280 days of age (40 weeks) in the three production systems is moderate (BFT x Cage and RAS x Cage) to high (BFT x RAS). These results suggest that high performance and genetically superior tilapias in RAS and BFT will not necessarily be in cages at older ages. In this situation, there are two options: (1) a breeding program is installed for each production system or (2) a selection index is used that combines the genetic merits for body weight in different production systems. NGTAqua chose the second option

Table 1. Genetic correlations between body weight traits evaluated at 280 days of age: body weight in biofloc technology (wBFT), recirculating aquaculture system (wRAS), and in cages (wCage) Adapted from Fernandes et al. (2019)

	wBFT	**wRAS**
wRAS	0.70	
wCAGE	0.50	0.50

for its breeding program. However, we focused on a strain that could be well adapted to the most intensive growth conditions in sustainable growth environments such as BFT and RAS. Thus, in the following three generations, families were evaluated only with more emphasis on these two systems, i.e. greater number of animals under evaluation and consequent selection intensity in BFT.

Similar to the first three generations, approximately 100 families were produced, and 3,000 candidate tilapia were evaluated in each of the subsequent three generations. The sixth generation of selected fish was produced and evaluated in 2019. Analysis of the genetic values throughout the selected generations (Figure 2) showed that the response to selection for body weight at 168 days in the BFT system was significant (30.32 ± 0.11 g/generation). Additionally, genetic improvement can significantly contribute to improve tilapia growth in this particular production system.

In breeding program planning, we established that selection would be based on the combination of the overall ranking (which receives the greatest emphasis) and within family selection so that at least 50 families should have been represented. This design was implemented to ensure that inbreeding was controlled. This strategy produced a mean coefficient of inbreeding of less than 3% in the population (Figure 3), a rate that will ensure a multiyear outlook with significant genetic gains per generation.

The results of this breeding program are important for academia, tilapia fingerling production companies, and the entire tilapia-related productive sector. The information can aid private or public companies in the design of other genetic improvement programs for tilapia and should ensure elevated productive and

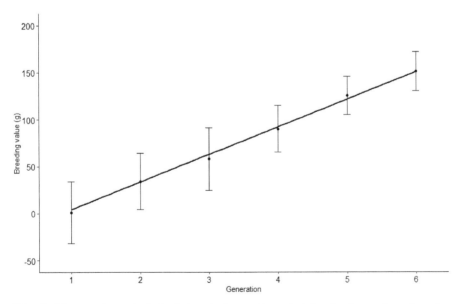

Fig. 2. Mean and standard deviation (points and vertical bars) of genetic values for body weight at 168 days of age (g) in the biofloc technology (BFT) and genetic trend (straight) over the first six generations of selection. The genetic tendency was statistically significant (30.32 ± 0.11 g/generation).

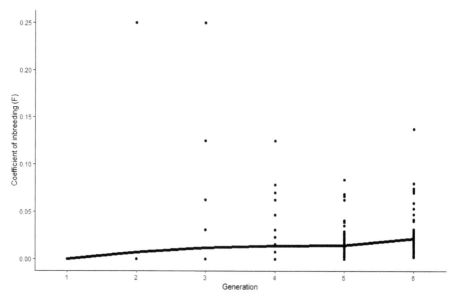

Fig. 3. Coefficient of inbreeding (%) over six generations of body weight selection at 168 days of age using biofloc technology(BFT) by the NGTAqua Nile tilapia breeding program.

economic efficiency of the tilapiculture chain. The zootechnical efficiency generated by the breeding program results in heavier animals that are more productive per unit of volume and with better feed conversion. These benefits contribute to reduce the environmental impact of the activity and allow companies involved in the production chain to have economic and environmental sustainability. The contribution to BFT, through the selection of improved animals for this system, should allow the expansion of its use by companies in the sector. Brazilian and worldwide tilapia production can be increased by utilizing a system that aligns environmental and economic sustainability.

4. Molecular genetics applied to Tilapia selective breeding programs

An important number of genetic improvement programs have been and continue to be implemented for tilapia across the world. They aim to enhance the growth rate and other economically important traits by means of selective breeding (Gjedrem and Rye 2018). For instance, one of the most widespread tilapia strains comes from the pioneering initiative Genetically Improved Farmed Tilapia (Lim and Webster 2006). This strain is currently being farmed on different continents, including Africa, Asia, and South America. The results from selective breeding in this species are encouraging, with a reported response to selection of up to a 15% per generation for faster growth after six generations of artificial selection (Ponzoni et al. 2011). Currently, most of the genetic improvement programs for Nile tilapia rely primarily on conventional pedigree-based selective breeding approaches. These tactics are well suited for traits that are directly measured on the selection candidates but have some limitations for traits that are evaluated via sib-testing. Given the advantages of tilapia aquaculture (e.g. rapid growth and good adaptability) and the existence of genetic improvement programs to support the production, there is an increasing interest in the application of genomic approaches to better exploit breeding methods, including mapping genomic variants that underlie genetic variation for desired traits and speeding up the selection response through the use of genomic selection.

4.1. Potential applications of genomics

Recent advances in genomic technologies, including high throughput-genotyping methods and next-generation sequencing, have allowed researchers to identify genetic variants involved in phenotypic variation for several traits in terrestrial animals (Georges et al. 2018) and aquaculture species (Lhorente et al. 2019). Molecular markers, e.g. single nucleotide polymorphisms (SNPs), can be used for several applications in livestock and aquaculture species, including strain and hybrid traceability, genetic variability and diversity assessment, parentage analyses, pedigree reconstruction, quantitative trait loci (QTL) and genome-wide association (GWA) analyses, marker assisted selection (MAS), and genomic predictions (Liu and Cordes 2004, Yáñez et al. 2014).

Data from a small number of molecular markers associated with QTL (i.e. genomic regions that span genes or genetic variants with an important effect on the trait of interest) might be employed in breeding programs through MAS, if they explain a high percentage of genetic variance for the trait. Additionally, information from thousands of SNP markers might be simultaneously implemented into routine genetic evaluations to estimate genomic-based breeding values (GEBVs), a method termed genomic selection (Meuwissen et al. 2001). These genomic approaches are particularly advantageous for the genetic improvement of traits that can be difficult or impossible to record directly on selection candidates (e.g. carcass quality and disease resistance traits; Sonesson and Meuwissen 2009, Villanueva et al. 2011, Taylor 2014).

Implementing MAS or GS requires previous knowledge on the genomic regions associated with the trait of interest. Therefore, studying the association between phenotypes and genotypes—and thus disentangling the genetic architecture of the trait through QTL mapping or genome-wide association studies (GWAS)—is typically the first step to incorporate genomic information to enhance selective breeding methodologies in livestock and aquaculture species. Increased knowledge about genes and causative variants that affect economically important traits in different aquaculture species is expected in the near future, expedited by the rapid adoption of genomic technologies and cumulative availability of genomic tools for farmed fish (Yáñez et al. 2015).

4.2. Genomic resources for Nile tilapia

The identification of the molecular basis of complex traits, at a whole-genome level, and the application of genomic selection (GS) for aquaculture species requires an extensive number of highly informative SNP markers that preferably segregate in multiple farmed populations. Therefore, the identification and validation of

SNP markers, densely distributed across the whole genome, will facilitate a better understanding of economically relevant traits in tilapia, with a high potential of accelerating the rate of genetic progress in this species. From a practical perspective, the utilization of dense SNP panels to assist breeding programs for tilapia, by means of GS, should have a marked impact in accelerating genetic gains for carcass quality traits (e.g. fillet yield, fat content and composition, and flesh quality, among others) and resistance against the main infectious and parasitic diseases that affect tilapia farming systems (e.g. *Streptococcus* spp., *Francisella* spp., viral nervous necrosis, and tilapia lake virus, among others). Indeed, given their nature, all these traits cannot be directly measured via selection candidates (Lhorente et al. 2019, Ødegård et al. 2014).

The discovery and validation of genome-wide SNP panels in tilapia have been greatly facilitated by the availability of a chromosome level-assembly for the reference genome of this species (Conte et al. 2017, 2019). For instance, these resources have allowed the development of two high-throughput SNP genotyping platforms for Nile tilapia. The first panel comprises a 58K Affymetrix (ThermoFisher) SNP array developed by means of whole-genome resequencing of 32 fish from a farmed population of the Genomar strain to discover the genetic variants included in the platform (Joshi et al. 2018). The second SNP panel was generated after a large-scale *de novo* variant discovery effort performed with whole-genome resequencing of more than 300 animals from three different commercial strains—two from Costa Rica and one from Brazil—information that was exploited for the discovery and validation of a 50K SNP panel posteriorly implemented in an Illumina bead chip for routine genotyping (Yáñez et al. 2020). These genomic tools, in the context of GWAS and GS, are being used to perform an increasing number of studies that aim to generate better knowledge on the species biology, identify the genetic basis of economically relevant traits, and evaluate genomic prediction approaches to hasten genetic progress in farmed Nile tilapia.

4.3. Application of genomics to Nile tilapia biology and populations

The 58K Affymetrix SNP array for Nile tilapia has been used to construct a high density linkage map for the species. The study also included the characterization of the differential patterns of recombination between sexes along the species genome (Joshi et al. 2018). This linkage map has facilitated the generation of an improved version of the Nile tilapia genome assembly (*O. niloticus* UMD NMBU; Conte et al. 2019). Furthermore, genome-wide linkage disequilibrium and genetic differentiation were recently assessed in farmed Nile tilapia from Latin America. The results indicated strong admixture and structure patterns among three breeding populations, with effective population sizes between 78 and 159 (Yoshida et al. 2019c). These results are a good example of how genomic information can be used to improve knowledge on tilapia biology and the demographic processed from current farmed stocks.

There is sexual dimorphism in tilapia: males exhibit a faster growth rate when compared to females. The generation of mono-sex populations (all-males in this case) is crucial for the profitability of tilapia aquaculture. Thus, sex determination in tilapia has been broadly studied to deepen knowledge of the mechanisms involved in the genetic differentiation of sex as a way to generate novel tools for assisting non-chemical sex reversal procedures.

Different factors, including genetic background and environmental variables (e.g. temperature), are reportedly involved in sex determination in tilapia (Baroiller et al. 2001, Cnaani et al. 2008, Palaiokostas et al. 2013, Eshel et al. 2014, Wessels et al. 2014). Several genomic regions involved with sex determination have been identified in this species, including regions in chromosomes 1, 3, 20, and 23 (Lee et al. 2003, 2005, Shirak et al. 2006, Eshel et al. 2010, 2012, Cnaani 2013, Palaiokostas et al. 2015, Baroiller and D'Cotta 2019, Cáceres et al. 2019). Additionally, several genes have been proposed to play a role in sex determination in the tilapia genus, including Wilms tumor suppressor protein 1b (wt1b), cytochrome P450 of family 19 subfamily A member 1 (cyp19a), anti-Müllerian hormone (Amh), and Double sex and Mab-3 related transcription factor 2 (Dmrt2; Shirak et al. 2006, Lee et al. 2007, Sun et al. 2014, Li et al. 2015, Cáceres et al. 2019).

4.4. GWAS and GS

High-throughput SNP genotyping platforms were recently used to identify QTL through GWAS and to assess genomic selection. For instance, a GWAS performed using a 50K SNP panel identified that harvest weight and fillet weight yield are traits primarily controlled by several loci that each exert a small effect. These data indicate a strong polygenic component for these two economically important traits in a commercial Nile tilapia population (Yoshida et al. 2019b). Furthermore, the implementation of genomic predictions using both real dense SNP data and imputed genotypes from low-density to higher-density SNP panels outperformed conventional pedigree-based genetic evaluations for growth and fillet yield (Yoshida et al. 2019b).

Together, these results indicate that genomic selection will quicken the genetic progress for these traits. The use of low-density panels combined with genotype imputation can decrease the costs for the practical implementation of genomic selection in Nile tilapia breeding schemes. SNP genotyping platforms have also been used for the genomic dissection of additive and non-additive genetic effects for body weight, fillet weight and conformational traits in Nile tilapia. Besides showing a lack of dominance, the data revealed the presence of detrimental effects of inbreeding over these commercially relevant traits (Joshi et al. 2019). Furthermore, univariate and multivariate genomic prediction approaches have been also tested using the same SNP array. The findings showed an increase in prediction accuracies when compared to conventional pedigree-based genetic evaluation methods for body weight, fillet yield, and fillet weight (Joshi et al. 2019). These results indicate the feasibility of multi-trait genomic evaluations for genetic improvement of these populations, which is typically used for the simultaneous improvement of all the traits included in the breeding objective.

In general, genomic selection has increased accuracy of selection for growth and fillet traits, when compared against ordinary pedigree-based genetic evaluations, in different Nile tilapia breeding populations. This phenomenon will lead to increased response to selection for these traits. However, neither GWAS nor GS methodologies have been tested for disease resistance traits in this species. Based on the experience obtained from other aquaculture species, in which genomic technologies have been rapidly adopted to enhance breeding programs, it is expected that in the near future these methods will be applied to resistance against a wide variety of pathogens. This undertaking should be similar to salmonid species, in which viral (Houston et al. 2012, Moen et al. 2015, Rodriguez et al. 2019, Yoshida et al. 2019a), bacterial (Correa et al. 2015, Vallejo et al. 2016, 2017a,b, 2018, Bangera et al. 2017, Barría et al. 2018, Yoshida et al. 2019a), and parasitic (Tsai et al. 2016, Correa et al. 2017a, b, Robledo et al. 2019) resistance traits have been disentangled and evaluated using GWAS and GS approaches.

There are studies that aimed to determine the genetic parameters for disease resistance traits in Nile tilapia. For instance, significant genetic variation has been detected for resistance against different bacterial infections in Nile tilapia, including *Flavobacterium columnare* (Wonmongkol et al. 2018), *Streptococcus iniae* (LaFrentz et al. 2016, Shoemaker et al. 2017), and *Streptococcus agalactiae* (Shoemaker et al. 2017, Suebsong et al. 2019). These results indicate the potential for improved survival by means of artificial selection in the studied tilapia populations. Further GWAS studies will allow researchers to determine the distribution of marker effects that explain these traits along the genome. Based on this information, animal breeders will be able to make the decision of implementing either MAS or GS programs, if the traits are explained by few loci of big effect or several loci of small effect each, respectively.

The multiple regions identified to play a role in sex determination in tilapia indicate that this trait is complex and can also be modulated from environmental stimulus. Therefore, we expect that epigenetic studies, using whole-genome approaches, will be crucial to better understand the causative variants involved in sex determination and differentiation across a variety of environmental conditions in this species. This knowledge will help to establish novel strategies for managing the sex ratio in tilapia aquaculture populations.

4.5. Genomic signatures of selection and domestication

The identification of genomic regions subjected to artificial selection and adaptation to captivity conditions during the domestication process can also be addressed by means of genomic information in aquaculture populations (López et al. 2015). For instance, more than a hundred putative selection signatures were identified in four farmed tilapia strains by using 1.4 million SNPs derived from resequencing 47 individual genomes. The gonadotropin-releasing hormone receptor, Wnt, and integrin signaling pathways, generally related with reproduction, development and growth traits, were under positive selection in all the analyzed tilapia lines (Xia et al. 2015). A recent study using a similar number of whole-genome resequencing derived SNPs and a higher sample size (>300 fish) identified several genes related to growth, development, cognition, and behavior that are under selection in three farmed Nile tilapia strains (Cádiz et al. 2020). These studies provide relevant information about the putative genes and pathways involved in artificial selection and early domestication in farmed tilapia.

5. Closing comments

Selective breeding in tilapia began in the 1990s with the GIFT program, and programs are currently distributed worldwide. Despite the high number and wide distribution of the programs, a large portion of farmed tilapia still come from unimproved populations, a factor that delays further development of the activity. This chapter, in addition to providing scientific and didactic important information to the academic field, may be useful to disseminate tilapia selective breeding to the productive sector by presenting genetic assessment methods in different environments and advancing genetic prediction models using the molecular tools currently available. Thus, aquaculture companies may be attracted to invest in breeding programs that will provide farmers with animals of high genetic potential and make tilapiculture an increasingly profitable and environmentally sustainable activity.

Acknowledgments

The authors thank Coordenação de Aperfeiçoamento de Pessoal de Nível Superior – Brasil (CAPES - Finance Code 001) and Conselho Nacional de Desenvolvimento Científico e Tecnológico - CNPq for their financial support.

References cited

Ahmad, I., A.M.B. Rani, A.K. Verma and M. Maqsood. 2017. Biofloc technology: An emerging avenue in aquatic animal healthcare and nutrition. Aquac. Int. 25: 1215–1226.

Avnimelech, Y. 2007. Feeding with microbial flocs by tilapia in minimal discharge bio-flocs technology ponds. Aquaculture 26: 140–147.

Avnimelech, Y. 2009. Biofloc Technology – A Practical Guide Book. Baton Rouge, Louisiana: The World Aquaculture Society.

Bangera, R., K. Correa, J.P. Lhorente, R. Figueroa and J.M. Yáñez. 2017. Genomic predictions can accelerate selection for resistance against *Piscirickettsia salmonis* in Atlantic salmon (*Salmo salar*). BMC Genomics 18: 2–12.

Baroiller, J. and H. D'Cotta. 2001. Environment and sex determination in farmed fish. Comparative. Comp. Biochem. Physiol., C: Comp. Pharmacol. 130: 399–409.

Baroiller, J.F. and H. D'Cotta. 2019. Sex Control in Aquaculture, vol. I, pp. 191–247, ed. 1, edited by Wang, H.-P., Piferrer, F., Chen, S.-L. and Shen, Z.-G. John Wiley & Sons Ltd, Hoboken, NJ.

Barría, A., K.A. Christensen, G.M. Yoshida, K. Correa, A. Jedlicki, J.P. Lhorente et al. 2018. Genomic predictions and genome-wide association study of resistance against *Piscirickettsia salmonis* in coho salmon (*Oncorhynchus kisutch*) using ddRAD sequencing. G3 – Genes Genom. Genet. 8: 1183–1194.

Cáceres, G., M.E. López, M.I. Cádiz, G.M. Yoshida, A. Jedlicki, R. Palma-Véjares, et al. 2019. Fine mapping using whole-genome sequencing confirms anti-Müllerian hormone as a major gene for sex determination in farmed Nile tilapia (*Oreochromis niloticus* L.) G3 – Genes Genom. Genet. 9: 3213–3223.

Cádiz, M.I., M.E. López, D. Díaz-Domínguez, G. Cáceres, G.M. Yoshida, D. Gomez-Uchida, et al. 2020, Whole genome re-sequencing reveals recent signatures of selection in three strains of farmed Nile tilapia (*Oreochromis niloticus*). Scientific Reports 10: 1-14.

Cnnani, A., B. Lee, N. Zilberman, C. Ozouf-Costaz, G. Hulata, M. Ron, et al. 2008. Genetics of sex determination in tilapine species. Sex Dev. 2: 43–54.

Cnaani, A. 2013. The tilapias' chromosomes influencing sex determination. Cytogenet Genome Res. 141: 195–205.

Conte, M., W. Gammerdinge, K. Bartie, D. Penman and T. Kocher. 2017. A high-quality assembly of the Nile Tilapia (*Oreochromis niloticus*) genome reveals the structure of two sex determination regions. BMC Genomics 18: 341.

Conte, M.A., R. Joshi, E.C. Moore, P. Nandamuri, W.J. Gammerdinger, R.B. Roberts, et al. 2019. Chromosome-scale assemblies reveal the structural evolution of African cichlid genomes. *Giga Science* 8: 1–20.

Correa, K., J.P. Lhorente, M.E. López, L. Bassini, S. Naswa, N. Deeb, et al. 2015. Genome-wide association analysis reveals loci associated with resistance against *Piscirickettsia salmonis* in two Atlantic salmon (*Salmo salar* L.) chromosomes. BMC Genomics 16: 854.

Correa, K., R. Bangera, R. Figueroa, J.P. Lhorente and J.M. Yáñez. 2017a. The use of genomic information increases the accuracy of breeding value predictions for sea louse (*Caligus rogercresseyi*) resistance in Atlantic salmon (*Salmo salar*). Genet. Sel. Evo. 49: 1–5.

Correa, K., J.P. Lhorente, L. Bassini, M.E. López, A. Di Genova, A. Maass, et al. 2017b. Genome wide association study for resistance to *Caligus rogercresseyi* in Atlantic salmon (*Salmo salar* L.) using a 50K SNP genotyping array. Aquaculture 472: 61–65.

Eshel, O., A. Shirak, J. Weller, T. Slossman, G. Hulata, A. Cnaani, et al. 2010. Fine-mapping of a locus on linkage group 23 for sex determination in Nile Tilapia (*Oreochromis niloticus*). Anim. Genet. 42: 222–224.

Eshel, O., A. Shirak, J. Weller, G. Hulata and M. Ron. 2012. Linkage and physical mapping of sex region on LG23 of Nile Tilapia (*Oreochromis niloticus*). G3 – Genes Genim. Genet. 2: 35–42.

Eshel, O., A. Shirak, L. Dor, M. Band, T. Zak, M. Markovich-Gordon, et al. 2014. Identification of male-specific Amh duplication, sexually differentially expressed genes and microRNAs at early embryonic development of Nile tilapia (*Oreochromis niloticus*). BMC Genomics 15: 774.

Fernandes, A.F.A., É.R. Alvarenga, G.F.O. Alves, L.G. Manduca, F.L.B. Toral, B.D. Valente, et al. 2019. Genotype by environment interaction across time for Nile tilapia, from juvenile to finishing stages, reared in different production systems. Aquaculture 513: 734429.

Garcia, A.L.S., C.A.L. de Oliveira, H.M. Karim, C. Sary, H. Todesco and R.P. Ribeiro. 2017. Genetic parameters for growth performance, fillet traits, and fat percentage of male Nile tilapia (*Oreochromis niloticus*). J. Appl. Genet. 58: 527–533.

Georges, M., C. Charlier and B. Hayes. 2018. Harnessing genomic information for livestock improvement. Nat. Rev. Genet. 20: 135–156.

Gjedrem, T. and M. Baranski. 2009. Selective Breeding in Aquaculture: An Introduction. Series: Reviews: Methods and Technologies in Fish Biology and Fisheries. V.10. [eds.]. Springer, London.

Gjedrem, T. and Rye, M. 2018. Selection response in fish and shellfish: A review. Rev. Aquacult. 10: 168–179.

Henderson, C.R. 1984. Applications of Linear Models in Animal Breeding. pp. 384. University of Guelph, Guelph.

Houston, R.D., J.W. Davey, S.C. Bishop, N.R. Lowe, J.C. Mota-Velasco, A. Hamilton, et al. 2012. Characterization of QTL-linked and genome-wide restriction site-associated DNA (RAD) markers in farmed Atlantic salmon. BMC Genomics 13: 244.

Joshi, R., M. Árnyasi, S. Lien, H.M. Gjøen, A.T. Alvarez and M. Kent. 2018. Development and validation of 58K SNP-array and high-density linkage map in Nile Tilapia (*O. niloticus*). Front. Genet. 9: 472.

Joshi, R., A. Skaarud, M. Vera, A.T. De Alvarez and J. Ødegård. 2019. Genomic prediction for commercial traits using univariate and multivariate approaches in Nile Tilapia (*Oreochromis niloticus*). Aquaculture 734641.

LaFrentz, B.R., C.A. Lozano, C.A. Shoemaker, J.C. García, D.H. Xu, M. Løvoll, et al. 2016. Controlled challenge experiment demonstrates substantial additive genetic variation in resistance of Nile tilapia (*Oreochromis niloticus*) to *Streptococcus iniae*. Aquaculture 458: 134–139.

Lee, B., D. Penman and T. Kocher. 2003. Identification of a sex-determining region in Nile tilapia (*Oreochromis niloticus*) using bulked segregant analysis. Anim. Genet. 34: 379–383.

Lee, B., W. Lee, J. Streelman, K. Carleton, A.E. Howe, G. Hulata, et al. 2005. A second-generation genetic linkage map of tilapia (*Oreochromis* spp.). Genetics 170: 237–244.

Lee, B. and T. Kocher. 2007. Exclusion of Wilms tumors (WT1b) and ovarian cytochrome P450 aromatase (CYP19A1) as candidates for sex determination genes in Nile tilapia (*Oreochromis niloticus*). Anim. Genet. 38: 85–86.

Lhorente, J.P., M. Araneda, R. Neira and J.M. Yáñez. 2019. Advances in genetic improvement for salmon and trout aquaculture: The Chilean situation and prospects. Rev. Aquacult. 11: 340–353.

Li, M., Y. Sun, J. Zhao, H. Shi, S. Zeng, K. Ye, et al. 2015. A tandem duplicate of anti-Müllerian hormone with a missense SNP on the Y chromosome is essential for male sex determination in Nile Tilapia, *Oreochromis niloticus*. PLOS Genetics 11(11): e1005678 https://doi.org/10.1371/journal.pgen.

Lim, C. and Webster, C.D. 2006. Tilapia: Biology, Culture, and Nutrition. Lim, C. and Webster, C. (Eds.). Harworth Press: Binghamton, NY, USA; p. 678.

Lind, C.E., C.E. Ponzoni, N.H. Nguyen and H.L. Khaw. 2012. Selective breeding in fish and conservation of genetic resources for aquaculture. Reprod. Domest. Anim. 47: 255–263.

Liu, Z. and J. Cordes. 2004. DNA marker technologies and their applications in aquaculture genetics. Aquaculture 238: 1–37.

López, M.E., R. Neira, J.M. Yáñez, W.S. Davidson and S. Fraser. 2015. Applications in the search for genomic selection signatures in fish. Front. Genet. 5: 1–12.

Masuda, Y. 2018. Introduction to BLUPF90 suite programs. Inc., University of Georgia. http://nce.ads.uga.edu/wiki/doku.php?id=documentation.

Meuwissen, T.H.E., B.J. Hayes and M.E. Goddard. 2001. Prediction of total genetic value using genome-wide dense marker maps. Genetics 157: 1819–1829.

Moen, T., J. Torgersen, N. Santi et al. 2015. A caderina epitelial determina a resistência ao vírus da necrose pancreática infecciosa no salmão do Atlântico. Genetics 200: 1313–1326.

Mrode, R.A. and R. Thompson. 2005. Linear models for the prediction of animal breeding values [eds.]. CABI, Boston, MA.

Ødegard, J., T. Moen, N. Santi, S.A. Korsvoll, S. Kjøglum, et al. 2014. Genomic prediction in an admixed population of Atlantic salmon (*Salmo salar*). Front. Genet. 5: 402.

Oliveira, C.A.L., R.P. Ribeiro, G.M. Yoshida, N.M. Kunita, G.R. Rizzato, S.N. de Oliveira, et al. 2016. Correlated changes in body shape after five generations of selection to improve growth rate in a breeding program for Nile tilapia *Oreochromis niloticus* in Brazil. J. Appl. Genet. 57: 487–493.

Palaiokostas, C., M. Bekaert, M. Khan, J. Taggart, K. Gharbi, B. Mcandrew et al. 2013. Mapping and validation of the major sex-determining region in Nile Tilapia (*Oreochromis niloticus* L.) Using Rad Sequencing. PLoS ONE 8: e68389.

Palaiokostas, C., M. Bekaert, M. Khan, J. Taggart, K. Gharbi, B. Mcandrew et al. 2015. A novel sex-determining QTL in Nile tilapia (*Oreochromis niloticus*). BMC Genomics 16: 171.

Ponzoni, R.W., H.L. Khaw, N.H. Nguyen and A. Hamzah. 2010. Inbreeding and effective population size in the Malaysian nucleus of the GIFT strain of Nile tilapia (*Oreochromis niloticus*). Aquaculture 302: 42–48.

Ponzoni, R.W., N.H. Nguyen, H.L. Khaw, A. Hamzah, K.R.A. Bakar and H.Y. Yee. 2011. Genetic improvement of Nile tilapia (*Oreochromis niloticus*) with special reference to the work conducted by the WorldFish Center with the GIFT strain. Rev. Aquacult. 3: 27–41.

Robledo, D., C. Palaiokostas, L. Bargelloni, P. Martínez and R.D. Houston, 2017. Applications of genotyping by sequencing in aquaculture breeding and genetics. Rev. Aquacult. 10(3): 670–682.

Rodríguez, F.H., R. Flores-mara, G.M. Yoshida and A. Barría. 2019. Genome-wide association analysis for resistance to infectious pancreatic necrosis virus identifies candidate genes involved in viral replication and immune response in rainbow trout (*Oncorhynchus mykiss*). G3 – Genes Genom. Genet. 9: 2897–2904.

Shirak, A., E. Seroussi, A. Cnaani, A. Howe, A.E. Domokhovsky, R.N. Zilberman, et al. 2006. Amh and Dmrta2 Genes Map to Tilapia (*Oreochromis* spp.) Linkage group 23 within quantitative trait locus regions for sex determination. Genetics 174: 1573–1581.

Sonesson, A.K. and T.H. Meuwissen. 2009. Testing strategies for genomic selection in aquaculture breeding programs. Genet. Sel. Evol. 174: 1573–1581.

Taylor, J.F. 2014. Implementation and accuracy of genomic selection. Aquaculture 420–421: S8–S14.

Thodesen, J., M. Rye, Y.X. Wang, K.S. Yang, H.B. Bentsen and T. Gedrem. 2011. Genetic improvement of tilapias in China: Genetic parameters and selection responses in growth of Nile tilapia (*Oreochromis niloticus*) after six generations of multi-trait selection for growth and fillet yield. Aquaculture 322–323: 51–64.

Tsai, H.Y., A. Hamilton, A.E. Tinch, D.R. Guy, J.E. Bron, J.B. Taggart, et al. 2016. Genomic prediction of host resistance to sea lice in farmed Atlantic salmon populations. Genet. Sel. Evol. 48: 47.

Turra, E.M., F.L.B. Toral, E.R. Alvarenga, F.S.S. Raidan, A.F.A. Fernandes, G.F.O. Alves, et al. 2016. Genotype × environment interaction for growth traits of Nile Tilapia in biofloc technology, recirculating water and cage systems. Aquaculture 460: 98–104.

Vallejo, R.L., T.D. Leeds, B.O. Fragomeni, G. Gao, A.G. Hernandez, et al. 2016. Evaluation of genome-enabled selection for bacterial cold water disease using progeny performance data in rainbow trout: Insights on genotyping methods and genomic prediction models. Front. Genet. 7: 96.

Vallejo, R.L., T.D. Leeds, G. Gao, J.E. Parsons, K.E. Martin, J.P. Evenhuis, et al. 2017a. Genomic selection models double the accuracy of predicted breeding values for bacterial cold water disease resistance compared to a traditional pedigree-based model in rainbow trout aquaculture. Genet. Sel. Evol. 49: 1–13.

Vallejo, R.L., S. Liu, G. Gao, B.O. Fragomeni, A.G. Hernandez, T.D. Leeds, et al. 2017b. Similar genetic architecture with shared and unique quantitative trait loci for bacterial cold water disease resistance in two rainbow trout breeding populations. Front. Genet. 8: 156.

Vallejo, R.L., R.M.O. Silva, J.P. Evenhuis, G. Gao, L. Sixin, J.E. Parsons, et al. 2018. Accurate genomic predictions for BCWD resistance in rainbow trout are achieved using low-density SNP panels: Evidence that long-range LD is a major contributing factor. J. Anim. Breed. Genet. 135: 263–274.

Villanueva, B., J. Fernández, L.A. García-Cortés, L. Varona, H.D. Daetwyler and M.A. Toro. 2011. Accuracy of genome-wide evaluation for disease resistance in aquaculture breeding programs. J. Anim. Sci. 89: 3433–3442.

Wessels, S., R. Sharifi, L. Luehmann, S. Rueangri, I. Krause, S. Pach, et al. 2014. Allelic variant in the anti-Müllerian hormone gene leads to autosomal and temperature-dependent sex reversal in a selected tilapia line. PLOS ONE 9: e104795.

Xia, H., J. Bai, Z. Meng, Z. Zhang, Y. Wang, L. Liu, et al. 2015. Signatures of selection in tilapia revealed by whole genome resequencing. Sci. Rep. 5: 14168.

Yáñez, J.M., R.D. Houston and S. Newman. 2014. Genetics and genomics of disease resistance in salmonid species. Front. Genet. 5: 415.

Yáñez, J.M., S. Newman and R.D. Houston. 2015. Genomics in aquaculture to better understand species biology and accelerate genetic progress. Front. Genet. 6: 128.

Yáñez, J.M., G. Yoshida, A. Barria, R. Palma-Véjares, D. Travisany, D. Díaz, et al. 2020. High-throughput single nucleotide polymorphism (SNP) discovery and validation through whole-genome resequencing in Nile tilapia (*Oreochromis niloticus*). Mar. Biotechnol. 22(1): 109-117.

Yoshida, G.M., J.M. Yáñez, S.A. de Queiroz and R. Carvalheiro. 2020. Mate selection provides similar genetic progress and average inbreeding than optimum contribution selection in the long-term. Aquaculture 526: 735376.

Yoshida, G.M., J.M. Yáñez, C.A.L. de Oliveira, R.P. Ribeiro, J.P. Lhorente, S.A. de Queiroz, et al. 2017. Mate selection in aquaculture breeding using differential evolution algorithm. Aquac. Res. 48(11): 5490-5497.

Yoshida, G.M., J.P. Lhorente, R. Carvalheiro and J.M. Yáñez. 2017. Bayesian genome-wide association analysis for body weight in farmed Atlantic salmon (*Salmo salar* L.). Animal Genetics 48: 698–703.

Yoshida, G.M., R. Carvalheiro, J.P. Lhorente, K. Correa, R. Figuero, R.D. Houston et al. 2018. Accuracy of genotype imputation and genomic predictions in a two-generation farmed Atlantic salmon population using high-density and low-density SNP panels. Aquaculture, 491: 147–154.

Yoshida, G.M., R. Carvalheiro, F.H. Rodríguez, J.P. Lhorente and J.M.Yáñez. 2019a. Single-step genomic evaluation improves accuracy of breeding value predictions for resistance to infectious pancreatic necrosis virus in rainbow trout. Genomics 111: 127–132.

Yoshida, G.M., J.P. Lhorente, K. Correa, J. Soto, D. Salas and J.M. Yáñez. 2019b. Genome-wide association study and cost-efficient genomic predictions for growth and fillet yield in Nile Tilapia (*Oreochromis niloticus*). G3 (Bethesda) 8; 9(8): 2597–2607.

Yoshida, G.M., A. Barria, K. Correa, G. Cáceres, A. Jedlicki, M.I. Cadiz, et al. 2019c. Genome-wide patterns of population structure and linkage disequilibrium in farmed Nile tilapia (Oreochromis niloticus). Front. Genet. 10: 745.

3

Structure and Physiology of the Gastrointestinal Tract

Sílvio Teixeira da Costa[1]* and Bernardo Baldisserotto[2]

[1] Department of Morphology, Universidade Federal de Santa Maria (UFSM), Santa Maria, RS, Brazil

[2] Department of Physiology and Pharmacology, Universidade Federal de Santa Maria, Santa Maria, RS, Brazil

1. Introduction

Teleost fish manifest numerous characteristics related to their morphology, physiology and biochemistry, allowing the colonization of different environments, as well as the exploration of different resources or food sources. Food selection by aquatic organisms is based on prey size and abundance, and on the ease with which it can be captured. According to Hunter (1981), prey selection by fish depends basically on size, although movement also plays an important role in the predator's ability to seize it. In many species, the prey selection pattern is also determined by the position, shape and size of the mouth. There is a close relationship between habitat, nutrition and the morphological organization of the digestive system, which manifests itself in adaptations that may have a temporary character, as well as modifications that impel a permanent character, interfering in the phylogenetic evolution of the species (Mokhtar 2017).

Anatomically, the major division of the alimentary canal in vertebrates is the mouth, buccal cavity, pharynx, esophagus, stomach, intestines, rectum and related organs or attachments to the digestive tract. In some fish, the digestive canal forms an almost continuous tube from the mouth to the anus, and may be best compartmentalized in the oral and pharyngeal cavities; some authors still characterize the rest of the tube as: *foregut* – including the esophagus and stomach; *midgut* – such as duodenum, jejunum and ileum; and posterior *intestine* – the large and straight intestine. Here we will treat each portion individually. However, it is perfectly visualized that the food channel forms lugs and is structurally and functionally subdivided into different parts and these compartments are separated by sphincters and valves. To McMillan and Harris (2018), the main histological layers observed in the description of the cavitary compartments of the vertebrate digestive tract are:

*Corresponding authors: silvio.teixeira.da.costa@gmail.com

Tunica mucosa: The internal mucous membrane consists of three layers:

(a) *Epithelium mucosae*, often simple epithelium, resting on a basement membrane;

(b) *Lamina propria mucosae*, the subepithelial layer consisting of a loose reticular tissue and areolar tissue connected with lymphoid nodules; and

(c) *Muscularis mucosae*, only two narrow bands of smooth musculature in circular and longitudinal arrangements.

Tunica submucosa: This layer of connective tissue contains larger blood vessels and lymphatic vessels and presents mucosal-associated lymphoid tissues (MALT). The autonomic plexus of *Meissner*, containing fibers and nerve ganglia associated with the autonomic system, is also observed. The secretory portions of the exocrine glands that contribute to the digestive events in the intestine can also be found in this layer.

Tunica muscularis: Generally, there are two layers of smooth muscle, one arranged in an internal circular arrangement and another one in longitudinal arrangement, with an interfascicular connective tissue between these two muscle layers. The *Auerbach's* myenteric plexus, which is related to the autonomic nervous system, is found immersed in this tissue.

Tunica serosa or tunica adventitia: The peripheral tunic may be a serous or an adventitious tunic. The tunic is considered serosa if it commonly presents a pavement epithelium associated with connective tissue, representing the surface of a visceral peritoneum, also qualified as mesothelium. Blood vessels, lymphatics and nerves transit in this conjunctiva that is also accompanied by adipose cells. The tunica is described as adventitia when it consists of areolar connective tissue verified in anatomical structures like the pharynx and the anterior esophagus.

2. Buccal cavity

The oral cavity of Nile tilapia (*Oreochromis niloticus*) opens anteriorly through a small terminal mouth delimited by the upper and lower lips. In the oral cavity, two regions are recognized: the roof (dorsal region) and the floor (ventral region). The roof of the oral cavity comprises the upper lip, upper band or arch of teeth, upper respiratory valve, palate, and epipharyngeal bones. The floor of the cavity comprises the lower lips, lower band or arch of the teeth, lower respiratory valve, tongue and hypopharyngeal bones (Maina 2000).

The labial surface, associated with the upper and lower arches, is endowed with a serrated calcified structure (like small teeth), showing a pair of similar upper and lower lines, serving to grind and gnaw food in sync with the teeth and the tongue. This set proceeds the selection, capture and conduction of food to the esophagus (Santos et al. 2019). The tongue is generally rigid and slightly mobile, contains no salivary glands, but exhibits taste buds (Elsheikh et al. 2012).

The oral cavity is shared by the respiratory and digestive systems and its digestive function is related to the selection, seizure and conduction of food to the esophagus. The structure of the oral cavity is essentially the same as that of the skin. The epidermis of the skin is represented by the stratified mucosal epithelium and *chorium* over a connective tissue submucosa (Ahsan-ul-Islam 1951). Generally, the epithelial cells lining the oral cavity have a circular cross-sectional profile and can have large intercellular spaces separating the cells. In some parts of its peripheral wall, however, epithelial cells are connected directly through zipper-like interdigitations (Maina 2000). In the margin of two to four epithelial cells, there is a crypt with almost round to irregular morphology, with variable dimensions and discernible at irregular intervals. These crypts present goblet cell openings and are often filled with drops of mucous secretions that normally do not protrude from the epithelial surface.

Three morphologically distinct types of taste buds can be identified. The taste buds type I are dispersed and found mainly in the upper lip, upper and lower tooth bands, and upper and lower breathing valves. Often, the base of a papilla may be deeper than neighboring epithelial cells, and between them is an almost round area where the sensory cell terminals inside the papilla extend to the surface. Terminals are tightly grouped in this area.

The type II taste buds are located particularly in the lower lip in the floor of the oral cavity and were also observed in the dorsal surface of the tongue and palate. However, there is no marked demarcation between the papillae epithelium and the surrounding epidermis. Type III taste buds do not protrude from the plane of neighboring epithelial cells. They are mainly seen in the squamous and cornified epithelium of the masticatory tract of the oral cavity and metabranchial area. This type is recognizable only as epithelial pores (Elsheikh et al. 2012).

3. Gill arches and rakers

Generally, the epithelial cells lining the buccal cavity have a circular profile in transverse section, being able to demonstrate wide intercellular spaces separating cells. The epithelial cells are connected directly through interdigitations similar to a zipper in some parts of their peripheral wall (Maina 2000).

Each gill arch supports two rows of gill filaments and two rows of gill traces, as shown in Figure 1. The branchial arch I differs from the other arches in that its external tracks are longer, forming the border with the operculum. In the inner row of rakers, it has the same structure as arcs II and III and the outer row of raker IV. The IV arch differs in being much shorter, laterally widened, and the elongated internal tracks overlapping each other without, or very few, inter-raker spaces. The internal tracks of the branchial arch IV form the limit with the lower pharyngeal mandible (Nnorthcott and Beveridge 1988).

All surfaces of the branchial arch and the tracks are covered by a squamous epithelium that has concentric surface micro-bridges and apparently without the crypts or openings for goblet cell secretion. In the branchial arches, small excrescences composed of 2-4 taste buds can be observed in smaller specimens. The anterior face of the arch is covered by rows of taste buds that connect the opposite tracks in the arch. As the fish develops, its gill traces become more prominent and with about 10-14 taste buds on each trail.

Gills may also contain a rear section, called a keel and an anterior main face. This anterior main face of the arch is covered by rows of taste buds which, in their adult phase, can reach 40 mm in length, displaying a smooth, continuous keel with the side of the gill arch. The two main edges of each gill raker show a single line of taste buds, where curiously their occurrence diminishes in fish arches larger than 40 mm in length. A few small goblet cells are observed only in association with the keel.

The center of the raker is filled with loose, vascularized connective tissue and contains a single gill element. The epithelium that covers the anterior face and sides of the branchial arch is considerably thinner, 8-12 µm, and contains only small mucous cells and few taste buds (Zayed and Mohamed 2004).

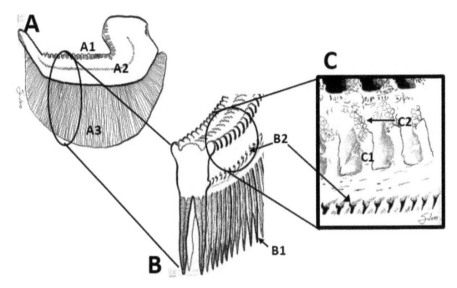

Fig. 1. A – Side view of the gill arch of Nile tilapia, *Oreochromis mossambicus*: A1 – Gill rakers; A2 – Gill arch; A3 – Gill filaments. **B** – Gill arch cross section: B1 – Gill filament; B2 – Microbranchiospines. **C** – Detail of the gill rakers: C1 – Gill rakers; C2 – Bud tastes. Figure by Sílvio Teixeira da Costa.

3.1. Gill rakers

The gill rakers form a net to capture plankton and particles suspended in the water. The Nile tilapia can retain particles as small as bacteria (Beveridge et al. 1989) and cyanobacteria (38 µm to 1.0 mm in diameter) using the gill rakers and mucus trapping. During pumps of the buccopharyngeal cavity in *O. niloticus*, most of the mucus (97%) either stayed in place on the gill arches or slid posteriorly along the arch surfaces (Sanderson et al. 1996). The occurrence of a brief reversal of flow (~80 ms duration) from posterior to anterior in the oropharyngeal cavity prior to every feeding pump (250–500 ms duration) in *O. aureus* increases filtration performance by lifting most of the mucus and enhancing the back-migration of particles from the gill rakers to the bulk flow region, thus reducing particle accumulation that can clog the filter (Smith and Sanderson 2008). The flow of water carrying the plankton and particles is parallel to the gill rakers, directing these particles and mucus toward the esophagus, and only the filtered water passes perpendicular to the gill rakers

(Sanderson et al. 2001). However, *O. esculentus* does not use the gill rakers and mucus entrapment as a mechanism of particle retention (Goodrich et al. 2000).

4. Esophagus

The esophagus is a short, swollen, membranous muscle tube that connects the pharynx to the initial portion of the stomach (Awaad et al. 2014). Its shape is cylindrical and rectilinear along its entire length and is located dorsally to the rostral portion of the liver.

In its histological organization, the esophagus is formed by the four concentric layers previously listed, and considered from the inside out: mucosa, submucosa, double-layered muscle (composed of internal longitudinal and external circular sublayers) and serous. The mucosa presents many longitudinal folds that extend along the esophageal tube along the entire length, giving the esophageal lumen a starry shape, contributing to its expansion capacity. These mucous folds are relatively thick and short and are represented by primary folds. They are lined with a squamous stratified epithelium composed of undifferentiated basal epithelial cells, followed by several layers of mucus-secreting cells superficially covered with 1-2 layers of low to flattened cuboidal cells (Ahsan-ul-Islam 1951).

Mucosal secretory cells are rounded, elongated and recognized in the epithelial lining of the esophagus. These rounded cells, larger in size, are concentrated near the base of the epithelium, close to the secondary folds, and represent most mucus-secreting cells. These cells are smaller in size and present in the superficial part of the epithelium in the primary folds. Both cell types produce a mucopolysaccharide secretion and are recognized by periodic shiff acid (PAS).

The mucus secreted by these cells can be attributed to the lack of salivary glands in fish and the mucin secreted by the esophageal epithelium acts as a compensation. In addition, mucin is important for the formation of a continuous lubricating blade along the entire esophageal wall, emulsifying food particles, favoring ionic absorption and protecting the esophageal mucosa against mechanical damage and microbial invasion (Albrecht et al. 2001). Likewise, mucous secretion participates in the enzymatic digestion of ingested foods and facilitates their transformation into chyme. Two types of mucus-secreting cells are observed in the esophageal mucosa, where two or more carbohydrate types can be produced. This secretory pattern oscillation may be due to different stages of maturation and/or age (Murray et al. 1996).

In the submucosa lamina propria, a dense and thick layer of intensely vascularized collagen fibers is observed below the epithelium. There is no mucosal muscle layer between these two sublayers. This connective aims to fill the nuclei of the mucosa and submucosa folds (Awaad et al. 2014). The muscular layer of the esophagus is composed of a thinner longitudinal inner layer and a thicker circular outer layer, both consisting of striated muscle fibers interconnected with loose connective tissue. The serous tunic is composed of two sublayers: sub-serous lamina, composed of richly vascularized loose connective; and serous lamina composed of the association of a simple epithelium with connective tissue, also called mesothelium (Ahsan-ul-Islam 1951, Awaad et al. 2014, Morrison and Wright 1999).

5. Stomach

Nile tilapia gastric morphology (Figure 2) reveals three anatomically distinct regions, topographically configuring a physical resemblance to the letter Y, where the arms correspond to the initial and terminal regions. Each communication with the main rod of the letter Y represents the blind sac of the middle region, gauging for separate openings. On microscopy, the stomach has four tunics or layers, like the digestive tract of the other vertebrate animals: a mucous tunic, submucosal tunic, muscular tunic and serous tunic.

The mucous tunic is also discriminated in the initial, middle and terminal regions. Each of these has characteristics similar to those of the cardiac, fundic and pyloric regions found in the stomach of monogastric mammals (Caceci et al. 1997, Morrison and Wright 1999). The stomach of this species is innervated by the vagal nerve (parasympathetic innervation), which arose from the medulla oblongata (Ohshimo and Uematsu 1996).

5.1 Initial gastric segment

The initial region, which is shaped by the junction with the esophagus, demonstrates an abrupt transition from the squamous epithelium to goblet cells of the esophagus to a simple columnar lining without goblet cells in the stomach. The mucosal surface in the initial region is smooth and devoid of gastric cavities or crypts. The epithelium of the initial portion is contiguous with that of the esophagus; however, its caudal portion merges with the superficial epithelium of the middle region, without a distinct demarcation line like *Margo plicatus*. This junctional zone is characterized by the presence of skeletal muscle fibers in the muscular tunic with invaginations of the superficial layer, where the mucous secretion epithelium is predominantly noticeable. This secretory epithelium, characterized in PAS stains, is observed as a set of cells contiguous to the columnar cells on the surface of these invaginations. It may take the form of simple glandular acini, invading the underlying lamina propria (Caceci et al. 1997).

On the free surface of the coating, the apex of the cells is characterized by a brush border. The glands in the lamina propria of the initial region are lined with tall cubic epithelium and opened by ducts in the shallow part of the mucosa. The glandular cell cytoplasm is clear and vacuolated. The mucous tunic is separated from the submucosa by a reasonably developed muscular sublayer of the mucosa consisting of longitudinal filaments of smooth muscle fibers.

The muscular tunic consists in part of striated muscle fibers contiguous with the striated fibers of the esophageal muscle tunic. Deep within the skeletal muscle layer, however, are two layers of smooth muscle fibers oriented in circular and longitudinal directions.

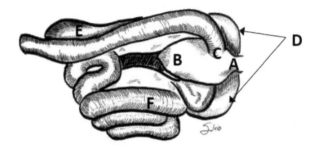

Fig. 2. Right lateral view of the viscera of the digestive system of Nile tilapia, *Oreochromis niloticus*: **A** – Stomach: esophageal portion; **B** – fundic portion; **C** – pyloric portion. **D** – Liver: left lobe (cranial); **E** – caudal lobe. Intestine: **F** – Gastric spiral loop. (Figure by Sílvio Teixeira da Costa).

5.2. Gastric intermediate segment or blind sac

The median region of the stomach or blind sac is larger than the initial and final regions of the gastric compartment. The lining is characterized by a simple cylindrical epithelium; however, in mucus-secreting areas, it becomes slightly more prominent. A denser excretion profile is observed in the first half of the middle region, that is, more adjacent to the junction with the initial region. The number and size of the mucus-secreting area reduces in the contour of the terminal region. The mucosal surface of the middle region shows recesses forming prominent gastric crypts similar to those observed in the fundic region of mammals. Khojasteh and Banan (2012) also noted that the tilapia stomach has secretion mechanisms for hormones such as ghrelin through diffuse neuroendocrine system (DNES) cells.

The glands almost completely occupy the lamina propria in the mucous membrane. They are of simple tubular morphology and open over the overlying epithelium in the first half (near the initial region), but directly into the gastric grooves near the junction with the terminal region. These glands are lined with truncated cells with acidophilic cytoplasm that show no alcianophilic or PAS-positive reaction, but the basal lamina and some cervical mucous cells are PAS-positive. Significant blood capillaries are visualized on the lamina propria of this region.

In the muscular tunic, there is an abrupt transition between striated and smooth mixed muscle fibers visualized in the initial region to a completely smooth muscle layer and organized into inner circular and longitudinal outer sublayers.

5.3. Final gastric segment

The terminal region is shorter and joins directly to the intestine. Softer folds and absence of gastric crypts are observed in the mucous membrane of the cranial part of this region. The lamina propria of the mucosa is more relevant than in the other two stomach regions and shows positive alcianophilic glands and PAS. The lining is typical and continuous to that found on the entire surface of the gastric mucosa, showing some goblet cells in their area of connection to the duodenum. The muscular tunic in the terminal region, as presented in the initial region, contains smooth and striated mixed muscle fibers. The submucosal tunic is unique throughout the length of the stomach and forms the nucleus of the gastric folds, composed of highly vascularized loose connective tissue, lymphatic vessels and nerves (Al-Hussaini 1949, Khojasteh and Banan 2012).

6. Intestine

The intestinal tract of Nile tilapia (Figure 2) follows a complex course that involves several turns and spirals arranged in an as yet undescribed form. From the cranial to the caudal portion, five regions are identified and designated as: hepatic loop, the proximal main spiral loop, the gastric loop, the distal main spiral loop, and the terminal segment (Smith et al. 2000).

The ileorectal valve of Nile tilapia is coarse and can measure up to 7 cm in a 32-35 cm specimen. It has a circular muscle ring, with a thickening of the inner circular layer, forming the rectum. Histologically, the rectum is quite similar to the midgut of other teleosts, with vacuolated rectal epithelial cells (Morrison and Wright 1999).

The mucous tunic usually consists of an epithelium that covers a loose connective tissue layer, also called lamina propria, vascularized and that presents nerves and leukocytes. The lamina propria may have a compact layer of dense connective tissue. This connective stratum with dense collagen layers acts to strengthen and preserve the intestinal wall. A marked presence of granular eosinophilic cells can also be noted in the lamina propria, as found in the intestine of many other species. These cells correspond to mammalian mast cells and have been widely described in mucous membranes of numerous tissues such as intestines, gills and skin. Below the lamina propria, a delicate and discrete mucous muscle sublayer is observed (Sklan et al. 2004, Smith et al. 2000, Khojasteh and Banan 2012).

The three sublayers of the mucous membrane are supported by a typical connective tissue submucosa, which has less cellularity than the lamina propria and contains blood vessels, an extensive network of lymphatic tissue, and nerve plexuses.

The lining of the intestinal mucosa is formed by a simple prismatic epithelium, accompanying the folds called intestinal villi. The villus is relatively narrow and dense in the hepatic loop; however, distal villi become wider and fewer villi are present per unit area. The number of villi in the terminal segment is approximately 50% of that observed near the hepatic loop and many of the villi are branched and with different statures (Sklan et al. 2004).

Histological treatment with special stains to indicate mitotic proliferation demonstrates that structures like Lieberkühn crypts are not observed, and that proliferative activity of enterocytes is concentrated in the intervillous region and the lower third of the villi in all intestinal segments. Scarce proliferative cells are observed in the upper portion of the villi of the entire intestinal tract. The use of stains such as PAS (mucopolysaccharides) and alcian blue (mucoproteins) demonstrate that mucus-secreting cells and mucoproteins, respectively, are located along the entire villus of the proximal small intestine. Thus, an increasing number of mucin-containing cells are visualized in the upper portion of the villi, and in the distal segments of the small intestine. Chalice-like mucous cells, lymphocytes, and enteroendocrine cells are scattered throughout the epithelium, and rodlet cells are also found in some teleost fish species, as well as in Nile tilapia (Khojasteh and Banan 2012, Sklan et al. 2004, Smith et al. 2000).

- *Enterocytes (absorptive columnar cells)*: They are usually tall and narrow, with elongated nuclei with central-basal polarization, mitochondria located in the apical and basal regions, a well-developed brush border and lamellar structures running parallel to the lateral plasma membrane. It should be noted that among unitary complexes, they generally do not verify the presence of interdigitations as seen between mammalian enterocytes. The brush edge contributes more than 90% to the total intestinal surface area and forms the digestive / absorptive interface with a functional microenvironment where enzymes involved in further food breakdown are located and where nutrient absorption and transport are present. A considerable number of enzymes such as alkaline phosphatase, disaccharidases, leucine aminopeptidase and tri- and dipeptidases are located on the brush edge membrane. The presence of these enzymes is demonstrated by enzymatic and immunohistochemical histochemical analysis. It is also demonstrated that the brush border condition, especially the size of the microvilli in the enterocytes, can vary significantly, denoting a close relationship between eating habits and the brush border structure of the enterocytes. The quantitative characteristic of the brush border is certainly dependent on the location of the enterocyte in different parts of the fish gut. In all, there is a fundamental similarity between the ultrastructure of the intestinal epithelium in blue tilapia, *Oreochromis aureus*, and other vertebrates (Frierson and Foltz 2004).
- *Goblet cells*: These cells are important elements in the postgastric mucosal lining and constitute the predominant mucous cell type in the gut of tilapia. Their nucleus has center-basal polarization and their cytoplasm widens and then contracts from an apical pore through which mucus is secreted. In transmission electron microscopy, mucin-containing secretion granules fill most cells, with varying sizes and occupying almost the entire cytoplasm apical center of the cells. The neutral secretions produced are usually sulfates and sialomucins that contain sialic acid. Acid mucous secretions are produced by carboxylated and sulfated muconconjugates. This secretion lubricates the ingested, undigested contents for peristaltic progression to the rectum and has a possible participation in osmoregulatory events (Khojasteh and Banan 2012). An increased number of goblet cells in the rectum may mean the need for greater mucosal protection and lubrication for fecal expulsion. The mucus produced by the goblet cells is known to protect the mucosa of the digestive tract. The sialic acid in the mucus still has an antimicrobial effect because it interferes with virus detection of the receptor and protects the mucosa against bacterial sialidase (Murray et al. 1996).
- *Diffuse Neuroendocrine System (DNES)*: Gastrointestinal endocrine cells are distributed in the mucosa of the gastrointestinal tract, synthesizing numerous types of gastrointestinal peptides or hormones. These cells act in the regulation of the physiological functions of the digestive tract and are observed by immunohistochemical analysis of the gastrointestinal tract mucosa and throughout the digestive system epithelium. They constitute the gastroenteropancreatic endocrine system (GEP) with the pancreas (Didinen et al. 2011).

The secretion of many endocrine cells of the gastrointestinal tract regulate digestion in tilapia. These chemicals (peptides and/or amines) are detected in the epithelial lining, glands, various connective tissues, nerve ganglia and intermuscular nervous plexus. The secretion of these peptides ensures mobility as well as the proper functioning of the nervous system, regulation of secretion through cellular interaction, cell proliferation and regulation of intestinal epithelium and contraction of smooth muscles (Holmgren and Olsson 2009).

- *Rodlet cells*: Mature Rodlet cells have an ovoid morphology, a polarized nucleus at the base, and are characterized by the presence of large rod-shaped cytoplasmic granules, from which they derive their name (Reite and Evensen 2006). They have functions related to pH regulation and control, participation in osmoregulatory events, as secretory cells observing antimicrobial effects, as units of transport of genetic material, and nonspecific immune cells. In tilapia, they are seen as an important presence in the intestinal lining as well as in the specialized parenchyma of the liver and pancreas (Manera and Dezfuli 2004).

7. Extramural organs of the digestive system

7.1. Hepatopancreas

The liver of Nile tilapia is large, showing only two lobes (Figure 3). The largest of them, the left lobe, spreads through most of the body cavity, marking the impression of the intestine on its visceral face. The liver is surrounded by a thin connective tissue capsule and has been shown to be cordonal, with hepatocytes arranged in almost linear arrangements from a central vein, similar to that found in several teleosts. The hepatic parenchyma consists of hepatocytes scattered in anastomosed arrangements arranged in pairs of cell layers, surrounded by sinusoid capillaries. The connective edges of the hepatic lobes do not appear clearly defined, as few triads are found in the liver, showing an absence of compartmentalization in the form of hepatic lobules, compromising the appearance of portal triads. This pattern is not unique to tilapia and is found in many other teleost species (Coimbra et al. 2007, Mori et al. 2019).

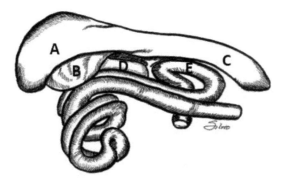

Fig. 3. Left lateral view of the viscera of the digestive system of Nile tilapia, *Orechromis niloticus*: **A** – Left lobe of the liver; **B** – Gall bladder; **C** – Posterior lobe of the liver; **D** – Visceral face of stomach; **E** – Gastric spiral loop. Figure by Sílvio Teixeira da Costa.

The bile ducts are usually found near the portal vein and exhibit a lining consisting of a simple cubic epithelium. A concentric layer of collagen and muscle fibers is observed under this epithelium.

Microscopy allows the identification of organized exocrine pancreatic tissue in the parenchymal mass of the liver (Figure 4). This pattern is given by its acinar arrangement as well as its diffuse distribution in the liver parenchyma. Intrahepatic exocrine pancreatic tissue is separated from the hepatocyte cords by a thin connective tissue septum, with pancreatic cells centered around the portal veins. The nuclei of these cells are also spherical and observed in cells of cuboid to cylindrical height characterizing the organ as a hepatopancreas (Nel et al. 1990).

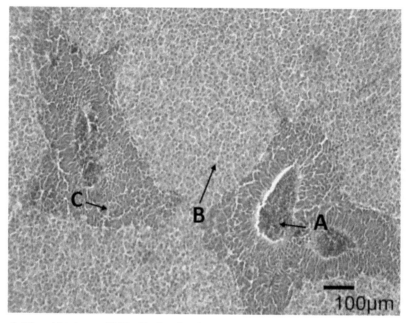

Fig. 4. Liver histology of Nile tilapia, *Orechromis niloticus.* **A** – Lobular center vein; **B** – Sinusoid capillary; **C** – Exocrine pancreas (accino). (Micrograph courtesy of Marcelo Leite da Veiga, Department of Morphology, UFSM)

Exocrine pancreatic cells are ultrastructurally characterized by the presence of secretion granules positioned at the apex of the cell, showing a well-developed rough endoplasmic reticulum with tubular-vesicular characteristics in circular arrangement in appearance. The gall bladder is well developed, demonstrating a rounded shape with right ventral-lateral positioning to the cranial and largest portion of the liver (Vicentini et al. 2009).

8. pH of gastrointestinal tract

The stomach content of Nile tilapia caught in the wild became more acidic towards the afternoon and most of the evening, when the amount of food in the stomach increases. The estimated stomach emptying was 0.36%/h (Getachew 1989). The lowest stomach pH, 1.54 in Mozambique tilapia (*O. mossambicus*) and 1.58 in *Tilapia rendalli* maintained in tanks and fed commercial diet, was observed around 12 h after feeding. The pH in the intestine of these species is slightly alkaline – proximal intestine: 7.0-8.0, distal intestine: 7.0-8.8 (Hlophe et al. 2014).

9. Digestive enzymes activity

Nile tilapia fed *ad libitum* showed higher pepsin-like activity in the stomach after noon up to 1 a.m., while in the intestine trypsin activity increased gradually from noon to 3 a.m. The activity of alkaline protease, amylase, chymotrypsin and lipase increased gradually from 4-5 p.m. to 3-5 a.m., as pepsin, but amylase activity decreased slightly after 4 h and further at midnight (Montoya-Mejía et al. 2016). The specific activities of amylase, lipase, total proteases, and trypsin of sex-reversed Nile tilapia were not affected by different feeding frequencies (one meal at 0600 h, two meals – 6 a.m. and noon or 6 a.m. and 6 p.m. – and three meals daily) (Thongprajukaew et al. 2017). Activity of acid protease and amylase also did not differ between Nile tilapia fed once a day at 11 a.m. or 11 p.m., but alkaline protease activity in the middle intestine showed daily rhythm with the achrophase at the beginning of the dark phase in both groups (Guerra-Santos et al. 2017). Digestive enzyme activities were also not affected in Nile tilapia maintained in in-pond raceways and at three different stocking densities: 0.28, 0.57 and 0.85 kg/m^3 (Wang et al. 2019).

The activity of γ-glutamyl transpeptidase and the disaccharidases maltase and sucrose activities were the highest in the mid-intestine of the hybrid *O. niloticus×O. aureus*, but alkaline phosphatase activity decreased from the proximal to the distal portion (Harpaz and Uni 1999). Amylase and the disaccharidase maltase activities in the intestine of Nile tilapia were not affected by different carbohydrates sources (wheat bran, cassava residue, ground corn, and broken rice) (Gominho-Rosa et al. 2015). Hybrids *O. mossambicus* × *O. aureus* that consumed a diet with 48% crude protein (CP) showed higher activity of maltase and lower activity of the proteolytic enzyme γ-glutamyl transpeptidase in the intestine compared to those that consumed a diet with 30% CP. The activity of alkaline phosphatase was higher in the proximal compared with distal intestinal section, and leucine amino peptidase was constant through the intestine and was not affected by protein levels in the diet (Hakim et al. 2006). Nile tilapia raised in a biofloc system showed higher amylase and lipase activities than those raised in clear water (Long et al. 2015). The activity of pepsine and trypsine increased and chymotrypsin decreased with the increase of dietary digestible protein (22 to 30%) in Nile tilapia raised in brackish (10 ppt) biofloc system (Durigon et al. 2019).

Amylase and lipase activities were also observed in the stomach of Mozambique tilapia and *T. rendalli*, as well as a low cellulase activity in the stomach and distal intestine of *T. rendalli*. Amylase and lipase activities in both species were higher in the proximal intestine, while protease activity was higher in the distal intestine (Hlophe et al. 2014). The cellulolytic activity in the intestine of Mozambique tilapia is due to the bacteria *Bacillus circulans* and *Bacillus megaterium* (Saha et al. 2006). Nile tilapia fed a diet containing 15% fish silage meal presented higher intestinal protease, trypsin, chymotrypsin, amylase and lipase activities than those fed with the same percentage of bovine blood meal or several vegetable products (Montoya-Mejía et al. 2017). Alkaline phosphatase was detected in the intestine of Nile tilapia (Hassaan et al. 2019). The intestinal brush border membrane of the hybrid tilapia (*O. niloticus*×*O. aureus*) has a low concentration of an enzyme capable of hydrolyzing phytic acid, and despite its low substrate affinity, it probably allows the effective use of phytate phosphorus by this hybrid (Lavorgna et al. 2003).

10. Nutrient transporters

Nile tilapia intestine presents a homogeneous expression of the Na^+-glucose cotransporter 1 (SGLT1, gene SLC5A1) and Na^+-myoinositol cotransporter 2 (SMIT2, gene SLC5A11) through its extension (Subramaniam et al. 2019). The expression of SGLT1 and facilitated glucose transporter 2 (GLUT2) increased during 1-3 h post feeding in GIFT Nile tilapia, decreased at 8 h post feeding and then returned to the levels observed in fasted (36 h) fish. These results indicated that intestinal uptake of glucose and its transport across the intestine to blood mainly occurred during 1-3 h post feeding (Chen et al. 2017). The insulin-responsive glucose transporter 4 (GLUT4) was also found in the intestine of this species (Li et al. 2019), but authors did not discuss its function in this organ.

Neutral or cationic amino acids are transported through the apical membrane (brush border membrane) of the enterocytes by the $B^{o,+}AT$ system (SLC7A9 and SLC3A1 genes), a Na^+-independent transporter. Neutral amino acids are also transported by the B^oAT system (SLC6A19 and SLC6A18 genes), Na^+-dependent transporters. The expression of the gene SLC6A19 has higher activity in the proximal and middle intestine (Nitzan et al. 2017), while SLC6A18 was mainly expressed in the middle and posterior intestine of Mozambique tilapia (Orozco et al. 2018). The expression of the SLC6A19 gene was not affected by salinity exposure despite being a Na^+-dependent transporter. In contrast, the expression of the SLC7A9 gene of the Na^+-independent transporter $B^{o,+}AT$ transporter increased with the increase of salinity, while the expression of the SLC3A1 gene decreased (Nitzan et al. 2017).

Peptides are transported through the apical membrane to the cytoplasm of enterocytes and hydrolyzed into free amino acids. There are two peptide transporters from the solute carrier 15 (SLC15) family (Verri et al. 2017), and they use the H^+ gradient at the apical membrane of enterocytes to mediate the absorption of small peptides, i.e. the uptake of peptides is coupled with H^+ entrance (Con et al. 2017, Wang et al. 2017). The peptide transporter 1 has two variants (PepT1a and PePT1b) (Con et al. 2017) that are high-capacity, low-affinity transporters whose expression decreased from proximal to distal intestine and the main carriers for the uptake of dietary peptides (Wang et al. 2017). Unexpectedly, because the driving force of PePT is the proton gradient, in Mozambique tilapia maximum activity was observed when PepT1 was exposed to alkaline pH, as found in the intestine. Probably this pH-dependent increase was due to an increase in the transporter capacity, while its affinity decreased (Con et al. 2017). In Mozambique tilapia submitted to short-term fasting, PePT1 expression did not change or even increased, but a down-regulation occurred with long-term food deprivation. The PePT1 expression increases following re-feeding, usually surpassing the levels of control group (Orozco et al. 2017). The peptide transporter 2 (PepT2) is a low-capacity, high-affinity transporter (Wang et al. 2017), found mainly in the medium and distal intestine, where it can absorb small peptides at lower substrate concentrations than the proximal intestine (Con et al. 2017). In Mozambique tilapia adapted to seawater, the expression of both PePT1 and PePT2 transporters shift to the distal portion of the intestine. Consequently, a more uniform dispersion of transporters' expression along the intestine was seen in seawater compared to the gradual decline verified in freshwater (Con et al. 2017). Dietary addition of 5% NaCl also increased the expression of PepT1a, PePT1b and PePT2 in the intestine of Nile tilapia (Hallali et al. 2018).

Acknowledgments

B. Baldisserotto is a recipient of a CNPq research fellowship.

References cited

Ahsan-ul-Islam. 1951. The comparative histology of the alimentary canal of certain fresh water teleost fishes. Proc. Indian Acad. Sci. B 33(6): 297–321.

Al-Hussaini, A.H. 1949. On the functional morphology of the alimentary tract of some fish in relation to differences in their feeding habits; anatomy and histology. Q. J. Microsc. Sci. 90(2): 109–139.

Albrecht, M.P., M.F.N. Ferreira and E.P. Caramaschi. 2001. Anatomical features and histology of the digestive tract of two related neotropical omnivorous fishes 8 (*Characiformes; Anostomidae*). J. Fish Biol. 58(2): 419–430.

Awaad, A.S., K.M. Usama and M.G. Tawfiek. 2014. Comparative histomorphological and histochemical studies on the oesophagus of Nile tilapia *Oreochromis niloticus* and African catfish *Clariasgariepinus*. J. Histol. 2014: 987041. doi: 10.1155/2014/987041.

Banan, K. and S. Mahdi. 2012. The morphology of the post-gastric alimentary canal in teleost fishes: A brief review. Int. J. Aquat.Sci. 3(2): 71–88.

Beveridge, M.C.M., M. Begum, G.N. Frerichs and S. Millar. 1989. The ingestion of bacteria in suspension by the tilapia *Oreochromis niloticus*. Aquaculture 81: 373–378.

Caceci, T., H.A. El-Habback, S.A. Smith and B.J. Smith. 1997. The stomach of *Oreochromis niloticus* has three regions. J. Fish Biol. 50: 939–952.

Chen, Y.J., T.Y. Zhang, H.Y. Chen, S.M. Lin, L. Luo and D.S. Wang. 2017. Simultaneous stimulation of glycolysis and gluconeogenesis by feeding in the anterior intestine of the omnivorous GIFT tilapia, *Oreochromis niloticus*. Biol. Open. 6: 818–824.

Coimbra, A.M., A. Figueiredo-Fernandes and M.A. Reis-Henriques. 2007. Nile tilapia (*Oreochromis niloticus*), liver morphology, CYP1A activity and thyroid hormones after endosulfan dietary exposure. Pest. Biochem. Physiol. 89: 230-236.

Con, P., T. Nitzan and A. Cnaani. 2017. Salinity-dependent shift in the localization of three peptide transporters along the intestine of the Mozambique tilapia (*Oreochromis mossambicus*). Front Physiol. 8. doi: 10.3389/fphys.2017.00008.

Didinen, B.I., A. Kubilay, O. Diler, S. Ekici, E.E. Onuk and A. Findik. 2011. First isolation of vagococcus salmoninarum from cultured rainbow trout (*Oncorhynchusmykiss*, Walbaum) broodstocks in Turkey. Bul. Eur. Ass. Fish Path. 31(6): 235–243.

Durigon, E.G., A.P.G. Almeida, G.T. Jeronimo, B. Baldisserotto and M.G.C. Emerenciano. 2019. Digestive enzymes and parasitology of Nile tilapia juveniles raised in brackish biofloc water and fed with different digestible protein and digestible energy levels. Aquaculture 506: 35–41.

Elsheikh, E.H., E.S. Nasr and A.M. Gamal. 2012. Ultrastructure and distribution of the taste buds in the buccal cavity in relation to the food and feeding habit of a herbivorous fish: *Oreochromis niloticus*. Tissue Cell 44: 164–169.

Frierson, E.W. and J.W. Foltz. 2004. Comparison and estimation of absorptive intestinal surface areas in two species of cichlid fish. T. Am. Fish. Soc. 121(4): 517–523.

Getachew, T. 1989. Stomach pH, feeding rhythm and ingestion rate in *Oreochromis niloticus* L. (Pisces, Cichlidae) in Lake Awasa, Ethiopia. Hydrobiologia 174: 43–48.

Gominho-Rosa, M.D., A.P.O. Rodrigues, B. Mattioni, A. Francisco, G. Moraes and D.M. Fracalossi. 2015. Comparison between the omnivorous jundiá catfish (*Rhamdia quelen*) and Nile tilapia (*Oreochromis niloticus*) on the utilization of dietary starch sources: Digestibility, enzyme activity and starch microstructure. Aquaculture 435: 92–99.

Goodrich, J.S., S.L. Sanderson, I.E. Batjakas and L.S. Kaufman. 2000. Branchial arches of suspension-feeding *Oreochromis esculentus*: Sieve or sticky filter? J. Fish Biol. 56(4): 858–875.

Guerra-Santos, B., J.F. Lopez-Olmeda, B.O. Mattos, A.B. Baiao, D.S.P. Pereira, F.J. Sanchez-Vazquez et al. 2017. Synchronization to light and mealtime of daily rhythms of locomotor activity, plasma glucose and digestive enzymes in the Nile tilapia (*Oreochromis niloticus*). Comp. Biochem. Phys. A. 204: 40–47.

Hallali, E., F. Kokou, T.K. Chourasia, T. Nitzan, P. Con, S. Harpaz et al. 2018. Dietary salt levels affect digestibility, intestinal gene expression, and the microbiome, in Nile tilapia (*Oreochromis niloticus*). Plos One. 13(8): e0202351.

Hakim, Y., Z. Uni, G. Hulata and S. Harpaz. 2006. Relationship between intestinal brush border enzymatic activity and growth rate in tilapias fed diets containing 30% or 48% protein. Aquaculture 257: 420–428.

Harpaz, S. and Z. Uni. 1999. Activity of intestinal mucosal brush border membrane enzymes in relation to the feeding habits of three aquaculture fish species. Comp. Biochem. Phys. A. 124: 155–160.

Hassaan, M.S., E.Y. Mohammady, M.R. Soaudy and A.A.S. Abdel Rahman. 2019. Exogenous xylanase improves growth, protein digestibility and digestive enzymes activities in Nile tilapia, *Oreochromis niloticus*, fed different ratios of fish meal to sunflower meal. Aquac. Nutr. 25: 841–853.

Hlophe, S.N., N.A.G. Moyo and I. Ncube. 2014. Postprandial changes in pH and enzyme activity from the stomach and intestines of *Tilapia rendalli* (Boulenger, 1897), *Oreochromis mossambicus* (Peters, 1852) and *Clarias gariepinus* (Burchell, 1822). J. Appl. Ichthyol. 30: 35–41.

Holmgren, S. and C. Olsson. 2009. The neuronal and endocrine regulation of gut function. pp. 468–499. *In*: Bernier, N.J., G.V.D. Kraak, A.P. Farrell and C.J. Brauner [eds]. Fish Neuroendocrinology. Academic Press, London.

Hunter, J.R. 1981. Feeding ecology and predation of marine fish larvae. pp. 34–77. *In*: Lasker, R. [ed]. Marine Fish Larvae: Morphology, Ecology and Relation to Fisheries.Washington Sea Grant Program, Seattle.

Kaiya, H., M. Kojima, H. Hosoda, L.G. Riley, T. Hirano, E.G. Grau et al. 2003. Identification of tilapia ghrelin and its effects on growth hormone and prolactin release in the tilapia, *Oreochromis mossambicus*. Comp. Biochem. Physiol. B Biochem. Mol. Biol. 135(3): 421–429.

Khojasteh, S. and M. Banan. 2012. The morphology of the post-gastric alimentary canal in teleost fishes: A brief review. Intern. J. Aquat. Sci. 3(2): 71–88.

Lavorgna, M.W., Y. Hafez, T. Handwerker and S.G. Hughes. 2003. Phosphorous digestibility and activity of intestinal phytase in hybrid tilapia, *Oreochromis niloticus* × *O. aureus*. J. Appl. Aquac. 14(1–2): 89–100.

Li, R.X., Q. Chen, H.Y. Liu, B.P. Tan, H.H. Dong, S.Y. Chi et al. 2019. Molecular characterization and expression analysis of glucose transporter 4 from *Trachinotus ovatus*, *Rachycentron canadums* and *Oreochromis niloticus* in response to different dietary carbohydrate-to-lipid ratios. Aquaculture 501: 430–440.

Long, L.N., J. Yang, Y. Li, C.W. Guan and F. Wu. 2015. Effect of biofloc technology on growth, digestive enzyme activity, hematology, and immune response of genetically improved farmed tilapia (*Oreochromis niloticus*). Aquaculture 448: 135–141.

Maina, J.N. 2000. The highly specialized secretory epithelium in the buccal cavity of the alkalinity adapted lake Magadi cichlid, *Oreochromis alcalicusgrahami* (Teleostei: Cichlidae): A scanning and transmission electron microscope study. J. Zool. 251(4): 427–438.

Manera, M. and B.S. Dezfuli. 2004. Rodlet cells in teleosts: A new insight into their nature and functions. J. Fish Biol. 65(3): 597–619.

McMillan, D.B. and R.J. Harris. 2018. An Atlas of Comparative Vertebrate Histology. Academic Press, London.

Mokhtar, D.M. 2017. Fish Histology from Cells to Organs. Apple Academic Press, New Jersey.

Montoya-Mejia, M., H. Rodriguez-Gonzalez and H. Nolasco-Soria. 2016. Circadian cycle of digestive enzyme production at fasting and feeding conditions in Nile tilapia, *Oreochromis niloticus* (Actinopterygii: Perciformes: Cichlidae). Acta Ichthyol. Piscat. 46: 163–170.

Montoya-Mejia, M., M. Garcia-Ulloa, A. Hernandez-Llamas, H. Nolasco-Soria and H. Rodriguez-Gonzalez. 2017. Digestibility, growth, blood chemistry, and enzyme activity of juvenile *Oreochromis niloticus* fed isocaloric diets containing animal and plant byproducts. Rev. Bras. Zootecn. 46: 873–882.

Mori, N.C., B.T. Michelotti, T.S. Pês, C.A. Bressan, F. Sutili, L.C. Kreutz et al. 2019. Citral as a dietary additive for *Centropomus undecimalis* juveniles: Redox, immune innate profiles, liver enzymes and histopathology. Aquaculture 501: 14–21.

Morrison, C.M. and J.R. Wright. 1999. A study of the histology of the digestive tract of the Nile tilapia. J. Fish Biol. 54: 597–606.

Murray, H.M., G.M. Wright and G.P. Goff. 1996. A comparative histological and histochemical study of the post-gastric alimentary canal from three species of pleuronectid, the Atlantic halibut, the yellowtail flounder and the winter flounder. J. Fish Biol. 48(2): 187–206.

Nel, M.M., J.H. Swanepoel and H.J. Geyer. 1990. Die histologie en ultrastruktuur van die hepatopankreas van die bloukurper (*Oreochromis mossambicus*). S. Afr. J. Sci. Technol. 9(1): 3–10.

Nitzan, T., P. Rozenberg and A. Cnaani. 2017. Differential expression of amino-acid transporters along the intestine of Mozambique tilapia (*Oreochromis mossambicus*) and the effect of water salinity and time after feeding. Aquaculture 472: 71–75.

Nnorthcott, M.E. and M.C.M. Beveridge. 1988. The development and structure of pharyngeal apparatus associated with filter feeding in tilapias (*Oreochromis niloticus*). J. Zool. 215(1): 133–149.

Ohshimo, S. and K. Uematsu. 1996. Histological study on the autonomic innervation of the stomach and ovaries of the Nile tilapia *Oreochromis niloticus*. Fish. Sci. 62(2): 196–206.

Orozco, Z.G.A., S. Soma, T. Kaneko and S. Watanabe. 2017. Effects of fasting and refeeding on gene expression of slc15a1a, a gene encoding an oligopeptide transporter (PepT1), in the intestine of Mozambique tilapia. Comp. Biochem. Phys. B. 203: 76–83.

Orozco, Z.G.A., S. Soma, T. Kaneko and S. Watanabe. 2018. Spatial mRNA expression and response to fasting and refeeding of neutral amino acid transporters slc6a18 and slc6a19a in the intestinal epithelium of Mozambique tilapia. Front Physiol. 9. doi: 10.3389/fphys.2018.00212.

Reite, O.B. and Ø. Evensen. 2006. Inflammatory cells of teleostean fish: A review focusing on mast cells/eosinophilic granule cells and rodlet cells. Fish Shell. Immun. 20: 192–208.

Saha, S., R.N. Roy, S.K. Sen and A.K. Ray. 2006. Characterization of cellulase-producing bacteria from the digestive tract of tilapia, *Oreochromis mossambica* (Peters) and grass carp, *Ctenopharyngodon idella* (Valenciennes). Aquacult. Res. 37: 380–388.

Sanderson, S.L., A.Y. Cheer, J.S. Goodrich, J.D. Graziano and W.T. Callan. 2001. Crossflow filtration in suspension-feeding fishes. Nature 412(6845): 439–441.

Sanderson, S.L., M.C. Stebar, K.L. Ackermann, S.H. Jones, I.E. Batjakas and L. Kaufman. 1996. Mucus entrapment of particles by a suspension-feeding tilapia (Pisces: Cichlidae). J. Exp. Biol. 199: 1743–1756.

Santos, E.L., G. Pereira Reis, I.M. Guimarães, V.C. Vasconcelos and J. Radünz Neto. 2019. Morfologia comparativa do trato digestório de tilápias do Nilo (*Oreochromis niloticus*) cultivadas em sistema semi-intensivo vs da pesca artesanal. J. Interdisc. Biociênc. 3(2): 19–24.

Sklan, D., T. Prag and I. Lupatsch. 2004. Structure and function of the small intestine of the tilapia *Oreochromis niloticus* × *Oreochromis aureus* (Teleostei, Cichlidae). Aquac. Res. 35(4): 350–357.

Smith, J.C. and S.L. Sanderson. 2008. Intra-oral flow patterns and speeds in a suspension-feeding fish with gill rakers removed versus intact. Biol. Bull-Us. 215: 309–318.

Smith, B.J., S.A. Smith, B. Tengjaroenkul and T.A. Lawrence. 2000. Gross morphology and topography of the adult intestinal tract of the tilapian fish, *Oreochromis niloticus* L. Cell Tissues Organs 166(3): 294–303.

Subramaniam, M., L.P. Weber and M.E. Loewen. 2019. Intestinal electrogenic sodium-dependent glucose absorption in tilapia and trout reveal species differences in SLC5A-associated kinetic segmental segregation. Am. J. Physiol-Reg I. 316: R222–R234.

Thongprajukaew, K., S. Kovitvadhi, U. Kovitvadhi and P. Prepramec. 2017. Effects of feeding frequency on growth performance and digestive enzyme activity of sex-reversed Nile tilapia, *Oreochromis niloticus* (Linnaeus, 1758). Agric. Nat. Res. 51: 292–298.

Verri, T., A. Barca, P. Pisani, B. Piccinni, C. Storelli and A. Romano. 2017. Di- and tripeptide transport in vertebrates: The contribution of teleost fish models. J. Comp. Physiol. B. 187: 395–462.

Vicentini, C.A., I.B. Franceschini-Vicentini, M.T.S. Bombonato, B. Bertolucci, S.G. Lima and A.S. Santos. 2009. Morphological study of the liver in the teleost *Oreochromis niloticus*. Intern. J. Morphol. 23(3): 211–216.

Wang, J., X. Yan, R. Lu, X. Meng and G. Nie. 2017. Peptide transporter 1 (PepT1) in fish: A review. Aquac. Fish. 2: 193–206.

Wang, Y.Y., P. Xu, Z.J. Nie, Q.J. Li, N.L. Shao and G.C. Xu. 2019. Growth, digestive enzymes activities, serum biochemical parameters and antioxidant status of juvenile genetically improved farmed tilapia (*Oreochromis niloticus*) reared at different stocking densities in in-pond raceway recirculating culture system. Aquac. Res. 50: 1338–1347.

Zayed, A.E. and S.A. Mohamed. 2004. Morphological study on the gills of two species of fresh water fishes: *Oreochromis niloticus* and *Clarias gariepinus*. Ann. Anat. 186(4): 295–304.

Current Improvements in Tilapia Nutrition

Wilson Massamitu Furuya[1], Priscila Vieira[2]*, Renan Rosa Paulino[2]
and Mariana Michelato[3]

[1] Department of Animal Science, Universidade Estadual de Ponta Grossa, PR, Brazil
[2] Department of Animal Science, Universidade Federal de Lavras, Lavras, MG, Brazil
[3] Ichthus Unlimited, San Diego, CA, USA

1. Introduction

Tilapia is one of the most valuable species of freshwater fish with a high impact economically and socially in many tropical and subtropical areas (FAO 2016). Currently, tilapia culture is mitigated by variable production in earth ponds, cages, circular tanks, raceways and biofloc systems. The continuous expansion of the tilapia industry has demanded an increasing need for more sustainable and economical practices, to identify alternative ingredients, and increase the nutritional value of aquafeeds. Over the last decade, a consistent number of research has been focused on the use of exogenous enzymes (Hassaan et al. 2019, Li et al. 2019a, Maas et al. 2019), prebiotics (Boonanuntanasarn et al. 2018b, Hamdan et al. 2016, Levy-Pereira et al. 2018, Van Doan et al. 2018b), probiotics (Abou-El-Atta et al. 2019, Nguyen et al. 2019a, Tan et al. 2019), essential oils (Vicente et al. 2019) and organic minerals (Nguyen et al. 2019a) in tilapia aquafeeds. Effects of prebiotics and probiotics improve growth performance, immune status, impact on gut health (Dawood et al. 2019, Elsabagh et al. 2018) and microbiota (Hassaan et al. 2018, Tan et al. 2019).

Fatty acid profile and sensory properties of the meat have been assessed to meet the demands of consumers' preferences. Globally, environmental issues continue to be a challenge, and water quality concerns have focused on reducing nitrogen and phosphorus excretions. The aim of this review is to update information on recent advances in tilapia nutrition, including dietary requirements and a wide variety of effects of prebiotics, probiotics, fatty acids, vitamins, and minerals on nutrient utilization, gut health and microbiota diversity. Additionally, recent developments in genomics and the use of insect meal and essential oils have also led to tilapia nutrition

---

*Corresponding author: priscila@ufla.br

advances; therefore, consideration for growth and reproductive performance, health status, and meat quality will be highlighted.

2. Dietary requirements

Nutrient requirements of tilapias were updated in the last decade and are summarized in Tables 1-5.

Table 1. Daily maintenance energy and protein requirements of tilapias[1]

Author	Equation[*]	Application[*]
Energy		
Trung et al. 2011	$25.9 \text{ kJ} \times BW^{0.80}$	Pre-growout
Saravanan et al. 2013	$57 \text{ kJ} \times BW^{0.8}$	Growout
Schrama et al. 2018	$56 \text{ to } 89 \text{ kJ} \times BW^{0.8}$	Growout
Lupatsch et al. 2010	$59.46 \text{ kJ} \times BW^{0.80}$	Breeder, female
Protein		
Trung et al. 2011	$0.45 \text{ g} \times BW(kg)^{0.67}$	Pre-growout
Lupatsch et al. 2010	$0.98 \text{ g} \times BW(kg)^{0.70}$	Breeder, female

[1] Fish size classes: as nursery, 1 to 30 g of body weight; pre-growout, 30 to 220 g of body weight; and growout, >220 g body weight (Chowdhury et al. 2013).
[*] BW, body weight as kg.

Tilapias are unable to meet maintenance requirements by filter-feeding plankton biomass (Dempster et al. 1995), and an artificial, well-balanced diet must be supplied to maximize performance and the health of the fish. Nutrient requirements are defined by the amount of nutrients that tilapia requires for maintenance, growth, reproduction, and health. Dietary requirements are determined for individual and recommendations for a population of fish. Tilapia requirements are described for carbohydrates, proteins (and amino acids), fats, minerals and vitamins. In addition to nutrients, dietary energy requirements are available in the literature. Recommendations are often determined as the requirements of an average animal in the population and are affected by several factors such as environmental temperature, stocking density, genetics and gender, health status, feeding management, and diet composition.

The energy and protein values of feeds and requirements are mainly expressed as the content of digestibles, while amino acids and minerals may be expressed as available values determined in tilapias. The energy and nutrients excreted in feces are measured by the total collection of feces in a conventional modified Guelph system by collecting samples using the marker technique. In recent years, there have been improvements in the apparent digestibility coefficients of energy, protein, amino acids (Davies et al. 2011, Vidal et al. 2017, 2015), and minerals (Guimarães et al. 2012, Moura et al. 2018), by utilizing common feed ingredients frequently used in commercial tilapia diets.

Protein and energy are the first nutritional requirements that should be considered when formulating aquafeeds, and they represent the most expensive and impacting

Table 2. Dietary energy, protein and linolenic acid requirements of tilapias
(Dry matter basis)

Author	Value	Application[1]
Digestible energy, MJ/kg		
Koch et al. 2017	12.6	Growout[a]
Orlando et al. 2017	15.1	Broodstock, female
Crude protein, g/kg		
Silva et al. 2019	420	Sex-reversal 11 to 20 d)
Silva et al. 2019	360	Sex-reversal(21 to 30 d)
Fernandes Junior et al. 2016[2]	350	Growout[a]
Liu et al. 2017a[3]	293	Growout[a]
El-Sayed and Kawanna 2008[4]	400	Broodstock
Pereira et al. 2019[5]	320	Broodstock
Lipids and fatty acids, g/kg		
Lipids		
Lim et al. 2011b	50-120	Nursey to growout
Linolenic acid, g/kg		
Chen et al. 2013[6]	4.5 to 6.4	Pre-growout
Nobrega et al., 2017[7]	7.0	Nursey to pre-growout
Linoleic acid		
Li et al. 2013	11.4	Nursery

[1] Fish size classes: sex-reversal. From birth to 0.4 g of body weight (FAO 2019) as nursery, 1 to 30 g of body weight; pre-growout, 30 to 220 g of body weight; and grow out >220 g body weight (Chowdhury et al. 2013).

[2] Nile tilapia reared in floating net cages placed in a reservoir considering body weight gain and fillet and fed diets containing 300 g/kg of digestible protein and 13.4 kcal/kg of digestible energy.

[3] Nile tilapia reared in cages placed in earth ponds with fertilized freshwater considering body weight gain as criteria response and fed diets containing 18.7 MJ/kg of gross energy.

[4] Nile tilapia placed in recycling system and fed diet containing 16.7 kcal/kg of gross energy.

[5] Nile tilapia placed in earth ponds and fed diets containing 14.6 kcal/kg of digestible energy.

[6] Nile tilapia reared under optimal temperature.

[7] Nile tilapia reared under sub-optimal temperature.

factors on the growth and reproductive performance (NRC 2011). Regardless of omnivorous food habits, high dietary carbohydrate has been associated with glucose intolerance in Nile tilapia, *Oreochromis niloticus,* suggesting that moderate dietary carbohydrate levels may be considered to improve the glucose tolerance (Boonanuntanasarn et al. 2018a, Liu et al. 2018). One way to overcome this problem involves the use of 6.1:1 carbohydrate to lipids ratio to optimize growth performance of grow-out Nile tilapia (Coutinho et al. 2018). In this context, lipids have been recommended in tilapia diets at 60 to 120 g/kg (Lim et al. 2011b).

Based on body weight (kg), a bioenergetic factorial modelling approach, the daily dietary energy (25.9 kJ $BW^{0.80}$ at 28 °C) and protein requirements (0.45 g $BW^{0.80}$)

Table 3. Dietary amino acids requirements of tilapias (Dry matter basis)

Author	Amino acid	g/kg	% Protein	Application[1]
Neu 2016)	Arginine	13.6	4.8	Pre-growout
Michelato et al. 2016c	Histidine	8.2	3.1	Pre-growout
Santiago and Lovell 1988	Isoleucine	8.3	3.1	Nursey
Gan et al. 2016	Leucine	12.5	4.3	Nursey
Michelato et al. 2016b	Lysine	14.6	5.8	Growout
Nguyen and Davis 2009	Methionine	8.1[2]	2.9	Nursey
Xiao et al. 2019	Phenylalanine	8.8[3]	3.0	Nursey to pre-growout
Michelato et al. 2016a	Threonine	11.5	4.0	Growout
Nguyen et al. 2019b	Tryptophan	3.1	1.0	Nursey to pre-growout
Xiao et al. 2018	Valine	11.5	4.1	Nursey to growout

[1] Fish size classified according to previously established as nursery, 1 to 30 g of body weight; pre-growout, 30 to 220 g of body weight; and growout >220 g of body weight (Chowdhury et al. 2013).
[2] Dietary cysteine at 0.4 g/kg of diet.
[3] Dietary tyrosine 9.8 g/kg of diet.

Table 4. Minerals requirements of tilapias (Dry matter basis)

Author	Mineral	Unit	Value	Application[1]
Shiau and Tseng 2007	Calcium	g/kg	3.5[2]	Nursey
Robinson et al. 1987	Calcium	g/kg	8.0[3]	Nursey
Yao et al. 2014	Phosphorus	g/kg	8.6[4]	Pre-growout
Carvalho et al. 2018	Phosphorus	g/kg	6.5[4]	Broodstock
Shiau and Ning 2003	Copper	mg/kg	4	Nursey
Shiau and Su 2003	Iron	mg/kg	85-160	Nursey
Lin et al. 2013	Magnesium	mg/kg	0.2[5]	Nursey
Lin et al. 2008	Manganese	mg/kg	7	Nursey
Shiau and Hsieh 2001	Potassium	g/kg	2-3	Nursey
Lee et al. 2016	Selenium	mg/kg	>1.06 and <2.06	Nursey
Shiau and Lu 2004	Sodium	g/kg	1.5[6]	Growout
Huang et al. 2015	Zinc	mg/kg	37.2[7]	Growout

[1] Fish size classified according to previously established (Chowdhury et al. 2013) as nursery, 1 to 30 g of body weight; pre-growout, 30 to 220 g of body weight; and growout >220 g of body weight.
[2] Calcium concentration in the water at 27.1–33.3 mg/L.
[3] Fish reared in calcium-free water
[4] Based on available phosphorus values.
[5] Fish reared in freshwater, and tilapia reared in seawater requires 0.02 mg/kg of magnesium.
[6] Fish reared in freshwater, and no dietary Na is required for tilapia reared in seawater.
[7] Zinc supplemented as zinc sulfate ($ZnSO_4 \cdot 7H_2O$) in purified casein-gelatin-dextrin. based diet.

Table 5. Dietary vitamins requirements of tilapias (Dry matter basis)

Author	Vitamin	Unit	Value	Application[1]
Guimarães et al. 2014	A	IU/kg	3910	Nursey to pre-growout
Shiau and Hwang 1993	D	IU/kg	375	Nursey
Shiau and Shiau 2001	E	IU/kg	50 to 120	Nursey
Lim et al. 2011a	B_1	mg/kg	3.5	Nursey to pre-growout
Lim et al. 1993	B_2	mg/kg	5	Nursey to pre-growout
Jiang et al. 2014	B_3	mg/kg	20.4	Pre-growout to growout
Roem et al. 1991	B_5	mg/kg	10	Nursey
Teixeira et al. 2011	B_6	mg/kg	10	Nursey to pre-growout
Shiau and Chin 1999	B_7	mg/kg	0.06	Nursey
Wu et al. 2016	B_9	mg/kg	0.4	Pre-growout to growout
Lovell and Limsuwan 1982	B_{12}		NR[2]	Nursey
Soliman et al. 1994	C	mg/kg	420	Nursey
Huang et al. 2016	C	mg/kg	115[3]	Pre-growout to growout
Barros et al. 2015	C	mg/kg	600[4]	Nursey
Sarmento et al. 2018	C	mg/kg	599 to 942	Broodstock
Baldisserato et al. 2019	Choline	mg/kg	800 to 1,200	Nursey
Shiau and Su 2005	Inositol	mg/kg	400	Nursey

[1] Fish size classified according to previously established as nursery, 1 to 30 g of body weight; pre-growout, 30 to 220 g of body weight; and growout, >220 g of body weight (Chowdhury et al. 2013).
[2] Not required
[3] Based on maximum growth
[4] Fish subjected to cold-induced stress or bacterial challenge

have been established for Nile tilapia fingerling-juvenile (Trung et al. 2011). The daily dietary methionine and lysine requirements for maintenance were estimated at 16.5 mg $BW^{0.7}$ and 68.8 mg $BW^{0.7}$, respectively, for adult Nile tilapia (He et al. 2013). In recent years, protein (Fernandes Junior et al. 2016), lysine (Michelato et al. 2016b), methionine and threonine (Michelato et al. 2016a) requirements have been considered to optimize fillet yield beyond growth and feed efficiency of Nile tilapia. The dietary digestible energy requirement for maintenance of female Nile tilapia breeder to optimize egg production has been found to be 59.46 kJ $BW^{0.80}$ per day, while digestible protein requirement has been established at 0.98 g $BW^{0.70}$ per day, where BW is the body weight of fish in kg (Lupatsch et al. 2010).

The roles of non-essential amino acids on growth performance have been focused on glutamine and/or arginine supplementation (Pereira et al. 2017). Likewise, dietary cysteine and tyrosine were found as important amino acids affecting skin color, by affecting the melanin synthesis in the red tilapia (Wang et al. 2018). In addition, taurine has been found to enhance growth and metabolic responses of larvae (Al-Feky et al. 2016) and juveniles (Michelato et al. 2018), also as important amino acids affecting reproductive performance of Nile tilapia bloodstock fed soybean meal-based diet (Al-Feky et al. 2014).

Different from energy, protein and amino acids, few studies have been carried out to update the nutritional requirements of minerals and vitamins for tilapias. However, dietary phosphorus for broodstock (Carvalho et al. 2018)and juveniles (Yao et al. 2014), and zinc (Huang et al. 2015), magnesium (Lin et al. 2013) and selenium (Lee et al. 2016) were recently established. Similarly, vitamin C requirements for broodstock (Sarmento et al. 2016) and pre-growout to growout fish were determined (Barros et al. 2015, Huang et al. 2016). In the same way, new data on vitamin A (Guimarães et al. 2014), vitamin B1 (Lim et al. 2011a) and vitamin B3 (Jiang et al. 2014), vitamin B6 (Teixeira et al. 2011) and vitamin B9 (Wu et al. 2016) were determined. For minerals, most studies have reported higher availability of organic rather than inorganic sources (Nguyen et al. 2019a). To date, nano forms of minerals have been recently evaluated in tilapia diets because of high availability and low toxicity (Dawood et al. 2020). Recently, the dietary optimum requirement of dietary linolenic acid for juvenile Nile tilapia reared under optimal temperature was found to be 4.5 to 6.4 g/kg (Chen et al. 2013). In addition, the dietary requirement of linolenic acid was updated—for juvenile Nile tilapia reared at sub-optimal temperature it was found to be 7.0 g/kg (Nobrega et al. 2017).

3. Broodstock nutrition

Nile tilapia is a continuous spawner throughout the year with short vitellogenetic periods, which makes it possible to improve spawning quality by modification of the nutritional quality of broodstock diets (Izquierdo et al. 2001). Broodstock diet may influence the maturation process of the gametes and energy storage in the form of the yolk (Mañanos et al. 2008), composed primarily of lipids and protein, wherein the dietary oil sources may influence fecundity (El-Sayed et al. 2005), and the protein may influence the tilapia spawning performance (El-Sayed and Kawanna 2008).

Broodstock nutrition is a quickly emerging yet controversial area of study because of its complex experimental conditions, particularly considering factors such as fish age, species, feed composition, breeding system, and reproductive management techniques (Bombardelli et al. 2017). Recent studies have indicated that tilapia broodstock require 300 to 400 g/kg of crude protein and 15 to 17 MJ/kg of digestible energy in the diet for optimized reproductive performance and hatchability (El-Sayed and Kawanna 2008, Oliveira et al. 2014, Orlando et al. 2017).

Furthermore, it is well known that dietary fatty acids markedly affect the reproductive performance of fish. Tilapia broodstock fed dietary soybean oil (high n-6 PUFA) or crude palm oil revealed an improvement in reproductive performance, compared to fish fed dietary fish oil (high in n-3 LC-PUFA) or linseed oil (Ng and Wang 2011). However, bloodstock reared at different water salinities require a source of dietary n-3-HUFA to enhance spawning performance, while it is not required for broodstock reared in freshwater (El-Sayed et al. 2005). Regarding the mineral nutrition of broodstock, the dietary requirements of available phosphorus was found to be 6.5 g/kg to enhance spawning performance and bone mineralization of female Nile tilapia (Carvalho et al. 2018). A number of questions regarding broodstock nutrition remains to be addressed. A more systematic research is required to explain the effects of additives and functional foods on reproductive performance.

4. Prebiotics, probiotics and microbiota

Tilapias are raised in a complex ecosystem where the microorganisms in the sediments, in the animal intestinal tract, and in the water interact with one another to influence the water quality and health of the fish (Fan et al. 2017). There is a growing interest in immunostimulatory feed supplements in global Tilapia aquaculture. Prebiotics and probiotics have been emerging as alternatives to substitute antibiotics and as an eco-friendly nutritional strategy to improve health and immunity, targeting synergistic enhancements in fish performance (Nayak 2010). In tilapias, probiotics have been shown to promote growth, stimulate the immune system and protect the host from diseases (Abou-El-Atta et al. 2019, Hamdan et al. 2016, Makled et al. 2019, Van Doan et al. 2018), modulate the intestinal histomorphology (Dawood et al. 2019, Elsabagh et al. 2018) and gastrointestinal ecosystem (Ferguson et al. 2010, Standen et al. 2015, Tan et al. 2019). Therefore, probiotics provide a potential alternative strategy to overcome adverse effects of in-feed antibiotics and drugs (Nayak 2010). However, factors like source, dose and duration of supplementation of probiotics may be considered to optimize the immunomodulatory effects (Hai 2015).

The synergistic effect of dietary probiotic *Lactobacillus plantarum* and whey protein concentrate on growth performance, antioxidant, immunity response and susceptibility to *Aeromonassobria* infection has been demonstrated in Nile tilapia (Abou-El-Atta et al. 2019). The combination of malic acid with *Bacillus subtilis* has revealed beneficial effects on growth, innate immunity and on gut health of Nile tilapia (Hassaan et al. 2018). In addition, in-diet mixture of *Bacillus subtilis*, *Bacillus licheniformis*, and *Bacillus pumilus* has confirmed beneficial effects on water quality, growth performance and intestinal morphology (Elsabagh et al. 2018), while probiotics used as water additive was demonstrated to improve growth and immune status of Nile tilapia (Zhou et al. 2010).

Prebiotics have been described as a functional additive, mainly focusing on benefits on growth performance, feed efficiency, digestive enzyme activities, health status, disease resistance and gut morphology and gut microbial composition (Guerreiro et al. 2018). These effects of prebiotics are mainly related to improved growth and feed efficiency because of enhanced nutrient availability due to changes in gut morphology, as described in Nile tilapia fed diet supplemented with inulin (Boonanuntanasarn et al. 2018a), orange peel-derived pectin (Van Doan et al. 2018b), β-glucans (Pilarski et al. 2017), fructo-oligosaccharides (Liu et al. 2017b), and mannan oligosaccharides (Levy-Pereira et al. 2018). Prebiotics influence gut immune system as well, and effects on microbiota have been addressed (Nawaz et al. 2018).

The gut microbiome has long been known to play important roles in tilapia health, and the establishment and maintenance of a microbial community structures affect the health and growth performance of tilapias. Factors include, among others, diet (Zhou et al. 2018), and probiotic and prebiotic administration (Haygood and Jha 2018). A recent strategy is the incorporation of antibiotic alternatives such as probiotics

(Ferguson et al. 2010, Standen et al. 2016) and additives such as organic acids (Koh et al. 2016) to modulate the gut microbiota in order to enhance gastrointestinal health and consequent growth performance and well-being of tilapias.

5. Fatty acid

Tilapia, like other vertebrates, cannot synthesize *de novo* n-3 or n-6 series of C18-polyunsaturated fatty acids (PUFA), such as α-linolenic acid (ALA, 18:3n-3) and linoleic acid (LNA, 18:2n-6), and consequently require a dietary supply of these essential fatty acids (Sargent et al. 2002). Dietary C18-PUFA are then converted by tilapia into the biologically active long-chain polyunsaturated fatty acids (LC-PUFA), namely, eicosapentaenoic acid (EPA, 20:5n-3), docosahexaenoic acid (DHA, 22:6n-3), and arachidonic acid (ARA, 20:4n-6). Marine fat fish, like Atlantic salmon, *Salmo salar*, has commonly been regarded as the main dietary source of abundant n-3 LC-PUFA. However, the content of n-6 fatty acids in freshwater fish like Nile tilapia is much greater than n-3 fatty acids (Liu et al. 2019). Vegetable oil sources added in freshwater fish diets is widely known to contain large amount of n-6 fatty acids, which increases the deposition of these fatty acids in the muscle. Therefore, several studies have previously shown that the fatty acid composition of the fillets in farmed fish may be modified by dietary oil sources (Ferreira et al. 2011).

Previous research showed that the content of n-3 PUFAs in Nile tilapia fillets increased in fish fed plant oils enriched with 18:3n-3 (Carbonera et al. 2014). However, the contents of n-3 LC-PUFA in fillets of tilapia fed vegetable oil diets are known to be lower than fish fed fish oil-based diets (Chen et al. 2018, Teoh and Ng 2016). Fish oil are considered the best source of n-3 LC-PUFA in aquafeed industry to promote improved meat quality; however, it is increasingly recognized as an environmentally unsustainable and economically unviable practice (Turchini and Francis 2009). Recently, linolenic acid supplementation was demonstrated to enhance cold tolerance in Nile tilapia raised under suboptimal temperature (Corrêa et al. 2017), also enhancing fatty acid profile in fish muscle. Thus, fatty acids' supplementation has been considered to improve beneficial fatty acid profile for consumers, also considering beneficial effects on fish health and meat quality.

6. Exogenous enzymes

Alternative vegetable sources have been evaluated in the substitution of fishmeal in tilapia diets. However, vegetable sources have high variability in antinutritional factors, particularly regarding non-starch polysaccharides (Haidar et al. 2016, Sinha et al. 2011) and phytic acids (Li et al. 2019a), which impair the nutritional value of the feed, contributes to increased waste production (Kokou and Fountoulaki 2018), and limits its inclusion in fish diets. The use of exogenous enzymes as feed additives to improve nutrient digestibility in vegetable-based feedstuffs has been increasing in fish nutrition, providing more energy and nutrients from feed for tilapias, allowing flexibility in feed formulation. In addition, exogenous enzymes have been demonstrating an important role in reducing the negative environmental impact of aquaculture by reducing waste excretion (Castillo and Gatlin III 2015).

Recently, combined supplementation of phytase, xylanase (Hassaan et al. 2019), protease (Li et al. 2016, Yang et al. 2019), cellulase (Yigit and Olmez 2011) and mannanase (Chen et al. 2016) has been assessed successfully in tilapia feeds, and the combination of phytase and xylanase (Maas et al. 2019), phytase and protease (Li et al. 2019b) and phytase, protease and xylanase (Adeoye et al. 2016) enzymes have demonstrated synergic effects to improve the utilization of dietary nutrients when compared to the isolated supplementation.

7. Insects meal

Edible insects as an alternative sustainable protein rich source for human food and animal feed are impressive in terms of lower water waste, low land use, large number of offspring per reproduction, and their ability to transform low-value organic side streams into high-value protein products (Makkar et al. 2014, Van Huis 2013). In general, insects have high protein content, essential amino acids, lipids, minerals and vitamins. Their nutritional composition may vary according to insect species, life stage, and rearing conditions (Finke 2002). Furthermore, insect meal has been shown to be promising as a nutritional and nutraceutical ingredient in animal feed (Barroso et al. 2014, Dietz and Liebert 2018).

Insects are a natural food source for marine and freshwater fish species, including Nile tilapia. Based on the apparent digestibility coefficients of various insect meals, *Tenebrio molitor* larvae meal has been demonstrated as suitable feed for Nile tilapia fingerlings. In addition, the presence of chitinolytic enzymes in the digestive system was described in Nile tilapia fingerlings. Despite this, high levels of chitin in insect meals may decrease the nutritional value by reducing the apparent digestibility coefficients (Fontes et al. 2019).

Dietary inclusion levels of insect meal varies from 18.5% for Black soldier, *Hermetia illucens* fly larvae (Dietz and Liebert 2018) to 25% for maggot meal (magmeal) Locusta migratoria for Nile tilapia. Insect meal has demonstrated a potential protein source in aquafeeds. However, production of insect meal needs to be processed in large amounts at a reasonable price and be available year-round. Furthermore, a number of questions regarding insect meals such as the nutritive value related to presence of chitin and fatty acids profile remain to be addressed in tilapia diets.

8. Plant extracts

The plant extracts have a variety of functions due to the presence of various active compounds like alkaloids, flavonoids, pigments, phenolics, terpenoids, steroids and essential oils. Plant extracts are known to possess properties such as appetite stimulation, growth promotion, antimicrobial, antiparasitic, antioxidant, and immunostimulant agents on *in-vitro* and *in-vivo* applications and therefore acting as functional foods. The use of plant extracts has been widely investigated as an alternative therapy to antibiotics in tilapia. A previous study has emphasized the inclusion of dietary dry oregano (1.5 g/kg) for improvement in the immune system and resistance of Nile tilapia against *Streptococcus agalactiae* infection (Espirito

Santo et al. 2019). Similarly, dietary supplementation of green tea, *Camellia sinensis* (5 g/kg) revealed improved growth performance, health and resistance of Nile tilapia in preventing Aeromoniosis (Abdel-Tawwab et al. 2010). Additional study attested Indian herb, *Solanum trilobatum,* as antioxidant, also resulting in enhanced serum lysozyme activity, and consequently lower mortality following fish challenged with *Aeromonas hydrophila* (Divyagnaneswari et al. 2007). In addition, Tilapia mossambica, *Oreochromis mossambicus* fed extract of the Chinese herb, *Toonasinensis* (4 or 8 µg/g) has demonstrated increased respiratory burst, lysozyme activity, and phagocytic cell activity (Wu et al. 2010). Plant extracts may act as growth promoters in tilapia feed by acting on the gastrointestinal tract secretions of digestive enzymes, accelerating gastric emptying, stimulating appetite, increasing mucus production and glucose uptake in the intestine (Freccia et al. 2014). Recently, allicin, a biological component of garlic clove extracts, was described to ameliorate deltamethrin-induced oxidative stress and might have some therapeutic properties to protect Nile tilapia from subacute deltamethrin toxicity (Abdel-Daim et al. 2015). In tilapia farming, garlic has been used for enhancing growth and resistance against challenge of infection with *Aeromonas hydrophila* (Aly and Mohamed 2010). In addition, antiparasitic activity against Trichodinosis and Gyrodactylosis has been reported (Abd El-Galil and Aboelhadid 2012). More recent evidence highlighted the use of clove basil oil in increasing the health status of fish (Brum et al. 2018), while rosemary, *Rosmarinus officinalis,* enhances growth performance, antioxidant activity, and innate immunity of Nile tilapia (Naiel et al. 2019). Plant extracts have demonstrated potential use in the aquafeeds industry. However, there is still considerable controversy concerning essential oils application in tilapia diets, particularly involving the standardization of the evaluated essential oils (crude extract vs isolated active ingredient). This standardization is recommended because the active principles may present synergistic or antagonistic effects and differences in edaphoclimatic characteristics.

9. Considerations and perspectives

Progress has been made in the last decade to better define the nutritional requirements to develop suitable and environmentally friendly feeds for tilapias.

There is an urgent research need to determine the requirements of tilapias above 1,000 g because of preferred market size. While optimizing growth and feed efficiency will remain important criteria in considering nutritional adequacy of diets for large tilapias, equally important will be the need for nutritional specifications that produce healthy fish with desirable market characteristics, especially modulating fatty acids' profile and enhancing carcass traits, particularly fillet yield of market size fish.

A continuous provision of adequate new biotechnological applications is necessary to sustain continuous improvements in production, particularly genomics, a valuable and relatively new tool that has been used to support findings in tilapia research. Thus, nutritional influence on gene expression has been widely evaluated in tilapias to elucidate how nutrients influence metabolism in terms of growth and health, in order to increase production in aquaculture. Mostly, nutrigenomics studies have focused on how carbohydrates, fatty acids, and amino acids regulate gene expression by modulating the activity of transcription factors or secretion of hormones, through qPCR, micro-array or RNA-seq analysis.

Although few studies have been conducted to evaluate gene expression modulation by vitamins (Kumer et al. 2012, Tang et al. 2013) in tilapias, the findings comprise a huge step forward by enhancing the understanding regarding immunological diseases related with nutritional status.

Recent research has quantified the broodstock requirements for energy, protein, fatty acids, phosphorus and vitamin C, and this will assume greater importance to produce healthy fish, also considering meat quality. The effects of dietary supplements like prebiotics, probiotics, and symbiotic have been evidenced, and has already been established in aquaculture practices especially as a promising alternative to chemicals and antibiotics.

The tilapia industry depends on a reliable supply of healthy fry and fingerlings to optimize grow-out performance on fish farms. Therefore, appropriate feed and feeding program may be considered to maximize growth performance of fish.

Sustainable and nutritive alternative source like insect meal has demonstrated potential to elaborate tilapia feeds. However, large-scale production may be considered and applied in commercial scale aquafeed production.

Acknowledgments

The authors wish to thank public and private research agencies and to all post-graduate students throughout the years. The authors are grateful for Dr. Yu Kawakami for English writing-review contribution.

References cited

Abd El-Galil, M.A.A. and S.M. Aboelhadid. 2012. Trials for the control of trichodinosis and gyrodactylosis in hatchery reared *Oreochromis niloticus* fries by using garlic. Vet. Parasitol. 185: 57–63.

Abdel-Daim, M.M., N.K.M. Abdelkhalek and A.M. Hassan. 2015. Antagonistic activity of dietary allicin against deltamethrin-induced oxidative damage in freshwater Nile tilapia *Oreochromis niloticus*. Ecotoxicol. Environ. Saf. 111: 146–152.

Abdel-Tawwab, M., M.H. Ahmad, M.E.A. Seden and S.F.M. Sakr. 2010. Use of green tea, *Camellia sinensis* L., in practical diet for growth and protection of Nile tilapia, *Oreochromis niloticus* (L.), against *Aeromonas hydrophila* infection. J. World Aquac. Soc. 41: 203–213.

Abou-El-Atta, M.E., M. Abdel-Tawwab, N. Abdel-Razek and T.M.N. Abbelhakim. 2019. Effects of dietary probiotic Lactobacillus plantarum and whey protein concentrate on the productive parameters, immunity response and susceptibility of Nile tilapia, *Oreochromis niloticus* (L.), to Aeromonas sobria infection. Aquac. Nutr. 1–11.1

Adeoye, A.A., R. Yomla, A. Jaramillo-Torres, A. Rodiles, D.L. Merrifield and S.J. Davies. 2016. Combined effects of exogenous enzymes and probiotic on Nile tilapia (*Oreochromis niloticus*) growth, intestinal morphology and microbiome. Aquaculture 463: 61–70.

Al-Feky, S.S.A., A.-F.M. El-Sayed and A.A. Ezzat. 2014. Dietary taurine improves reproductive performance of Nile tilapia (*Oreochromis niloticus*) broodstock. Aquac. Nutr. 22: 392–399.

Al-Feky, S.S.A., A.-F.M. El-Sayed and A.A. Ezzat. 2016. Dietary taurine enhances growth and feed utilization in larval Nile tilapia (*Oreochromis niloticus*) fed soybean meal-based diets. Aquac. Nutr. 22: 457–464.

Aly, S.M. and M.F. Mohamed. 2010. Echinacea purpurea and Allium sativum as immunostimulants in fish culture using Nile tilapia (*Oreochromis niloticus*). J. Anim. Physiol. Anim. Nutr. (Berl). 94: e31–e39.

Barros, M.M., D.R. Falcon, R.O. Orsi, L.E. Pezzato, A.C. Fernandes Junior, A. Fernandes Junior, et al. 2015. Immunomodulatory effects of dietary β-glucan and vitamin C in Nile tilapia, *Oreochromis niloticus* L., subjected to cold-induced stress or bacterial challenge. J. World Aquac. Soc. 46: 363–380.

Barroso, F.G., C.de. Haro, M.-J. Sánchez-Muros, E. Venegas, A. Martínez-Sánchez and C. Pérez-Bañón. 2014. The potential of various insect species for use as food for fish. Aquaculture 422: 193–201.

Bombardelli, R.A., E.S. dos Reis Goes, S.M. de Negreiros Sousa, M.A. Syperreck, M.D. Goes, A.C. de O. Pedreira, et al. 2017. Growth and reproduction of female Nile tilapia fed diets containing different levels of protein and energy. Aquaculture 479: 817–823.

Boonanuntanasarn, S., A. Jangprai, S. Kumkhong, E. Plagnes-Juan, V. Veron, C. Burel, et al. 2018a. Adaptation of Nile tilapia (*Oreochromis niloticus*) to different levels of dietary carbohydrates: New insights from a long term nutritional study. Aquaculture 496: 58–65.

Boonanuntanasarn, S., N. Tiengtam, T. Pitaksong, P. Piromyou and T. Teaumroong 2018b. Effects of dietary inulin and Jerusalem artichoke (*Helianthus tuberosus*) on intestinal microbiota community and morphology of Nile tilapia (*Oreochromis niloticus*) fingerlings. Aquac. Nutr. 24: 712–722.

Brum, A., S.A. Pereira, L. Cardoso, E.C. Chagas, F.C.M. Chaves, J.L.P. Mouriño, et al. 2018. Blood biochemical parameters and melanomacrophage centers in Nile tilapia fed essential oils of clove basil and ginger. Fish Shellfish Immunol. 74: 444–449.

Carbonera, F., G.E. Bonafe, C.A. Margins, P.F. Montanher, R.P. Pibeiro, L.C. Figueiredo, et al. 2014. Effect of dietary replacement of sunflower oil with perilla oil on the absolute fatty acid composition in Nile tilapia (GIFT). Food Chem. 148: 230–234.

Carvalho, P.L.P.F., J.F.A. Koch, F.T. Cintra, A.C. Fernandes Júnior, M.M.P. Sartori, M.M. Barros, et al. 2018. Available phosphorus as a reproductive performance enhancer for female Nile tilapia. Aquaculture 486: 202–209.

Castillo, S. and D.M. Gatlin III. 2015. Dietary supplementation of exogenous carbohydrase enzymes in fish nutrition: A review. Aquaculture 435: 286–292.

Chen, C., B. Sun, X. Li, P. Li, W. Guan, Y. Bi, et al. 2013. N-3 essential fatty acids in Nile tilapia, *Oreochromis niloticus*: Quantification of optimum requirement of dietary linolenic acid in juvenile fish. Aquaculture 416: 99–104.

Chen, C., W. Guan, Q. Xie, G. Chen, X. He, H. Zhang, et al. 2018. N-3 essential fatty acids in Nile tilapia, *Oreochromis niloticus*: Bioconverting LNA to DHA is relatively efficient and the LC-PUFA biosynthetic pathway is substrate limited in juvenile fish. Aquaculture 495: 513–522.

Chen, W., S. Lin, F. Li and S. Mao. 2016. Effects of dietary mannanase on growth, metabolism and non-specific immunity of Tilapia (*Oreochromis niloticus*). Aquac. Res. 4: 2835–2843.

Chowdhury, M.A.K., S. Siddiqui, H. Katheline and D.P. Bureau. 2013. Bioenergetics-based factorial model to determine feed requirement and waste output of tilapia produced under commercial conditions. Aquaculture 410–411: 138–147.

Corrêa, C.F., R.O. Nobrega, B. Mattioni, J.M. Block and D.M. Fracalossi. 2017. Dietary lipid sources affect the performance of Nile tilapia at optimal and cold, suboptimal temperatures. Aquac. Nutr. 23: 1016–1026.

Coutinho, J.J.O., L.M. Neira, C.G. Sandre, J.I. Costa, M.I.E.G. Martins, M.C. Portella, et al. 2018. Carbohydrate-to-lipid ratio in extruded diets for Nile tilapia farmed in net cages. Aquaculture 497: 520–525.

Davies, S.J., A.A. Abdel-Warith and A. Gouveia. 2011. Digestibility characteristics of selected feed ingredients for developing bespoke diets for Nile tilapia culture in Europe and North America. J. World Aquac. Soc. 43: 338–398.

Dawood, M.A.O., N.M. Eweedah, E.M. Moustafa and M.G. Shahin. 2019. Effects of feeding regimen of dietary *Aspergillus oryzae* on the growth performance, intestinal morphometry and blood profile of Nile tilapia (*Oreochromis niloticus*). Aquac. Nutr. 25: 1063–1072.

Dawood, M.A.O., M. Zommara, N.M. Eweedah and A.I. Helal. 2020. The evaluation of growth performance, blood health, oxidative status and immune-related gene expression in Nile tilapia (*Oreochromis niloticus*) fed dietary nanoselenium spheres produced by lactic acid bacteria. Aquaculture 515: 734571.

Dempster, P., D.J. Baird and M.C.M. Beveridge. 1995. Can fish survive by filter-feeding on microparticles? Energy balance in tilapia grazing on algal suspensions. J. Fish Biol. 47: 7–17.

Dietz, C. and F. Liebert. 2018. Does graded substitution of soy protein concentrate by an insect meal respond on growth and N-utilization in Nile tilapia (*Oreochromis niloticus*)? Aquac. Reports 12: 43–48.

Divyagnaneswari, M., D. Christybapita and R.D. Michael. 2007. Enhancement of nonspecific immunity and disease resistance in *Oreochromis mossambicus* by *Solanum trilobatum* leaf fractions. Fish Shellfish Immunol. 23: 249–259.

El-Sayed, A.-F.M., C.R. Mansour and A.A. Ezzat. 2005. Effects of dietary lipid source on spawning performance of Nile tilapia (*Oreochromis niloticus*) broodstock reared at different water salinities. Aquaculture 248: 187–196.

El-Sayed, A.-F.M. and M. Kawanna. 2008. Effects of dietary protein and energy levels on spawning performance of Nile tilapia (*Oreochromis niloticus*) broodstock in a recycling system. Aquaculture 280: 179–184.

Elsabagh, M., R. Mohamed, E. Moustafa, A. Hamza, F. Farrag, O. Decamp, et al. 2018. Assessing the impact of *Bacillus* strains mixture probiotic on water quality, growth performance, blood profile and intestinal morphology of Nile tilapia, *Oreochromis niloticus*. Aquac. Nutr. 24: 1613–1622.

Espirito Santo, A.H., T.S. Brito, L.L. Brandão, G.C. Tavares, M.P. Leibowitz, S.A. Prado, et al. 2019. Dietary supplementation of dry oregano leaves increases the innate immunity and resistance of Nile tilapia against *Streptococcus agalactiae* infection. J. World Aquac. Soc. 1–19.

Fan, L., J. Chen, S. Meng, C. Song, L. Qiu, G. Hu, et al. 2017. Characterization of microbial communities in intensive GIFT tilapia (*Oreochromis niloticus*) pond systems during the peak period of breeding. Aquac. Res. 48: 459–472.

FAO, 2016. (Food and Agriculture Organization). The State of World Fisheries and Aquaculture. FAO, Rome.

FAO, 2019. Cultured Aquatic Species Information Programme. *Oreochromis niloticus* (Linnaeus, 1758), FAO. Rome.

Ferguson, R.M.W., D.L. Merrifield, G.M. Harper, M.D. Rawling, S. Mustafa, S. Picchietti, et al. 2010. The effect of *Pediococcus acidilactici* on the gut microbiota and immune status of on-growing red tilapia (*Oreochromis niloticus*). J. Applied Microbiol. 109: 851–862.

Fernandes Junior, A.C., P.L.P.F.C. Carvalho, L.E. Pezzato, J.F.A. Koch, C.P. Teixeira, F.T. Cintra, et al. 2016. The effect of digestible protein to digestible energy ratio and choline supplementation on growth, hematological parameters, liver steatosis and size-sorting stress response in Nile tilapia under field condition. Aquaculture 456: 83–93.

Ferreira, M.W., F.G. Araujo, D.V. Costa, P.V. Rosa, J.C.P. Figueiredo and L.D.S Murgas. 2011. Influence of dietary oil sources on muscle composition and plasma lipoprotein concentrations in Nile tilapia, *Oreochromis niloticus*. J. World Aquac. Soc. 42: 24–33.

Finke, M.D. 2002. Complete nutrient composition of commercially raised invertebrates used as food for insectivores. Zoo Biol. 21: 269–285.

Fontes, T.V., K.R.B. Oliveira, I.L.G. Almeida, T.M. Orlando, P.B. Rodrigues, D.B. Costa, et al. 2019. Digestibility of insect meals for Nile tilapia fingerlings. Animals 9: 181.

Freccia, A., S.M. de N. Sousa, F. Meurer, A.J. Butzge, J.K. Mewes and R.A. Bombardelli. 2014. Essential oils in the initial phase of broodstock diets of Nile tilapia. Rev. Bras. Zootec. 43: 1–7.

Gan, L., L.L. Zhou, X.X. Li and Y.R. Yue. 2016. Dietary leucine requirement of juvenile Nile tilapia, *Oreochromis niloticus*. Aquac. Nutr. 22: 1010–1046.

Guerreiro, I., A. Oliva-Teles and P. Enes. 2018. Prebiotics as functional ingredients: Focus on Mediterranean fish aquaculture. Rev. Aquac. 10: 800–832.

Guimarães, I.G., C. Lim, M. Yildirim-Aksoy, M.H. Li and P.H. Klesius. 2014. Effects of dietary levels of vitamin A on growth, hematology, immune response and resistance of Nile tilapia (*Oreochromis niloticus*) to *Streptococcus iniae*. Anim. Feed Sci. Technol. 188: 126–136.

Guimarães, I.G., L.E. Pezzato, M.M. Barros and R.N. Fernandes. 2012. Apparent nutrient digestibility and mineral availability of protein-rich ingredients in extruded diets for Nile tilapia. Rev. Bras. Zootec. 41: 1801–1808.

Guimarães, I.G., L.E. Pezzatto, M.M. Barros and L. Tachibana. 2008. Nutrient AA digestibility of cereal grain products and by-products in extruded diets for Nile tilapia. J. Word Aquculture Soc. 39: 781–789.

Hai, N.V. 2015. Research findings from the use of probiotics in tilapia aquaculture: A review. Fish Shellfish Immunol. 45: 592–597.

Haidar, M.N., M. Petie, L.T.N. Heinsbroek, J.A.J. Verreth and J.W. Schrama. 2016. The effect of type of carbohydrate (starch vs. nonstarch polysaccharides) on nutrients digestibility, energy retention and maintenance requirements in Nile tilapia. Aquaculture 463: 241–247.

Hamdan, A.M., A.F.M. El-Sayed and M.M. Mahmoud. 2016. Effects of a novel marine probiotic, Lactobacillus plantarum AH 78, on growth performance and immune response of Nile tilapia (*Oreochromis niloticus*). J. Applied Microbiol. 4: 1061–1073.

Hassaan, M.S., M.A. Soltan, S. Jarmotowicz and H.S. Abdo. 2018. Combined effects of dietary malic acid and *Bacillus subtilis* on growth, gut microbiota and blood parameters of Nile tilapia (*Oreochromis niloticus*). Aquac. Nutr. 24: 83–93.

Hassaan, M.S., E.Y. Mohammady, M.R. Soaudy and A.A.S. Abdel Rahman. 2019. Exogenous xylanase improves growth, protein digestibility and digestive enzymes activities in Nile tilapia, *Oreochromis niloticus*, fed different ratios of fish meal to sunflower meal. Aquac. Nutr. 25: 841–853.

Haygood, A.M. and R. Jha. 2018. Strategies to modulate the intestinal microbiota of tilapia (*Oreochromis* sp.) in aquaculture: A review. Rev. Aquac. 10: 320–333.

He, J.Y., L.X. Tian, A. Lemme, W. Gao, H.J. Yang, J. Niu, et al. 2013. Methionine and lysine requirements for maintenance and efficiency of utilization for growth of two sizes of tilapia (*Oreochromis niloticus*). Aquac. Nutr. 19: 629–640.

Huang, F., M. Jiang, H. Wen, F. Wu, W. Liu, J. Tian, et al. 2016. Dietary vitamin C requirement of genetically improved farmed tilapia, *Oreochromis niloticus*. Aquac. Res. 47: 689–697.

Huang, F., M. Jiang, H. Wen, F. Wu, W. Liu, J. Tian, et al. 2015. Dietary zinc requirement of adult Nile tilapia (*Oreochromis niloticus*) fed semi-purified diets, and effects on tissue mineral composition and antioxidant responses. Aquaculture 439: 53–59.

Izquierdo, M.S., H. Fernández-Palacios and A.G.J. Tacon. 2001. Effect of broodstock nutrition on reproductive performance of fish. Aquaculture 197: 25–42.

Jiang, M., F. Huang, H. Wen, C. Yang, F. Wu, W. Liu, et al. 2014. Dietary niacin requirement of GIFT tilapia, *Oreochromis niloticus*. J. World Aquac. Soc. 45: 333–341.

Koch, J.F.A., M.M. Barros, C.P. Teixeira, P.L.P.F. Carvalho, A.C. Fernandes Junior, F.T. Cintra, et al. 2017. Protein-to-energy ratio of 21.43 g MJ^{-1} improves growth performance of Nile tilapia at the final rearing stage under commercially intensive rearing conditions. Aquac. Nutr. 23: 560–570.

Koh, C.B., N. Romano, A.S. Zahrah and W.K. Ng. 2016. Effects of a dietary organic acids blend and oxytetracycline on the growth, nutrient utilization and total cultivable gut microbiota of the red hybrid tilapia, *Oreochromis* sp., and resistance to *Streptococcus agalactiae*. Aquac. Res. 47: 357–369.

Kokou, F. and E. Fountoulaki. 2018. Aquaculture waste production associated with antinutrient presence in common fish feed plant ingredients. Aquaculture 495: 295–310.

Kumer, P., S. Rodrigue and G.W. Vandenberg. 2012. Influences of dietary biotin and avidin on growth, survival, deficiency syndrome and hepatic gene expression of juvenile Nile tilapia *Oreochromis niloticus*. Fish Physiol. Biochem. 38: 1183–1193.

Lee, S., R.W. Nambi, S. Won, K. Katya and S.C. Bai. 2016. Dietary selenium requirement and toxicity levels in juvenile Nile tilapia, *Oreochromis niloticus*. Aquaculture 464: 153–158.

Levy-Pereira, N., G.S. Yasui, M.V. Cardozo, J. Dias Neto, T.H.V. Farias, R. Sakabe, et al. 2018. Immunostimulation and increase of intestinal lactic acid bacteria with dietary mannan-oligosaccharide in Nile tilapia juveniles. Rev. Bras. Zootec. 47: e20170006.

Li, E., C. Lim, P.H. Klesius and T.L. Welker. 2013. Growth, body fatty acid composition, immune response, and resistance to *Streptococcus iniae* of hybrid tilapia, *Oreochromis niloticus × Oreochromis aureus*, fed diets containing various levels of linoleic and linolenic acids. J. World Aquac. Soc. 44: 42–55.

Li, X.Q., X.Q. Chai, D.Y. Liu, M.A. Kabir, M.A. Chowdhury and X.J. Leng. 2016. Effects of temperature and feed processing on protease activity and dietary protease on growths of white shrimp, *Litopenaeus vannamei*, and tilapia, *Oreochromis niloticus × O. aureus*. Aquac. Nutr. 22: 1283–1292.

Li, X.Q., X.Q. Zhang, M.A.K. Chowdhury, Y. Zhang and X.J. Leng. 2019a. Dietary phytase and protease improved growth and nutrient utilization in tilapia (*Oreochromis niloticus × Oreochromis aureus*) fed low phosphorus and fishmeal-free diets. Aquac. Nutr. 25: 46–55.

Li, X.Q., X.Q. Zhang, M.A.K. Chowdhury, Y. Zhang and X.J. Leng. 2019b. Dietary phytase and protease improved growth and nutrient utilization in tilapia (*Oreochromis niloticus × Oreochromis aureus*) fed low phosphorus and fishmeal-free diets. Aquac. Nutr. 25: 46–55.

Lim, C., M. Yildirim-Aksoy, M.M. Barros and P. Klesius. 2011a. Thiamin requirement of Nile tilapia, *Oreochromis niloticus*. J. World Aquac. Soc. 42: 824–833.

Lim, C., M. Yildirim-Aksoy and P. Klesius. 2011b. Lipids in aquaculture lipid and fatty acid requirements of tilapias. N. Am. J. Aquac. 73: 188–193.

Lin, Y.H., S.M. Lin and S.Y. Shiau. 2008. Dietary manganese requirements of juvenile tilapia, *Oreochromis niloticus × O . aureus*. Aquaculture 284: 207–210.

Lin, Y.H., C.Y. Ku and S.Y. Shiau. 2013. Estimation of dietary magnesium requirements of juvenile tilapia, *Oreochromis niloticus × Oreochromis aureus*, reared in freshwater and seawater. Aquaculture 380–383: 47–51.

Liu, H.Y., Q. Chen, B.P. Tan, X.H. Dong, S.Y. Chi, Q.H. Yang, et al. 2018. Effects of dietary carbohydrate levels on growth, glucose tolerance, glucose homeostasis and GLUT4 gene expression in *Tilapia nilotica*. Aquac. Res. 49: 3735–3745.

Liu, W., M. Jiang, J.P. Wu, F. Wu, J. Tian, C.G. Yang, et al. 2017a. Dietary protein level affects the growth performance of large male genetically improved farmed tilapia, *Oreochromis niloticus*, reared in fertilized freshwater cages. J. World Aquac. Soc. 48: 718–728.

Liu, W., W. Wang, C. Ran, S. He, Y. Yang and Z. Zhou. 2017b. Effects of dietary scFOS and lactobacilli on survival, growth, and disease resistance of hybrid tilapia. Aquaculture 470: 50–55.

Liu, Y., J.G. Jiao, S. Gao, L.J. Ning, S.M. Limbu, F. Qiao, et al. 2019. Dietary oils modify lipid molecules and nutritional value of fillet in Nile tilapia: A deep lipidomics analysis. Food Chem. 277: 515–523.

Lupatsch, I., R. Deshev and I. Magen. 2010. Energy and protein demands for optimal egg production including maintenance requirements of female tilapia *Oreochromis niloticus*. Aquac. Res. 41: 763–769.

Maas, R.M., M.C.J. Verdegem and J.W. Schrama. 2019. Effect of non-starch polysaccharide composition and enzyme supplementation on growth performance and nutrient digestibility in Nile tilapia (*Oreochromis niloticus*). Aquac. Nutr. 25: 622–632.

Makkar, H.P.S., G. Tran, V. Heuzé and P. Ankers. 2014. State-of-the-art on use of insects as animal feed. Anim. Feed Sci. Technol. 197: 563–583.

Makled, S.O., A.M. Hamdan and A.F. El-Sayed. 2019. Effects of dietary supplementation of a marine thermotolerant bacterium, *Bacillus paralicheniformis* SO-1, on growth performance and immune responses of Nile tilapia, *Oreochromis niloticus*. Aquac. Nutr. 25: 817–827.

Mañanos, E., N. Duncan and C. Mylonas. 2008. Reproduction and control of ovulation, spermiation and spawning in cultured fish. pp. 3–80. *In*: Cabrita, E., Robles, V. and Herraez, P. (Eds.), Methods in Reproductive Aquaculture: Marine and Freshwater Species. CRC Press, Taylor and Francis Group, Boca Raton.

Michelato, M., W.M. Furuya and D.M. Gatlin. 2018. Metabolic responses of Nile tilapia *Oreochromis niloticus* to methionine and taurine supplementation. Aquaculture 485: 66–72.

Michelato, M., L.V.O. Vidal, T.O. Xavier, T.S. Graciano, L.B. Moura, V.R.B. Furuya, et al. 2016a. Dietary threonine requirement to optimize protein retention and fillet production of fast-growing Nile tilapia. Aquac. Nutr. 22: 759–766.

Michelato, M., L.V.O. Vidal, T.O. Xavier, L.B. Moura, F.L.A. Almeida, V.B. Pedrosa, et al. 2016b. Dietary lysine requirement to enhance muscle development and fillet yield of finishing Nile tilapia. Aquaculture 457: 124–130.

Moura, L.B., T.O. Xavier, D.A.V. Campelo, M. Michelato, F.L.A. Almeida, L.V.O. Vidal, et al. 2018. Availability of minerals in rendered meat and bone meal for Nile tilapia: Preliminary observations. Aquac. Nutr. 24: 991–997.

Naiel, M.A.E., N.E.M. Ismael and S.A. Shehata. 2019. Ameliorative effect of diets supplemented with rosemary (*Rosmarinus officinalis*) on aflatoxin B1 toxicity in terms of the performance, liver histopathology, immunity and antioxidant activity of Nile tilapia (*Oreochromis niloticus*). Aquaculture 511: 734264.

Nawaz, A., A. Bakhsh javaid, S. Irshad, S.H. Hoseinifar and H. Xiong. 2018. The functionality of prebiotics as immunostimulant: Evidences from trials on terrestrial and aquatic animals. Fish Shellfish Immunol. 76: 272–278.

Nayak, S.K. 2010. Role of gastrointestinal microbiota in fish. Aquac. Res. 41: 1553–1573.

Neu, D. 2016. Growth performance, biochemical responses, and skeletal muscle development of juvenile Nile tilapia, *Oreochromis niloticus*, fed with increasing levels of arginine. J. World Aquac. Soc. 47: 248–259.

Ng, W.-K. and Y. Wang. 2011. Inclusion of crude palm oil in the broodstock diets of female Nile tilapia, *Oreochromis niloticus*, resulted in enhanced reproductive performance compared to brood fish fed diets with added fish oil or linseed oil. Aquaculture 314: 122–131.

Nguyen, L., F. Kubitza, S.M.R. Salem, R. Terry and D.A. Davis. 2019a. Comparison of organic and inorganic microminerals in all plant diets for Nile tilapia *Oreochromis niloticus*. Aquaculture 498: 297–304.

Nguyen, L., S.M.R. Salem, G.P. Salze, H. Dinh and D.A. Davis. 2019b. Tryptophan requirement in semi-purified diets of juvenile Nile tilapia *Oreochromis niloticus*. Aquaculture 502: 258–267.

Nguyen, T.N. and D.A. Davis. 2009. Re-evaluation of total sulphur amino acid requirement and determination of replacement value of cystine for methionine in semi-purified diets of juvenile Nile tilapia, *Oreochromis niloticus*. Aquac. Reports. 15: 247–253.

Nobrega, R.O., C.F. Corrêa, B. Mattioni and D.M. Fracalossi. 2017. Dietary α-linolenic for juvenile Nile tilapia at cold suboptimal temperature. Aquaculture 471: 66–71. https://doi.org/10.1016/j.aquaculture.2016.12.026

NRC, 2011. Nutrient requirements of fish and shrimp. National Academy Press, Washington.

Oliveira, M.M., T. Ribeiro, T.M. Orlando, D.G.S. Oliveira, M.M. Drumond, R.T.F. Freitas, et al. 2014. Effects crude protein levels on female Nile tilapia (*Oreochromis niloticus*) reproductive performance parameters. Anim. Reprod. Sci. 150: 62–69.

Orlando, T.M., M.M. Oliveira, R.R. Paulino, A. Carvalho, I.B. Allaman and P.V. Rosa. 2017. Reproductive performance of female Nile tilapia (*Oreochromis niloticus*) fed diets with different digestible energy levels. Rev. Bras. Zootec. 46: 1–7.

Pereira, M.M., M.M. Evangelista, E. Gisbert and E. Romagosa. 2019. Nile tilapia broodfish fed high-protein diets: Digestive enzymes in eggs and larvae. Aquac. Res. 50: 2181–2190. https://doi.org/10.1111/are.14098

Pereira, R.T., P.V. Rosa and D.M. Gatlin III. 2017. Glutamine and arginine in diets for Nile tilapia: Effects on growth, innate immune response, plasma amino acid profiles and whole-body composition. Aquaculture 473: 135–144.

Pilarski, F., C.A.F. Oliveira, F.P.B.D. Souza and F.S. Zanuzzo. 2017. Different β-glucans improve the growth performance and bacterial resistance in Nile tilapia. Fish Shellfish Immunol. 70: 25–29.

Ridha, M. and T.I.S. Azad. 2012. Preliminary evaluation of growth performance and immune response of Nile tilapia *Oreochromis niloticus* supplemented with two putative probiotic bacteria. Aquac. Res. 43: 843–852.

Robinson, E.H., D. Labomascus, P.B. Brown and T.L. Linton. 1987. Dietary calcium and phosphorus requirements of *Oreochromis aureus* reared in calcium-free water. Aquaculture 64: 267–276.

Santiago, C.B and R.T. Lovell. 1988. Amino acid requirements for growth of Nile tilapia. J. Nutr. 118: 1540–1546.

Saravanan, S., I. Geurden, Z.G.A. Orozco, S.J. Kaushik, J.A.J. Verreth and J.W. Schrama. 2013. Dietary electrolyte balance affects the nutrient digestibility and maintenance energy expenditure of Nile tilapia. Br. J. Nutr. 110: 1948–1957.

Sargent, J.R., D.R. Ocher and J.G. Bell. 2002. The lipids. pp. 181–257. *In*: Halver, J.E. and Hardy, R.W. (Eds.), Fish Nutrition. Academic Press, San Diego.

Sarmento, N.L.A.F., E.F.F. Martins, D.C. Costa, W.S. Silva, C.C. Mattioli and R.K. Luz. 2016. Effects of supplemental dietary vitamin C on quality of semen from Nile tilapia (*Oreochromis niloticus*) breeders. Reprod. Domest. Anim. 1–9.

Schrama, J.W., M.N. Haidar, I. Geurden, L.T.N. Heinsbroek and S.J. Kaushik. 2018. Energy efficiency of digestible protein, fat and carbohydrate utilisation for growth in rainbow trout and Nile tilapia. Br. J. Nutr. 119: 782–791.

Shiau, S.-S. and J.-H. Hsieh. 2001. Quantifying the dietary potassium requirement of juvenile hybrid tilapia (*Oreochromis niloticus* × *O. aureus*). Br. J. Nutr. 85: 213–218. https://doi.org/10.1079/BJN2000245

Shiau, S.-Y and L.-W. Su. 2003. Ferric citrate is half as effective as ferrous sulfate in meeting the iron requirement of juvenile tilapia, *Oreochromis niloticus* × *O. aureus*. J. Nutr. 133: 483–488.

Shiau, S.Y. and Y.C. Ning. 2003. Estimation of dietary copper requirements of juvenile tilapia, *Oreochromis niloticus* × *O. aureus*. Anim. Sci. 77: 287–292.

Shiau, S.Y and L.S. Lu. 2004. Dietary sodium requirement determined for juvenile hybrid tilapia (*Oreochromis niloticus* × *O. aureus*) reared in fresh water and seawater. Br. J. Nutr. 91: 585–590.

Shiau, S.Y. and H.C. Tseng. 2007. Dietary calcium requirements of juvenile tilapia, *Oreochromis niloticus* × *O. aureus*, reared in fresh water. Aquac. Nutr. 13: 298–303.

Silva, W.S., L.S. Costa, N.C.S. Costa, W.M. Santos, P.A.P. Ribeiro and R.K. Luz. 2019. Gene expression, enzyme activity and performance of Nile tilapia larvae fed with diets of different CP levels. Animal 13: 1376–1384.

Sinha, A.K., V. Kumar, H.P.S. Makkar, G. De. Boeck and K. Becker. 2011. Non-starch polysaccharide in Fish Nutrition: An overview non-starch polysaccharides and their role in fish nutrition – A review. Food Chem. 127: 1409–1426.

Standen, B.T., D.L. Peggs, M.D. Rawling, A. Foey, S.J. Davies, G.A. Santos, et al. 2016. Dietary administration of a commercial mixed-species probiotic improves growth performance and modulates the intestinal immunity of tilapia, *Oreochromis niloticus*. Fish Shellfish Immunol. 49: 427–435.

Standen, B.T., A. Rodiles, D.L. Peggs, S.J. Davies, G.A. Santos and D.L. Merrifield. 2015. Modulation of the intestinal microbiota and morphology of tilapia, *Oreochromis niloticus*, following the application of a multi-species probiotic. Appl. Microbiol. Biotechnol. 99: 8403–8417.

Tan, H.Y., S.W. Chen and S.Y. Hu. 2019. Improvements in the growth performance, immunity, disease resistance, and gut microbiota by the probiotic *Rummeliibacillus stabekisii* in Nile tilapia (*Oreochromis niloticus*). Fish Shellfish Immunol. 92: 265–275.

Tang, X., M. Xu, Z. Li, Q. Pan and J. Fu. 2013. Effects of vitamin E on expressions of eight microRNAs in the liver of Nile tilapia (*Oreochromis niloticus*). Fish Shellfish Immunol. 34: 1470–1475.

Teixeira, C.P., M.M. Barros, L.E. Pezzato, Jr. A.C. Fernandes, J.F.A. Koch and C.R. Padovani. 2011. Growth performance of Nile tilapia, *Oreochromis niloticus*, fed diets containing levels of pyridoxine and haematological response under heat stress. Aquac. Res. 43: 1080–1088.

Teoh, C.Y. and W.K. Ng. 2016. The implications of substituting dietary fish oil with vegetable oils on the growth performance, fillet fatty acid profile and modulation of the fatty acid elongase, desaturase and oxidation activities of red hybrid tilapia, *Oreochromis* sp. Aquaculture 465: 311–322.

Trung, D.V., N. Thi, N. Tat and B. Glencross. 2011. Development of a nutritional model to define the energy and protein requirements of tilapia, *Oreochromis niloticus*. Aquaculture 320: 69–75.

Turchini, G.M. and D.S. Francis. 2009. Fatty acid metabolism (desaturation, elongation and *b*-oxidation) in rainbow trout fed fish oil- or linseed oil-based diets. Br. J. Nutr. 102: 69–81.

Van Doan, H., S.H. Hoseinifar, C. Chanongnuch, A. Kanpiengjai, K. Unban, V.V. Kim, et al. 2018a. Host-associated probiotics boosted mucosal and serum immunity, disease resistance and growth performance of Nile tilapia (*Oreochromis niloticus*). Aquaculture 491: 94–100.

Van Doan, H., S.H. Hoseinifar, P. Elumalai, S. Tongsiri, C. Chitmanat, S. Jaturasitha, et al. 2018b. Effects of orange peels derived pectin on innate immune response, disease resistance and growth performance of Nile tilapia (*Oreochromis niloticus*) cultured under indoor biofloc system. Fish Shellfish Immunol. 80: 56–62.

Van Huis, A. 2013. Potential of insects as food and feed in assuring food security. Annu. Rev. Entomol. 58: 563–583. https://doi.org/10.1146/annurev-ento-120811-153704

Vicente, I.S.T., L.F. Fleuri, P.L.P.F. Carvalho, M.G. Guimarães, R.F. Naliato, H.C. Müller, et al. 2019. Orange peel fragment improves antioxidant capacity and haematological profile of Nile tilapia subjected to heat/dissolved oxygen-induced stress. Aquac. Res. 50: 80–92.

Vidal, L.V.O., T.O. Xavier, M. Michelato, E.N. Martins, L.E. Pezzato and W.M. Furuya. 2015. Apparent protein and energy digestibility and amino acid availability of corn and co-products in extruded diets for Nile tilapia, *Oreochromis niloticus*. J. World Aquac. Soc. 46: 183–190.

Vidal, L.V.O., T.O. Xavier, L.B. Moura, T.S. Graciano, E.N. Martins and W.M. Furuya. 2017. Apparent digestibility of soybean coproducts in extruded diets for Nile tilapia, *Oreochromis niloticus*. Aquac. Nutr. 23: 228–235.

Wang, L.M., W.B. Zhu, J. Yang, L.H. Miao, J.J. Dong, F.B. Song, et al. 2018. Effects of dietary cystine and tyrosine on melanogenesis pathways involved in skin color differentiation of Malaysian red tilapia. Aquaculture 490: 149–155.

Wu, C.C., C.H. Liu, Y.P. Chang and S.L. Hsieh. 2010. Effects of hot-water extract of Toona sinensis on immune response and resistance to *Aeromonas hydrophila* in *Oreochromis mossambicus*. Fish Shellfish Immunol. 29: 258–263.

Wu, J.P., F. Wu, M. Jiang, H. Wen, Q.W. Wei, W. Liu, et al. 2016. Dietary folic acid (FA) requirement of a genetically improved Nile tilapia *Oreochromis niloticus* (Linnaeus, 1758). J. Appl. Ichthyol. 32: 1155–1160.

Xiao, W., D.Y. Li, J.L. Zhu, Z.Y. Zou, Y.R. Yue and H. Yang. 2018. Dietary valine requirement of juvenile Nile tilapia, *Oreochromis niloticus*. Aquac. Nutr. 24: 315–323.

Xiao, W., Z. Zou, D. Li, J. Zhu, Y. Yue and H. Yang. 2019. Effect of dietary phenylalanine level on growth performance, body composition, and biochemical parameters in plasma of juvenile hybrid tilapia, *Oreochromis niloticus* × *Oreochromis aureus*. Aquac. Nutr. 51: 1–15.

Yang, H., X. Li, G. Liang, Z. Xu and X.J. Leng. 2019. Cork and guar gum supplementation enhanced the buoyancy of faeces, and protease supplementation alleviated the negative effects of dietary cork on growth and intestinal health of tilapia, *Oreochromis niloticus* × *O. aureus*. Aquac. Nutr. 1–11.

Yao, Y.F., M. Jiang, H. Wen, F. Wu, W. Liu, J. Tian, et al. 2014. Dietary phosphorus requirement of GIFT strain of Nile tilapia *Oreochromis niloticus* reared in freshwater. Aquac. Nutr. 20: 273–280.

Yigit, N.O. and M. Olmez. 2011. Effects of cellulase addition to canola meal in tilapia (*Oreochromis niloticus* L.) diets. Aquac. Nutr. 17: e494–e500.

Zhou, X., Z. Tian, Y. Wang and W. Li. 2010. Effect of treatment with probiotics as water additives on tilapia (*Oreochromis niloticus*) growth performance and immune response. Fish Physiol. Biochem. 36: 501–509.

Zhou, Z., E. Ringø, R.E. Olsen and S.K. Song. 2018. Dietary effects of soybean products on gut microbiota and immunity of aquatic animals: A review. Aquac. Nutr. 24: 644–665.

Feeding Practices and Their Determinants in Tilapia

**Rodrigo Fortes-Silva[1,3]*, Igo Gomes Guimaraes[2] and
Edenilce de Fátima Ferreira Martíns[3]**

[1] Department of Animal Science, Laboratory of Aquaculture, University of Viçosa (UFV),
36570-000 Viçosa, Brazil

[2] Department of Animal Science, Laboratory of Aquaculture, University of Goiás (UFG),
36570-000 Jataí, Brazil

[3] Center of Agricultural Sciences, Environmental and Biological (CCAAB), Laboratory of
Feeding Behavior and Fish Nutrition (Aqua/UFRB), Federal University of Bahia (UFRB),
44380-000 Cruz das Almas, Brazil

1. Introduction

Food can represent about 40-70% of the total cost of aquaculture production. The knowledge of feeding behavior in tilapia could aid in formulation and development of appropriate aquafeeds (Fortes-Silva et al. 2016, Matthew et al. 2017). Numerous studies have been conducted in recent years on the use of alternative plant ingredients in tilapia feed; however, there is a lack of knowledge about feeding practices when considering food composition changes. Freshwater fish such as tilapia also exhibit food preferences. On the other hand, the farming of tilapia is practiced in many ways, and for each aquaculture system the food practice is different. In short, the purpose of this chapter is thus to review how food affects feeding practice in tilapia. Also, it addresses the aspects that tilapia feeding is facing considering practical production systems.

2. Food acceptance and anti-nutritional factors

The use of increased amounts of plant protein and oil sources in aquafeed can mean sustainable aquaculture (Joblin 2015). However, the acceptance of plant products may be an issue for some fish species, even for omnivores like Nile tilapia. Aversive responses to potentially toxic or harmful bitter molecules can reduce fish plant meal intake. Both dietary acceptance and bioavailability of nutrients may be related to

*Corresponding author: fortes-silva@ufv.br, fortesrs@yahoo.com.br

substances deriving from vegetable meal, known as anti-nutritional factors. This is treated as a natural animal defense. The taste barrier blocks the entry of any unpleasant, undesirable or harmful foods, and protects animals against metabolic stress and true health and life hazards (Kasumyan 2019). In this way, limitations of the inclusion levels of plant ingredients appear for most fish species, including tilapia, due to the presence of these substances (Adeoye et al. 2016). Some examples of anti-nutritional substances found in aquafeed are protease inhibitors, phytates, glucosinolates, saponins tannins, lectins, oligosaccharides and non-starch polysaccharides, phytoestrogens, alkaloids, antigenic compounds, gossypols, cyanogens, mimosine, cyclopropenoid fatty acids, canavanine, antivitamins and phorbol esters (Francis et al. 2001). The factors present in moringa (*Moringa oleifera*) leaves can negatively affect tilapia growth (Dongmeza et al. 2006). These results may be related to the acceptability of diets that are negatively affected by increasing the inclusion of plant materials and their taste characteristics. Therefore, dietary acceptability tests can provideplenty of benefits for tilapia breeding. A new diet calculation is done by considering that its composition is commonly achieved by performing time-consuming growth trials, which may negatively impact the welfare of many fish if feed is not accepted (Carlberg et al. 2015). When exogenous sodium phytate is added to the diet, Nile tilapia prefer the phytase diet with 1500 IU phytase kg^{-1} of diet (Fortes-Silva et al. 2010b) (Table 1). Phytase is an enzyme that has proven capable of not only hydrolyzing nondigestible phytate, but also of improving protein digestibility and mineral availability for several fish species.

3. Feeding stimulants

Surely the effect of feeding stimulate can be influenced by regulating food intake with consequences in metabolic or physiologic responses, including behavioral output. In behavioral terms, during buccal handling fish decide if feed is rejected or swallowed (Olsén and Lundh 2016). The ontogenetic development of Nile tilapia sense organs reflects adaptations to early life in a parent's buccal cavity (Kawamura and Washiyama 1989). However, Nile tilapia lack external taste buds on head, body and fins (Marusov and Kasumyan 2017), which may be present in other fish species. Generally speaking, the taste system, especially its functionality, is regarded as being stable throughout fish's lifetime (Kasumyan 2019), which can lead fish to display similar feeding behavior during their lifetime. Some feeding stimulants like betaine can stimulate the intestinal protease activity of liver protease and intestinal amylase activities in Nile tilapia, strain GIFT or neuropeptide Y (NPY) mRNA expression in brain by tryptophan stimulus or still ghrelin mRNA expression in the stomach by dimethylthetin (Zou et al. 2017) (Table 2). The consequence of this complex regulatory network can lead tilapia to change their feeding behavior.

Although tilapia generally accept a wide variety of plant diets, the evaluation of feeding stimulants as a flavor additive in practical diets might be useful (Table 2). Eating certain diets can change to favor growth performance. For example, the weight gain in Nile tilapia fed diets with a choline:betaine ratio of 10:90 was heavier than those tilapia fed diets with choline:betaine ratios of 100:0 and 85:15 (Kasper et al. 2002). However, food stimulants can act in different ways. For example, food consumption, growth and lipid deposition increased in the Nile tilapia that received n-decanoic (ghrelin-C10), but not n-octanoic (ghrelin-C8), and both are thought to stimulate feeding (Riley et al. 2005). According to these authors, forms of ghrelin may act through different receptors in tilapia.

As regards several feeding stimulants used in aquiculture, silage supplements are generally more studied in tilapia. Supplementation of krill meal at 1.5% as an attractant or stimulant in diets based on soybean proteins may lead to increased feed intake, growth performance and feed utilization in Nile tilapia (Gaber et al. 2007). These results apparently agree with the following study on shrimp waste as a powerful appetite stimulant. According to Leal et al. (2010), shrimp protein hydrolysate is a promising protein feedstuff or a feeding stimulant, and can account for as much as 6% of Nile tilapia diets, in which no adverse effects on growth and nutrient utilization are noted. In fact, peptides and free amino acids can act as stimulants of fish. According to Carr et al. (1996), the major tissue components in fish and invertebrates (e.g. glycine and alanine) correlate with compounds that have been previously shown to stimulate feeding behavior in several fish species. In Nile tilapia, Testes meal used as a feed additive when added to both fish meal and plant protein meal diets increases not only feed intake, but also metabolic efficiency (Lee et al. 2013). Moreover, these stimulant feeds seem to provide more than just flavor, and have even improved the nutritional quality of diets. Diets with added protein shrimp silage have been reported to be well accepted by Nile tilapia, which avidly ate feed during the experiment, and this led to improved growth at dietary inclusion levels as high as 15% (Plascencia-Jatomea et al. 2002). Rainbow trout viscera silage cannot only act as a feeding stimulant, but can also improve the non-specific immunity of Mozambique tilapia (*Oreochromis mossambicus*), thanks to its essential amino acids profile (Goosen et al. 2014).

4. Nutrient intake target

All these metabolic and blood responses seem associated with tilapia's acceptance of, or aversion to, food. Several behavioral studies have tested tilapia's ability to regulate the intake of specific nutrients (Table 1). The preferential intake of linseed and fish oil diets could be due to their better palatability, or could most likely be due to their better nutritional value (Fortes-Silva et al. 2010a). A study conducted with self-demand feeders has concluded that tilapia can adapt to protein dilution/ removal to sustain energy intake to reach a nutrient intake target as protein (41.7%), lipid (34.8%) and carbohydrate (23.5%) (Fortes-Silva and Sánchez-Vázquez 2012). This nutrient intake target was expressed as a percentage of diet after taking the total macronutrient to be 100%. Beyond orosensory control, recent studies have supported the existence of taste receptors and signaling elements in the gastrointestinal tract of fish, which suggests that sensory properties of diet can also have functional effects other than oral taste sensations and palatability (Morais et al. 2016).

The first study that employed the macronutrient encapsulation method was conducted in 2011, which demonstrated tilapia's ability to show a nutritional target for protein, carbohydrate and lipid consumption. According to Fortes-Silva et al. (2011), Nile tilapia can self-compose a balanced diet, and is also able to regulate energy intake without the oral properties of diet. This study raised questions about the post-ingestive effects of diet in tilapia. In the following years, another study challenged tilapia's ability to regulate the consumption of essential amino acids. Using self-feeders, Nile tilapia were able to self-supplement a deficient diet by selecting a synchronized combination of l-tryptophane, l-methionine and l-threonine that matched their nutritional requirements (Fortes-Silva et al. 2012).

This could be due to them eating an unbalanced diet, which could lead to metabolic consequences that imply fish developing a particular behavior that involved preferring certain nutrients (Fortes-Silva et al. 2016). Simpson and Raubenheimer (1996) proposed three rules to explain forage behavior and the nutrient search of animals: (1) learning from positive associations as reminder clues that lead to places where food is rich in a specific nutrient; (2) learning from aversion so as to avoid locations associated with toxic or nutrient-poor foods; (3) nonassociative responses like simply moving to find new nutrients in the environment. By taking all this into account, the geometric approach to address multidimensional nutrition has been suggested to evaluate the nutrient intake target in fish (Simpson and Raubenheimer 2001) (Figures 1A, B, C). When there is only one possible nutrient that can be ingested (see the dashed black line in Figures 1A, B), fish only have the option of increasing or decreasing their intake to meet that nutrient requirement. However, when fish are allowed to eat more than one food with different nutritional compositions, they may regulate their intake to defend an intake target (Figure 1C). This approach considers

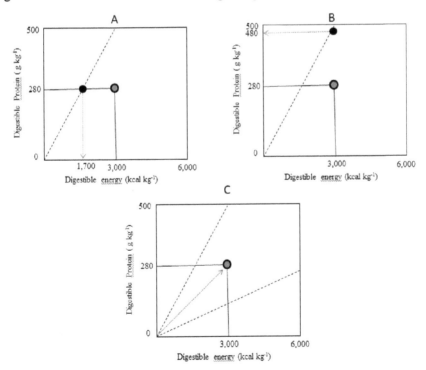

Fig. 1. Nutritional scenery of digestible protein and energy ration based on the intake target concept adapted from Simpson and Raubenheimer (2001).

animal behavior to be a guide for studies on intake diets. However, it is noteworthy that several research methodologies exist for assessing nutrient intake and animal requirements, such as dose-response or factorial methods (Sakomura and Rostagno 2007).

Digestible energy refers to non-protein energy sources such as carbohydrate and lipid. The black dotted lines in each figure represent food rails and the black circular symbols depict the hypothetical points of nutrient intake reached by fish. The red circle refers to the intake target based on Nile tilapia's nutritional requirement. Red arrows refer to ingestion. A: Fish eat enough protein, but energy is low; B: fish eat enough energy, but with excess protein; C: by considering two hypothetical diets with different nutritional values, fish can regulate their intake to obtain the necessary nutrients.

Table 1. Diet selection based on compromise rule approach in tilapia

Substances	Class	Intake target	Method used for diet preference	References
Protein (P), carbohydrate (C), lipid (L)[1]	Macronutrient	41.7% P, 34.8% C and 23.5% L	Self-feeder	Fortes-Silva et al. (2012)
Protein (P), carbohydrate (C), lipid (L)[1]	Macronutrient	45.4% P, 32.2% C, and 22.4%	Encapsulated diet "without orosensory characteristics"	Fortes-Silva and Sánchez-Vázquez (2012)
Soybean, linseed and fish oil	Lipid (fatty acids)	Linseed and fish oil	Self-feeder	
Essential amino acids	Micronutrient	Tryptophane, l-methionine, l-threonine	Self-feeder and encapsulated diet "without orosensory characteristics"	Fortes-Silva et al. (2012)
Phytase	Enzyme	1500 IU kg^{-1}	Self-feeder	Fortes-Silva et al. (2010b)

Nutrient intake target

[1] The relative macronutrient selection was expressed as the percentage of diet, considering the total of P, C and F as 100%.

5. Feeding management in fry

The quantity and quality of food supply for post-larvae is extremely important during larviculture because it is directly related to growth and survival. An adequate feeding frequency stimulates larvae to seek food at predetermined times and contributes to weight gain while reducing feed waste and production costs. It also improves water quality and makes it easier to observe the health status of fish through changes in feeding activity (Sanches and Hayashi 2001, Carneiro and Mikos 2005, Santos et al. 2015).

Table 2. Ingredient preference, feed attractant and diet selection based on compromise rule approach in tilapia

Substances	Effects	Concentration	References
Ingredients preferences			
Wheat meal, soybean meal, cottonseed meal, cassava meal, cassava scrapings, sunflower meal and corn meal	Low attraction and palatability	.	Pereira-da-Silva and Pezzato (2000)
Sugar-cane yeast and corn gluten meal	Medium attraction and palatability	.	Pereira-da-Silva and Pezzato (2000)
Integral lyophilized egg, silk worm meal, fish meal, meat meal, shrimp meal	High attraction and palatability	.	Pereira-da-Silva and Pezzato (2000)
Feed attractant and stimulant used in plant-based diets			
Krill meal	Feed intake, growth performance, and feed utilization	1.5% in the diets	Gaber (2007)
Choline:betaine	Consumption	40:60 ratio (choline concentration: 3 g kg^{-1} and dietary phosphatidylcholine concentration was 15 g kg^{-1} dry diet)	Kasper et al. (2002)
Dried basil leaves	Palatability index and reduction in feed waste percent from 33.48%	2%	El-Dakar et al. (2008)
Shrimp head hydrolysate	Consumption, weight, weight gain, mean daily weight gain, specific growth ratio and feed conversion ratio	10 and 15%	Plascencia-Jatomea et al. (2002)
Tryptophan	Neuropeptide Y (NPY) mRNA expression in brain	1.8 g kg^{-1}	Zou et al. (2017)
Dimethylthetin	Ghrelin mRNA expression in stomach	0.6 g kg^{-1}	Zou et al. (2017)
Betaine	Feed intake	5 g kg^{-1}	Luo et al. (2010)
Taurine	Feed consumed and feed conversion	10.0 and 15.0 g kg^{-1}	Al-Feky et al. (2015)

The main source of energy and nutrition for larvae is the yolk sac. The first exogenous feeding brings a need for increased feeding frequency, which is related to the species, the age of larvae and water quality and temperature (Hayashi et al. 2004, Soares et al. 2007). Dividing a diet into several daily meals improves digestibility, which favors growth and greater survival (Johnston et al. 2003). Thus, the amount of diet and/or frequency of feeding influences diet utilization when diet is directly applied to the water and any portion not consumed will be dissolved and lixiviated. However, gastrointestinal transit and/or gastric emptying can be altered, which modifies the action of digestive enzymes (Riche et al. 2004).

On the other hand, many fish larvae and fingerlings, such as those of tilapia, require live food at the onset of exogenous feeding. Thus, the use of the primary productivity (plankton) can increase larval development and growth in a semi-intensive system. In order to efficiently utilize primary productivity, it is necessary to maintain an appropriate biomass of phytoplankton and zooplankton.

Fertilization of ponds is an effective way to utilize solar energy and natural food organisms in fish culture (Boyd and Tucker 1998). Ponds with high fertilizer inputs have greater nutrient concentrations in the water and greater primary production and fish production (Diana et al. 1991b).

The feeding rate of tilapia is mainly affected by plankton community composition and ability of different zooplankton to escape (Drenner et al. 1982, Geiger 1983, Grover et al. 1989). The biomass ingested by tilapia, in general, mainly consists of copepods, rotifers and a few protozoa, which are important protein sources for post-larval growth.

Rotifers in particular are a good food source for tilapia due to their size and high reproductive rate in cultural media (Diana et al. 1991a, Shalloof and Khalifa 2009). Zooplankton diets generally have higher digestibility than diets of artificial feed (Wilcox et al. 2006). Changes in prey choice occur according to mouth size, age and size of post-larva (Jackson 2011), as well as zooplankton prey size, color, movement and locomotion (Gemmel and Buskey 2011, Gemmel et al. 2013).

As larvae grow, storage density needs to be adjusted with regard to food supply and feeding frequency (Table 3).

6. Importance of feeding practices with regard to sex reversal

Sex reversal techniques using feeding practices are widely used for the control of sex in fish farming. Male populations are preferred in tilapia aquaculture because they have higher growth rates, prevent uncontrolled reproduction (Beardmore et al. 2001), are less aggressive and have a more uniform size (El-Sayed 2006, Meurer et al. 2008). The metabolic energy of males is channeled towards growth, and they benefit from anabolism-enhancing androgens (Tran-Duy et al. 2008, Angienda et al. 2010).

The synthetic androgen 17 α-methyltestosterone (MT) has been added to diets for sexual reversion, which has the advantage of being easily excreted soon after hormonal treatment (Pompa and Green 1990), and thus no residue on the meat (Hoga et al. 2018).

Table 3. Food management with practical diet for fry raised in green water (plankton) tanks (Adapted from Kubitza, 2006)

Cultivation Phase	Fish weight (g)	Final biomass (g/m²)	Stocking density (fish/m²)	% Protein of feed	Granulometry of feed	Feed delivery (% weight/day)	Feeding frequency (day)
Nursery 1	0.3 to 5	400	80	40–36	Powder	6 to 4	3
Nursery 2	5 to 30	600	20	36-32	2–3 mm	6 to 3	2 to 3
Grow-out	30 to 150	600	4	28-32	3–4 mm	4 to 2	2
Grow-out	150 to 180	600 to 800	4-6	32	3–4 mm	4 to 2	2
Fattening	150 to 800	800	1	28	4-6 mm	2 to 1	1 to 2
Fattening	150 to 800	800 to 1.500	1-2	32	4-6 mm	2 to 1	1 to 2
Fattening	150 to 800	1.500 to 2.400	2-3	32	4-6 mm	2 to 1	1 to 2

Sex-reversal treatment involves giving tilapia larvae a powdered feed at the first feeding, while they are still sexually undifferentiated. Such a diet contains 30 – 60 mg of 17 α-methyltestosterone/kg (El-Sayed 2006). It is necessary then to feed post-larvae a large amount of food at a high rate, which causes post-larvae to consume less plankton and more diet with hormone.

Feeding post-larvae 17 α-methyltestosterone for about 38 days (Phelps and Pompa 2000) will suppress aromatase expression (Moret al. 2001), leading to sex differentiation to a monosex population (Bhandari et al. 2006).

Increased feeding frequency can make masculinization more efficient, as it promotes a constant supply of the hormone in the bloodstream for a longer period of time (Meurer et al. 2012).

Sexual reversion to a monosexual male population can be achieved using a feeding frequency of at least six times a day for a period of 30 days, which results in 98% male individuals (Mainardes-Pinto et al. 2000).

For more details on this topic, see Chapter 11 of the present book.

7. Semi-intensive tilapia production

This system is widely spread among developing countries from Southeast Asia, Africa and Latin America. It is based on natural food production of earthen ponds generally accomplished by fertilization with organic or inorganic fertilizers. Although this system is usually employed by small-scale farmers, a great variability on feed management practices has been reported since it is affected by adoption of technology to increase productivity.

One of the main issues observed on farms using this system has been the over use of supplemental feeding. Although there has been a consensus in the literature for reducing the use of high quality commercial supplemental feeding in these system for decades, there will still be some farmers using high quality feeds in their production which is not cost-effective.

Semi-intensive culture could be more convenient and cost effective for small-scale farmers than intensive farming systems (El-Sayed 2006), depending of fertilization practices. (El-Sayed 2006).

7.1 Correct use of fertilization and supplemental feeding

In general, it is assumed that tilapia produced in semi-intensive production systems rely completely on natural food produced through fertilization. Although it is difficult to determine the exact point in the production cycle to start the use of supplemental feeding, it is common sense that at least in early stages the natural food is sufficient to meet the requirements for larvae and fingerlings (Table 4). According to previous studies, tilapia production in fertilized ponds was more efficient when raised only on fertilizers up to 100-150 g. Afterwards, it is recommended to feed the fish with supplemental feed at 50% satiation. Therefore, delaying the start of using commercial feeds in this system sharply reduces feed cost without affecting fish yield (Diana et al. 1996).

Table 4. Results of different fertilization schemes observed in the literature for Nile tilapia

Size (g)	Stocking density (fish m²)	Fertilization scheme	Optimum feeding timing	Reference
15	3	Urea (60 kg/ha/wk) + triple super phosphate (34 kg/ha/wk)	When fish reaches 100-150 g	Diana et al. (1996)
25 g	3	Similar to the previous	80 days after stocking	Lin et al. (1997)
0.1 g	4	Ammonia (28 kg/ha/wk) + ammonium phosphate (5.6 kg/ha/wk)	75 days after stocking	Brown et al. (2000)
13.8 g	3	Chicken liter (750 kg/ha) + TSP (100 kg/ha) + urea (20 kg/ha) – biweekly	42 days after stocking	Abdelghany et al. (2002)

Another issue associated with this system is the use of high feeding rates in fertilized ponds. A report from FAO has indicated that most of the tilapia farms use semi-intensive production systems overfeed the fish which affect the economic return of tilapia farming (FAO 2013). Additionally, the appropriate use of supplemental feeds in this system will reduce the environmental impact of the farm and improve economic return.

7.2 The use of mixed-feeding approach

The mixed-feeding approach is a feeding strategy where the fish is fed on a high protein diet for a certain time and followed by feeding with a low protein diet. This method has been tested for several fish species and the general outcome is an increase in feed utilization and a reduced production cost. This approach is based on the assumption that when a fish is fed a high protein diet through the production cycle, the feed utilization could be reduced over time. In fact, this is a feeding strategy that has been used in other farm animals' management for several decades, such as the feed management of dietary calcium in dairy cattle, and is based on the physiological adaptation of nutrient transporter's expression at the enterocytes following adequate and inadequate nutrient contents. Using this approach with altered low protein diet (usually 10% lower than the optimal requirement), the feed cost could be reduced and fish performance maintained.

Up till now, the available literature on tilapia has reported no significant decline in growth performance of tilapia fed on mixed-feeding approach with an improvement in feed and protein utilization (Table 5). The best results obtained for tilapia reported feeding the fish with a high protein diet for two days followed by a low protein diet for three days during 60 days.

Table 5. Summary of the main studies on mixed-feeding for Nile tilapia

Size (g)	Protein levels	Design	Outcomes	Reference
1.92	25% → 18%	1-3 d on 25% & 1-4 d on 18% for 7 weeks	Better results on 2d on 25% & 1d on 18%	Santiago and Laron (2002)
20-250 g	40% → 25%	Feeding schedule similar to the previous	Better broodstock performance on 3d on 40% & 2d on 25%	Santiago and Laron (2002)
13.6 g	33% → 22%	7 mixed feeding schedules + 2 controls for 60 days	Better performance and yields on 2d on 33% & 3d on 22%	Patel and Yakupitiyage (2003)
14.5 g	31% → 24%	6d31; 5d31/1d24; 4d31/2d24; 3d31/3d24 for 90 days	Feeding schedule on 3d in each dietary CP level	El-Husseiny et al. (2008)

8. Feed management of tilapia in intensive systems

Feed is responsible for the highest economic inputs in intensive tilapia farm operations. Moreover, strategies to reduce feed cost along the production cycle have been one of the main aims in feeding strategies research in fish.

Most of the feeding strategies are based on physiological alteration in the way fish uses the nutrients and the capacity of fish species to depict a compensatory growth after feed deprivation. By using these strategies, farm operations may be more economically efficient and reduce feed waste and environmental impact of aquaculture.

In intensive production systems, such as RAS, net cages and earthen ponds usually use the same basic feeding strategy to deliver the correct amount of feed for tilapia. Although a debate still exists in the literature about the proper feeding level used in tilapia culture (satiation vs restricted feeding to % of BW), a consensus in the results observed in various studies is that when feeding tilapia at satiation (or ad libitum), a higher feeding frequency should be used while when feeding tilapia to a restricted % of BW 2-6 meals per day should be provided according to developmental stage of tilapia (Table 6). This is in line with the natural feeding habit and structure of the digestive tract of tilapia. As an opportunistic species, tilapia has adapted to feed several times a day in different regions of the aquatic environment. Moreover, because of the limited capacity of the stomach of tilapia, more frequent feeding would be appropriate for them.

A rule of thumb when feeding fish is not feed the fish in excess to their requirement. Higher feeding levels might lead to feed waste, reduced feed digestibility and efficiency, and finally increased environmental impact. Thus, most of tilapia farms nowadays use a certain level of feed restriction throughout the production cycle.

Table 6. Recommended feeding chart for Nile tilapia raised intensively (Adapted from Kubitza 2011)

Average weight (g)	30-32 °C		25-29 °C		20-24 °C		16-19 °C	
	Rate (% BW)	Frequency (times/day)	Rate (% BW)	Frequency (times/day)	Rate (% BW)	Frequency (times/day)	Rate (% BW)	Frequency (times/day)
25	3.6	3	4.5	3	3.6	2	2.7	1
50	3.0	3	3.7	3	3.0	2	2.2	1
75	2.7	3	3.4	3	2.7	2	2.0	1
100	2.5	3	3.2	3	2.5	2	1.9	1
150	2.4	2	3.0	2	2.4	1	1.8	1
200	2.2	2	2.8	2	2.2	1	1.7	1
250	2.0	2	2.5	2	2.0	1	1.5	1
300	1.8	2	2.3	2	1.8	1	1.4	1
400	1.6	2	2.0	2	1.6	1	1.2	1
500	1.4	2	1.7	2	1.4	1	1.0	1
>600	1.1	2	1.4	2	1.1	1	0.8	1

9. Exploring the compensatory growth in Nile tilapia raised in intensive production systems

Compensatory growth is a period of accelerated growth that follows growth-limiting conditions once non-limiting conditions are renewed. Characteristic features of compensatory growth include increased food-intake, accelerated mitosis and enhanced rate of food utilization. Most animals, especially fish, are capable of compensatory growth (He et al. 2015). Therefore, this biological feature has been used in aquaculture to formulate feeding strategies to increase fish yields and reduce the amount of feed during the production cycle. In summary, CG in fish is generally a response of enhanced feed intake (hyperphagia) or/and improved feed and nutrient utilization efficiency during refeeding period (Ali et al. 2003, 2016, Gaylord and Gatlin III 2000, Mohanta et al. 2017). The molecular and metabolic mechanisms that govern this physiological process include the growth hormone (GH), growth hormone receptor (GHR), insulin-like growth factor (IGF) I and II, growth hormone-releasing hormone (GHRH), leptins, growth hormone inhibiting hormone (GHIH), myostatin (MSTN), myogenic regulatory factors (MRFs), and many other endocrine regulators and associated physiological effects of their release.

Although there are conflicting results in the effectiveness of CG in tilapia (Table 7), a common feature observed in most of the studies is the hyperphagia state soon after feed deprivation without improving feed efficiency. Hyperphagia state and associated CG seems to be time-limited and dependent on fish size, previous feeding history, duration of starvation and refeeding phases, and nutritional history. This might explain the great variation in the results observed in the literature. Additionally, the available literature showed that changes in muscle growth after food deprivation in tilapia are limited. Tilapia consistently showed muscle atrophy during starvation, while in the refeeding period muscle growth restarted similarly to the control groups

Table 7. Summary of the results on compensatory growth studies with Nile tilapia

Size (g)	Starvation period	Refeeding period	Response	Reference
25-30 g	1-3 weeks	10 weeks	Limited CG Down-regulation of MRFs Hyperphagia	Nebo et al. (2017)
6.6 g	1, 2 and 4 wks	2, 4, 8 wks	Partial CG Hyperphagia	Wang et al. (2009)
0.6 g	0, 5 and 10 days	42, 37, 32 days	Fish starved for 5d showed complete CG Changes in muscle growth-related genes Hyperphagia	Nebo et al. (2013)
5 g	0,1, 2 and 4 wks	13, 12, 11 & 9 wks	Complete CG Hyperphagia	Gao and Lee (2012)
11.8 g	2, 3, 4 or 7 days through 80 days	105 days	Partial CG Hyperphagia	Gao et al. (2015)

which were fed throughout the trials. Additionally, no effect on IGF-1 and GH expression has been consistently observed in tilapia at refeeding stage.

Acknowledgments

The present research was partially funded by projects "PVE-CNPq No. 401416/2014-3; Produtividade em Pesquisa No. 305734/2016-4 and 303203/2020-0; FAPESB No. APP0067/2016" grants to Dr. Rodrigo Fortes.

References cited

Abdelghany, A.E., M.S. Ayyat and M.H. Ahmad. 2002. Appropriate timing of supplemental feeding for production of Nile tilapia, silver carp, and common carp in fertilized polyculture ponds. J. World Aquac. Soc. 33: 307–315.

Adeoye, A.A., A. Jaramillo-Torres, S.W. Fox, D.L. Merrifield and S.J. Davies. 2016. Supplementation of formulated diets for tilapia (*Oreochromis niloticus*) with selected exogenous enzymes: Overall performance and effects on intestinal histology and microbiota. Anim. Feed Sci. Technol. 215: 133–143.

Al-Feky, S.S.A., A.F.M. El-Sayed and A.A. Ezzat. 2015. Dietary taurine enhances growth and feed utilization in larval Nile tilapia (*Oreochromis niloticus*) fed soybean meal-based diets. Aquacult. Nutr. 22: 457–464.

Ali, M., A. Nicieza and R.J, Wootton. 2003. Compensatory growth in fishes: A response to growth depression. Fish Fish. 4: 147–190.

Ali, T.E.S., S. Martínez-Llorens, A.V. Moñino, M.J. Cerdá and A. Tomás-Vidal. 2016. Effects of weekly feeding frequency and previous ration restriction on the compensatory growth and body composition of Nile tilapia fingerlings. Egypt. J. Aquat. Res. 42: 357–363.

Angienda, P.O., B.O. Aketch and E.N. Waindi. 2010. Development of all-male fingerlings by heat treatment and the genetic mechanism of heat induced sex determination in Nile tilapia (*Oreochromis niloticus* L.). Int. J. Biol. Sci. 6: 38–43.

Beardmore, J. A., G.C. Mair and R.I. Lewis. 2001. Monosex male production in finfish as exemplified by tilapia: Applications, problems, and prospects. Aquaculture 197: 283–301.

Bhandari, R.K., M. Nakamura, T. Kobayashi and Y. Nagahama. 2006. Suppression of steroidogenic enzyme expression during androgen-induced sex reversal in Nile tilapia (*Oreochromis niloticus*). Gen. Comp. Endocrinol. 145: 20–24.

Boyd, C.E. and C.S. Tucker. 1998. Pond Aquaculture Water Quality Management. [Ed.]. Kluwer Academic Publishers. Boston.

Brown, C.L., R.B. Bolivar, E.T. Jimenez and J. Szyper. 2000. Timing of the onset of supplemental feeding of Nile tilapia (*Oreochromis niloticus*) in ponds. Eighteenth Annual Technical Report 4: 33–35.

Carlberg, H., K. Cheng, T. Lundh and E. Brännäs. 2015. Using self-selection to evaluate the acceptance of a new diet formulation by farmed fish. Appl. Anim. Behav. Sci. 171: 226–232.

Carneiro, P.C.F and J.D. Mikos. 2005. Freqüência alimentar e crescimento de alevinos de jundiá, *Rhamdia quelen*. Cienc. Rural. 35: 187–191.

Carr, W.E.S., J.C. Netherton III, R.A. Gleeson and C.D. Derby. 1996. Stimulants of feeding behavior in fish: Analyses of tissues of diverse marine organisms. Biol. Bull. 190: 149–160.

Diana, J.S., C.K. Lin and Y. Yi. 1996. Timing of supplemental feeding for tilapia production. J. World Aquacult. Soc. 27: 410–419.

Diana, J.S., D.J. Dettweiler and C.K. Lin. 1991a. Effect of Nile tilapia (*Oreochromis niloticus*) on the ecosystem of aquaculture ponds, and its significance to the trophic cascade hypothesis. Can. J. Fish Aquat. Sci. 48: 183–190.

Diana, J.S., C.K. Lin and P.J. Schneeberger. 1991b. Relationships among nutrient inputs, water nutrient concentrations, primary production, and yield of *Oreochromis niloticus* in ponds. Aquaculture 92: 323–341.

Dongmeza, E., P. Siddhuraju, G. Francis and K. Becker. 2006. Effects of dehydrated methanol extracts of moringa (*Moringa oleifera* Lam.) leaves and three of its fractions on growth performance and feed nutrient assimilation in Nile tilapia (*Oreochromis niloticus* (L.)). Aquaculture 261: 407–422.

Drenner, R.W., F. Jr. de Noyelles and D. Kettle. 1982. Selective impact of filter-feeding gizzard shad on zooplankton community structure. Limnol. Oceanogr. 27: 965–968.

El-Dakar, A., G. Hassanien, S. Gad and S. Sakr. 2008. Use of dried basil leaves as a feeding attractant for hybrid tilapia, *Oreochromis niloticus* × *Oreochromis aureus*, Fingerlings. Mediterr. Aquac. J. 1: 35–44.

El-Husseiny, O.M., G. El Din, M. Abdul-Aziz and R.S. Mabroke. 2008. Effect of mixed protein schedules combined with choline and betaine on the growth performance of Nile tilapia (*Oreochromis niloticus*). Aquac. Res. 39: 291–300.

El-Sayed, A.F.M. 2006. Tilapia Culture [Ed.]. CABI Publishing Series.

Fortes-Silva, R., F.J. Martínez, M. Villarroel and F.J. Sánchez-Vázquez. 2010a. Daily feeding patterns and self-selection of dietary oil in Nile tilapia. Aquacult. Res. 42: 157–160.

Fortes-Silva, R., F.J. Martínez, M. Villarroel and F.J. Sánchez-Vázquez. 2010b. Daily rhythms of locomotor activity, feeding behavior and dietary selection in Nile tilapia (*Oreochromis niloticus*). Comp. Biochem. Physiol. A Mol. Integr. Physiol. 156: 445–450.

Fortes-Silva, R., F.J. Martínez and F.J. Sánchez-Vázquez. 2011. Macronutrient selection in Nile tilapia fed gelatin capsules and challenged with protein dilution/restriction. Physiol. Behav. 102: 356–360.

Fortes-Silva, R. and F.J. Sánchez-Vázquez. 2012. Use of self-feeders to evaluate macronutrient self-selection and energy intake regulation in Nile tilapia. Aquaculture 326–329: 168–172.

Fortes-Silva, R., P.V. Rosa, S. Zamora and F.J. Sánchez-Vázquez. 2012. Dietary self-selection of protein-unbalanced diets supplemented with three essential amino acids in Nile tilapia. Physiol. Behav. 105: 639–644.

Fortes-Silva, R., A. Kitagawa and F.J. Sánchez Vázquez. 2016. Dietary self-selection in fish: A new approach to studying fish nutrition and feeding behavior. Rev. Fish. Biol. Fisher. 26: 39–51.

Francis, G., H.P.S. Makkar and K. Becker. 2001. Antinutritional factors present in plant-derived alternate fish feed ingredients and their effects in fish. Aquaculture 199: 197–227.

Gaber, M.M.A. 2007. The effect of different levels of krill meal supplementation of soybean-based diets on feed intake, digestibility, and chemical composition of juvenile Nile tilapia (*Oreochromis niloticus*, L). J. World Aquacult. Soc. 36: 346–353.

Gao, Y. and J.-Y. Lee. 2012. Compensatory responses of Nile tilapia *Oreochromis niloticus* under different feed-deprivation regimes. Fish Aquat. Sci. 15: 305–311.

Gao, Y., Z. Wang, J.-W. Hur and J.-Y. Lee. 2015. Body composition and compensatory growth in Nile tilapia *Oreochromis niloticus* under different feeding intervals. Chin. J. Oceanol. Limn. 33: 945–956.

Gaylord, I.G. and D.M. Gatlin III. 2000. Assessment of compensatory growth in channel catfish *Ictalurus punctatus* R. and associated changes in body condition indices. J. World Aquacult. Soc. 31: 326–336.

Geiger, J.G. 1983. A review of pond zooplankton production and fertilization for the culture of larval and fingerling striped bass. Aquaculture 35: 353–369.

Gemmell, B.J. and E.J. Buskey. 2011. The transition from nauplii to copepodites: Susceptibility of developing copepods to fish predators. J. Plankton Res. 33: 1773–1777.

Gemmell, B.J., D. Adhikari and E.K. Longmire. (2013). Volumetric quantification of fluid flow reveals fish's use of hydrodynamic stealth to capture evasive prey. J. R. Soc. Interface. 11: 1–8.

Goosen, N.J., L.F. De Wet and J.F. Görgens. 2014. Rainbow trout silage as immune stimulant and feed ingredient in diets for Mozambique tilapia (*Oreochromis mossambicus*). Aquacult. Res. 47: 329–340.

Grover, J.J., B.L. Olla, M. O'Brien and R.I. Wicklund. 1989. Food habits of Florida red tilapia fry in manured seawater pools in the Bahamas. Prog. Fish-Cult. 51: 152–156.

Hayashi, C., F. Meurer, W.R. Boscolo, C.H.F. Lacerda and L.C.B. Kavata. 2004. Freqüência de arraçoamento para alevinos de lambari do rabo-amarelo (*Astyanax bimaculatus*). R. Bras. Zootec. 33: 21–26.

He, L., Y. Pei, Y. Jiang, Y. Li, L. Liao, Z. Zhu, et al. 2015. Global gene expression patterns of grass carp following compensatory growth. BMC Genomics 16: 184.

Hoga, C.A., F.L. Almeida and F.G.R. Reyes. 2018. A review on the use of hormones in fish farming: Analytical methods to determine their residues. CyTA – J. Food. 16: 679–691.

Jackson, J.M. 2011. Larval clownfish *Amphiprion ocellaris* predatory success and selectivity when preying on the Calanoid Coppepod *Parvocalanus crassirostris*. Thesis of Master. University of Hawai'i.

Jobling, M. 2015. Fish nutrition research: Past, present and future. Aquacult. Int. 24: 767–786.

Johnston, G., H. Kaiser, T. Hecht and L. Oellermann. 2003. Effect of ration size and feeding frequency on growth size distribution and survival of juvenile Clownfish, *Amphiprion percula*. J. Appl. Ichthyol. 19: 40–43.

Kasper, C.S., M.R. White and P.B. Brown. 2002. Betaine can replace choline in diets for juvenile Nile tilapia, *Oreochromis niloticus*. Aquaculture 205: 119–126.

Kasumyan, A.O. 2019. The taste system in fishes and the effects of environmental variables. J. Fish Biol. 95: 155–178.

Kawamura, G. and N. Washiyama. 1989. Ontogenetic changes in behavior and sense organ morphogenesis in large mouth bass and tilapia nilotica. Trans. Am. Fish. Soc. 118: 203–213.

Kubitza, F. 2006. Questões freqüentes dos produtores sobre a qualidade dos alevinos de tilápia. Panorama da Aquicultura 16: 14–23.

Leal, A.L.G., P.F. de Castro, J.P.V. de Lima, E. de Souza Correia and R. de Souza Bezerra. 2009. Use of shrimp protein hydrolysate in Nile tilapia (*Oreochromis niloticus*, L.) feeds. Aquac. Int. 18: 635–646.

Lee, K.J., S. Rahimnejad, M.S. Powell, F.T. Barrows, S. Smiley, P.J. Bechtel, et al. 2013. Salmon testes meal as a functional feed additive in fish meal and plant protein-based diets for rainbow trout (*Oncorhynchus mykiss* Walbaum) and Nile tilapia (*Oreochromis niloticus* L.) fry. Aquacult. Res. 46: 1590–1596.

Lin, C.K., Y. Yi and J.S. Diana. 1997. The effects of pond management strategies on nutrient budgets. pp. 19–24. *In*: Thailand. Fourteenth Annual Technical Report. [Ed.]. Pond Dynamics/Aquaculture CRSP, Oregon State University, Corvallis, Oregon, USA.

Luo, Z., X.-Y. Tan, X.-Y. Liu and H. Wen. 2010. Effect of dietary betaine levels on growth performance and hepatic intermediary metabolism of GIFT strain of Nile tilapia *Oreochromis niloticus* reared in freshwater. Aquacult. Nutr. 17: 361–367.

Mainardes-Pinto, C.S.R., N. Fenerich-Verani, B.E.S. Campos and A.L. Silva. 2000. Masculinização da Tilápia do Nilo, *Oreochromis niloticus*, utilizando diferentes rações e diferentes doses de 17α-metiltestosterona. R. Bras. Zootec. 29: 654–659.

Marusov, E.A. and A.O. Kasumyan. 2017. Feeding behavior and responsivity to food odors in Nile tilapia *Oreochromis niloticus* (Cichlidae) after chronic polisensory deprivation. J. Ichthyol. 57: 747–752.

Matthew, M.T., O. Simon-Olok, N. Eric, K. Nicholas, K. Richard and M.W. Wilson. 2017. Feeding regimes for Singida tilapia (*Oreochromis esculentus*, Graham 1928) under controlled conditions: Acceptance and utilization of natural feeds compared to dry rations. Int. J. Fish. Aquac. 5: 420–424.

Meurer, F., C. Hayashi, M.M. Costa, A.S. Mascioli, L.M. Saragiotto and A. Freccia. 2008. Levedura como probiótico na reversão sexual da tilápia-do-Nilo. Ver. Bras. Saúde Prod. Anim. 9: 804–812.

Meurer, F., R.A. Bombardelli, P.S. Paixão, L.C.R. Silva and L.D. Santos. 2012. Feeding frequency on growth and male percentage during sexual reversion phase of Nile tilapia. Rev. Bras. Saúde Prod. Anim. 13: 1133–1142.

Mohanta, K., S. Rath, K. Nayak, C. Pradhan, T. Mohanty and S. Giri. 2017. Effect of restricted feeding and refeeding on compensatory growth, nutrient utilization and gain, production performance and whole body composition of carp cultured in earthen pond. Aquac. Nutr. 23: 460–469.

Mor, G., M. Eliza, J. Song, B. Wiita, S. Chen and F. Naftolin. 2001. 17alpha-methyl testosterone is a competitive inhibitor of aromatase activity in Jar choriocarcinoma cells and macrophage-like THP-1 cells in culture. J. Steroid Biochem. Mol. Biol. 79: 239–246.

Morais, S. 2016. The physiology of taste in fish: Potential implications for feeding stimulation and gut chemical sensing. Rev. Fish. Sci. Aquac. 25: 133–149.

Nebo, C., K. Overturf, A. Brezas, M. Dal-Pai-Silva and M.C. Portella. 2017. Alteration in expression of atrogenes and IGF-1 induced by fasting in Nile tilapia *Oreochromis niloticus* juveniles. Int. Aquat. Res. 9: 361–372.

Nebo, C., M.C. Portella, F.R. Carani, F.L. de Almeida, C.R. Padovani, R.F. Carvalho, et al. 2013. Short periods of fasting followed by refeeding change the expression of muscle growth-related genes in juvenile Nile tilapia (*Oreochromis niloticus*). Comp. Biochem Phys. B. 164: 268–274.

Olsén, K.H. and T. Lundh. 2016. Feeding stimulants in an omnivorous species, crucian carp *Carassius carassius* (Linnaeus 1758). Aquacult. Rep. 4: 66–73.

Patel, A.B. and A. Yakupitiyage. 2003. Mixed feeding schedules in semi-intensive pond culture of Nile tilapia, *Oreochromis niloticus*, L.: Is it necessary to have two diets of differing protein contents? Aquac. Res. 34: 1343–1352.

Pereira-Da-Silva, E.M. and L.E. Pezzato. 2000. Response of Nile tilapia (*Oreochromis niloticus*) to the attraction and palatability of the used ingredients in the feeding of fishes. Rev. Bras. Zootec. 29: 1273–1280.

Phelps, R.P. and T.J. Pompa. 2000. Sex reversal of tilapia. pp. 34–59. *In*: Costa-Pierce, B.A. and J.E. Rakocy [eds.]. Tilapia Aquaculture in the Americas. The World Aquaculture Society, Baton Rouge, LA, United States.

Plascencia-Jatomea, M., M.A. Olvera-Novoa, J.L. Arredondo-Figueroa, G.M. Hall and K. Shirai. 2002. Feasibility of fishmeal replacement by shrimp head silage protein hydrolysate in Nile tilapia (*Oreochromis niloticus* L.) diets. J. Sci. Food Agric. 82: 753–759.

Popma, T.J., and B.W. Green. 1990. Aquaculture Production Manual: Sex reversal of tilapia in earthen ponds. Research and Development Series, No. 35. International Center for Aquaculture, Auburn University, Alabama.

Riche, M., D.I. Haley, M. Oetker, S. Garbrecht and D.L. Garling. 2004. Effect of feeding frequency on gastric evacuation and the return of appetite in tilapia *Oreochromis niloticus* (L.). Aquaculture 234: 657–673.

Riley, L.G., B.K. Fox, H. Kaiya, T. Hirano and E.G. Grau. 2005. Long-term treatment of ghrelin stimulates feeding, fat deposition, and alters the GH/IGF-I axis in the tilapia, *Oreochromis mossambicus*. Gen. Comp. Endocr. 142: 234–240.

Sakomura, N.K. and H.S. Rostagno. 2007. Métodos de pesquisa em nutrição de monogástricos. Funep, Jaboticabal. 283.

Sanches, L.E.F. and C. Hayashi. 2001. Effect of feeding frequency on Nile tilapia, *Oreochromis niloticus* (L.) fries performance during sex reversal in hapas. Acta Scientiarum 23: 871–876.

Santiago, C.B. and M.A. Laron. 2002. Growth and fry production of Nile tilapia, *Oreochromis niloticus* (L.), on different feeding schedules. Aquac. Res. 33: 129–136.

Santos, M.M., J.A. Calumby, P.A. Coelho Filho, E.C. Soares and A.L. Gentelini. 2015. Nível de arraçoamento e frequência alimentar no desempenho de alevinos de tilápia-do-nilo. Bol. Inst. Pesca. 41: 387–395.

Shalloof, K.A.S. and N. Khalifa. 2009. Stomach contents and feeding habits of *Oreochromis niloticus* (L.) from Abu-Zabal Lakes, Egypt. World Appl. Sci J. 6: 01–05.

Simpson, S.J. and D. Raubenheimer. 1996. Feeding behaviour, sensory physiology and nutrient feedback: A unifying model. Entomol. Exp. Appl. 80: 55–64.

Simpson, S.J. and D. Raubenheimer. 2001. A framework for the study of macronutrient intake in fish. Aquacult. Res. 32: 421–432.

Soares, E.C., M. Pereira-Filho, R. Roubach and C.S. Silva. 2007. Condicionamento alimentar no desempenho zootécnico do tucunaré. Braz. J. Fish Eneg. 2: 35–48.

State of world fisheries and aquaculture (FAO). 2013. Contributing to food security and nutrition for all. Food and Agriculture Organization of the United Nations (FAO), Rome. http://www.fao.org/3/a-i5555e.pdf.

Tran-Duy, A., J.W. Schrama, A.A. van Dam and J.A.J. Verreth. 2008. Effects of oxygen concentration and body weight on maximum feed intake, growth and hematological parameters of Nile tilapia, *Oreochromis niloticus*. Aquaculture 275: 152–162.

Wang, Y., C. Li, J.G. Qin and H. Han. 2009. Cyclical feed deprivation and refeeding fails to enhance compensatory growth in Nile tilapia, *Oreochromis niloticus* L. Aquac. Res. 40: 204–210.

Wilcox, J.A., P.L. Tracy and N.H. Marcus. 2006. Improving live feeds: Effect of a mixed diet of Copepod Nauplii (*Acartiatonsa*) and Rotifers on the survival and growth of first-feeding larvae of the southern Flounder, *Paralichthys lethostigma*. J. World Aquacult. Soc. 37: 113–120.

Zou, Q., Y. Huang, J. Cao, H. Zhao, G. Wang, Y. Li, et al. 2017. Effects of four feeding stimulants in high plant-based diets on feed intake, growth performance, serum biochemical parameters, digestive enzyme activities and appetite-related genes expression of juvenile GIFT tilapia (*Oreochromis* sp.). Aquacult. Nutr. 23: 1076–1085.

6

Osmoregulation in Tilapia: Environmental Factors and Internal Mechanisms

Carlos Eduardo Copatti[1]* and Bernardo Baldisserotto[2]

[1] Biology Institute, Universidade Federal da Bahia (UFBA), Av. Adhemar de Barros, 668, Salvador, BA, 40170-115, Brazil

[2] Department of Physiology and Pharmacology, Universidade Federal de Santa Maria, Santa Maria, RS, 97105-900, Brazil

1. Introduction

The tilapias (*Oreochromis* spp., *Coptodon* spp. and *Sarotherodon* spp.) inhabit diverse freshwater (FW) habitats and are considered in general euryhaline species. The ability to accurately quantify osmolality and co-ordinate a response of appropriate magnitude over a range of stress levels suggests that they may have evolved novel and optimized osmoregulatory mechanisms (Wang et al. 2009). The gills of FW-adapted Mozambique tilapia possess three types of ionocytes (or mitochondria-rich cells) (Dymowska et al. 2012, Furukawa et al. 2014): type I, which expresses the Na^+/K^+-ATPase (NKA) in the basolateral membrane (Inokuchi et al. 2009) and a K^+ channel (ROMK) in the apical membrane (Furukawa et al. 2014); type II, which expresses the Na^+/Cl^- cotransporter (NCC) in the apical membrane and the electrogenic Na^+/HCO_3^- cotransporter (NBC), NKA and a Cl^- channel (CIC-3) in the basolateral membrane; Type III, which expresses the Na^+/H^+ exchanger 3 (NHE3) and ROMK in the apical membrane and NKA in the basolateral membrane (Inokuchi et al. 2009, Dymowska et al. 2012, Furukawa et al. 2014). In these tilapias, NHE3 expressed in ionocytes is involved in both FW adaptation and acid–base regulation, in which Na^+ is absorbed from external environments in exchange for H^+ secretion (Watanabe et al. 2008), whereas NCC absorbs external Na^+ and Cl^- simultaneously (Inokuchi et al. 2009) (Figure 1). The H^+-ATPase (v-type) is also found in the gills of Mozambique tilapia (Ruíz-Jarabo et al. 2017), but it was not determined yet in which type of ionocyte.

*Corresponding author: carloseduardocopatti@yahoo.com.br

The type III ionocytes of the gills of Mozambique tilapia apparently also contain an apical epithelial Ca^{2+} channel as well as basolateral Ca^{2+}-ATPase 2 and Na^+/Ca^{2+}exchanger 1b (Lin et al. 2016). The Ca^{2+}-ATPase and Na^+/Ca^{2+} cotransporter transport Ca^{2+} from the cells into the plasma (Baldisserotto 2003), thereby reducing its intracellular concentration, which facilitates the entry of Ca^{2+} from the medium into the ionocytes through the Ca^{2+} channel (Baldisserotto et al. 2004) (Figure 1). Intestinal Ca^{2+} uptake occurs through membrane of enterocytes by the presence of a basolateral Ca^{2+}-ATPase (Klaren et al. 1997). Absorption of monovalent ions occurs through all the gastrointestinal tract by several transporters: NKA in the basolateral membrane and NCC, Na^+/$2Cl^-$/K^+ cotransporter (NKCC) (Li et al. 2014). The enterocytes (mainly in anterior and medium intestine) also present NBC, the Cl^-/HCO_3^- co-transporter, and a H+-ATPase (v-type) in the apical membrane (Ruíz-Jarabo et al. 2017). The apical cystic fibrosis transmembrane conductance regulator channel (*cftr*) is important for Cl- absorption through the basolateral membrane (Li et al. 2014) (Figure 2).

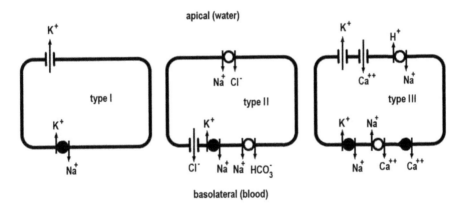

Fig. 1. Schematic model of gill ionocytes in freshwater-adapted Mozambique tilapia. Full circles: transporters that need energy to execute their function. Empty circles: transporters that do not need energy directly to execute their function.

2. Water pH

The suitability of tilapias for aquaculture is related to their ability to tolerate a wide range of environmental conditions. Tilapias remain in continuous contact with environmental water, and are constantly challenged to maintain plasma ions concentration within the defined range necessary for proper cell function.

Besides osmotic stresses, tilapias often face changes in ambient water pH, which lead to fluctuation of their blood pH as well as osmolality (Furukawa et al. 2011). This problem transpires in places where water acidification or alkalinisation may occur, resulting in the decline of fish populations, both in the environment and in fish farms (Copatti et al. 2019). So, water pH influences the stress response and fish development.

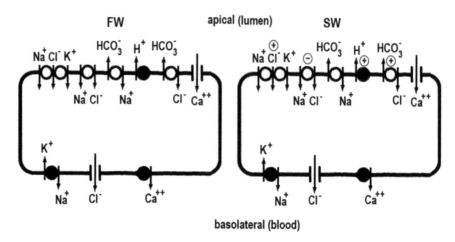

Fig. 2. Schematic model of ionocytes in the intestine of fresh- (left) and seawater-adapted (right) Mozambique tilapia. Full circles: transporters that need energy to execute their function. Empty circles: transporters that do not need energy directly to execute their function. ⊙ : indicates higher activity compared to freshwater-adapted fish. ⊙ : indicates lower activity compared to freshwater-adapted fish.

Acidic water tends to promote ionic imbalance due to the high concentrations of H^+, which causes the inhibition of Na^+, K^+ and Ca^{2+} influxes (Aride et al. 2007) as well as ionic losses by disruption of tight junctions in the gill cells (Baldisserotto 2003). This results in disorders in blood cells, plasma proteins and the volume of fish fluids, which can trigger reduction in growth and reproductive responses or even death from circulatory insufficiency (Wood 1989). Lemos et al. (2018) verified that high H^+ levels at pH 5.5 react directly with calcium carbonate ($CaCO_3$), reducing alkalinity values below 30 mg $CaCO_3$ L^{-1}. Under these conditions, water buffering capacity can be reduced, triggering high fluctuations of water pH levels and, consequently, disturbances in fish acid–base balance (Bhatnagar and Devi 2013).

The exposure of Nile tilapia to acidic pH under high stocking density contributed to elevation of plasma cortisol levels (Lemos et al. 2018). Although cortisol is involved in the stress response, this remarkably versatile hormone also has a well-established role in the endocrine control of osmoregulation (Mommsen et al. 1999). High plasma cortisol levels in individuals exposed to acidic pH may be an osmoregulatory adaptation against ionic disturbances in these environments (Kumai et al. 2012), since cortisol stimulates an increase in the functional area of ionocytes (Lin et al. 2015) and decreases gill permeability by tightening the tight junctions (Chasiotis et al. 2012) to maintain ionic balance.

The size distribution of branchial ionocytes changed drastically during acclimation to acidified FW in Mozambique tilapia. The mean ionocytes size was 1.5-fold larger in the fish exposed to water pH 3.5 for 7 days compared to normal FW (pH 7.2) (Furukawa et al. 2010). A structural damage of cells, which may result in cell death by necrosis, can occur in acute acidification (Wendelaar Bonga et al. 1990). The gradual water acidification down to pH 4.0 can be tolerated (Ginneken et al. 1997) or could even contribute to the Nile tilapia juvenile growth performance (Rebouças et al. 2015). For example, the reduction of plasma osmolality and sodium uptake rates is less severe under gradual water acidification in Mozambique tilapia (Wendelaar Bonga et al. 1984, Flik et al. 1989). According to Furukawa et al. (2010), somatolactin changed ion-transport functions of ionocytes to correct plasma osmotic balance disturbed by acid exposure in Mozambique tilapia. In addition, plasma osmolality drastically decreased until two days after transfer to acidified FW but was restored to normal levels after one week of acclimation (Furukawa et al. 2010), which demonstrated the excellent acid tolerance of tilapia.

This adjustment to acidic waters occurs probably via prolactin (PRL), which controls diffusional sodium losses across gill surface (Flik et al. 1989). This hormone, as well as growth hormone and somatolactin, can contribute to acid-base balance when tilapias are exposed to acidic waters (Furukawa et al. 2010).

Mozambique tilapia has some additional mechanisms described for adaptation to acidic waters. Under acidity, this species increases the release of α-melanophore stimulating hormone, which darkens the skin of the animal, and also changes its pattern of ionic regulation. This adjustment is usually successful if the pH reduction is slow (Van der Salm et al. 2005). Acclimation to acidic water (pH 4.0) increased the mRNA expression of NHE3 and NCC, where apical-NHE3 ionocytes were enlarged, and frequently formed multicellular complexes with developed deep apical openings at pH 4.0 (Furukawa et al. 2011).

Finally, four transitory stages are distinguished in the ionocytes cycle in tilapias under gradual acidification: accessory or replacement cells, immature, mature, and degenerating (apoptotic) cells (Wendelaar Bonga et al. 1990). According to Mai et al. (2010), the ATM-p53 signal pathway is activated in response to DNA damage under acute acidic exposure (pH 5.3) and acts as a mediator of cell cycle arrest, apoptosis or cell death in Nile tilapia. Under normal conditions, ATM-p53 levels are maintained at a low state by virtue of the extremely short-half-life of the polypeptide (Zakaria et al. 2009). In summary, both osmoregulatory and acid–base regulatory mechanisms are activated in interaction in tilapias adapted to acidic waters.

High pH levels (>9.0) can harm fish by denaturing cellular membranes (Zahangir et al. 2015). The main problem of exposure to alkalinisation pH is the increase of ammonia concentration in the body fluids due to inhibition of un-ionized (NH_3) ammonia efflux (Wilkie and Wood 1996). This inhibition occurs because of the rise in pH at the external boundary layer of the gills, which reduces the conversion of NH_3 to its ionized form (NH_4^+) due to the low concentration of H^+ available to react with NH_3 and transform it into NH_4^+ and inhibits the Rhesus protein-based Na^+/NH_4^+ exchange complex (Wright and Wood 2009, 2012, Copatti et al. 2015). There is also a decrease in water CO_2 that creates a higher blood-water gradient, which promotes branchial losses (Wilkie and Wood 1996, Wood 2001). In redbelly tilapia (*Coptodon* (*Tilapia*) *zillii*) exposed to higher ammonia levels, the resultant respiratory alkalosis increases blood pH levels (El-Shafey 1998).

Additionally, exposures to very alkaline pH waters may reduce Na^+ and Cl^- influxes by the gills (Wilkie et al. 1999). *Alcolapia* (*Tilapia*) *graham*, which lives in Lake Magadi (pH 10.0), has low branchial permeability to both Na^+ and Cl^- (Eddy et al. 1981).

3. Water hardness

Hardness can be defined as the sum of concentration of the divalent cations, particularly Ca^{2+} and Mg^{2+} in water. Calcium is crucial for osmoregulation and maintenance of many physiological activities, such as of heart, muscle and nerve function (Flik et al. 1995). Fish, which live in aquatic environments with inconsistent Ca^{2+} levels, have to maintain their body fluid Ca^{2+} homeostasis through an efficient Ca^{2+} regulation mechanism (Flik et al. 1995). Mozambique tilapia maintained at low Ca^{2+} levels (0.2 mmol Ca^{2+} L^{-1}) presented higher Ca^{2+} influx and efflux rates than those kept at 0.8 mmol $Ca^{2+}L^{-1}$, allowing fish that lived in low Ca^{2+} levels to maintain higher Ca^{2+} plasma levels, although they presented lower Ca^{2+} levels in their hard tissues (Flik et al. 1986a). The increase of waterborne Ca^{2+} decreased the expression of the epithelial Ca^{2+} channel and Ca^{2+} influx in larvae and gills of adults of Mozambique tilapia, while the expression of the Ca^{2+}-ATPase 2 and Na^+/Ca^{2+} exchanger 1b was not affected (Lin et al. 2016).

The extracellular calcium-sensing receptor (CaSR) serves an important detector function in fish (Takei and Loretz 2006). This receptor is a component of the complex teleost skeletal system, like skeletal bone and cartilage, epidermis and red muscle. It is localized mainly in the ion-transporting ionocytes of gill, in ion- and nutrient-transporting epithelia of middle and posterior intestine, suggesting its potential role in osmoregulation, as described in Mozambique tilapia (Loretz et al. 2004, 2009, 2012). CaSR is involved in Ca^{2+} homeostasis at the levels of integrative endocrine signaling through PRL (a hypercalcemic factor), stanniocalcin (STC, a hypocalcemic factor), somatolactin (SL, a proposed hypercalcemic factor) and calcitonin hormones (Takei and Loretz 2006, Loretz et al. 2009). It was also verified that somatostatin inhibits PRL release through a membrane receptor-coupled mechanism and similarly reduces intracellular free Ca^{2+} (Hyde et al. 2004).

In fish, growth hormone (GH) acts as a growth-promoting hormone, but it also appears to be involved in osmoregulation. In Mozambique tilapia, GH has specific calcitropic effects (Flik et al. 1993). In tilapia, Ca^{2+} uptake from the water via the gills is pivotal for growth and Ca^{2+} homeostasis (Flik et al. 1986b). Apparently, GH contributes to improve Ca^{2+} storage in the body of tilapia and decreases Ca^{2+} efflux, which may be related to the increase of ionocytes density in the opercular epithelium (Flik et al. 1993). The effect of GH on Ca^{2+} balance of tilapia is clearly different from that of PRL. The PRL does not stimulate growth, but stimulates Ca^{2+} influx and reduces Ca^{2+} efflux and, by doing so, induces hypercalcemia and increases bone Ca^{2+} density (Flik et al. 1986b). In Mozambique tilapia, GH can exert a negative control over the interrenals, because this hormone influences the reduction of cortisol levels (Flik et al. 1993).

Cortisol may exert its hypercalcemic function through glucocorticoid and/ or mineralocorticoid receptor and low water hardness stimulates Ca^{2+} uptake and expression of epithelial Ca^{2+} channels and a cortisol-synthesis enzyme in Mozambique tilapia (Lin et al. 2016). Cortisol inhibits tilapia PRL release through rapid reductions in intracellular free Ca^{2+} that likely involve an attenuation of Ca^{2+} entry through L-type voltage-gated Ca^{2+} channels (Hyde et al. 2004). Cortisol also promotes Na^+ excretion and thereby reduces the rise in blood osmolality that occurs when animals move to hyperosmotic environments (McCormick 2001).

Apparently, one of the survival mechanisms of species living in waters of low hardness is a high affinity for Ca^{2+}, which would contribute to closing paracellular junctions in the gills (Gonzalez et al. 1998), consequently contributing to osmoregulatory homeostasis and survival of fish by limiting diffusive branchial ion loss to the water by paracellular route (Chasiotis et al. 2012). Exposure to high waterborne Ca^{2+} levels led to hypercalcemia and both high waterborne Mg^{2+} and Ca^{2+} concentrations reduced the osmotic water permeability of the gills in Mozambique tilapia (Wendelaar Bonga et al. 1983). In addition, yolk sac larvae and swim-up fry survival was found to increase with the increase in water hardness in Nile tilapia (Bart et al. 2013).

The relationship between water pH and hardness also substantially influences fish physiology (Copatti et al. 2019). The protective effect of high water hardness occurs because the high ion loss is reduced by increasing the water hardness in acidic or alkaline pH, which could cause an improvement in the osmoregulation of the fish in these water pH values (Copatti et al. 2011). In acidic waters, the excess H^+ ions compete with waterborne Ca^{2+} and Na^+, inhibiting their capture by the fish, due to an opening of tight junctions of gill epithelia, increasing ion loss by a paracellular route (Wood 2001). To survive in this type of environment, it is possible that tilapias control the efflux of ions through the high affinity of the Ca^{2+} ions to these paracellular junctions in the gills, acting as a barrier to ion output (Baldisserotto 2003). At a low Ca^{2+} content of the water, pH 4.0 can be tolerated by tilapias, but the water acidification needs to be gradual (Ginneken et al. 1997). The increased ion loss during acid exposure are thought to be largely associated with the disruption of paracellular tight junctions (Kwong and Perry 2013), which apparently is caused by Ca^{2+} displacement from these junctions (Kwong et al. 2014). Low pH condition may induce ROS formation, mitochondrial dysfunction, and Ca^{2+} dysregulation in blood cells of tilapias (Mai et al. 2010). This could be avoided by increasing the water hardness as this would help to regulate Ca^{2+} levels in the organism. Acidic environments could also lead to an increase in cytosolic Ca^{2+} levels because Ca^{2+} is released from the mitochondria into the cytoplasm for lipoperoxidation of the mitochondrial membrane (Wang et al. 2009).

4. Salinity

The native distribution of Mozambique tilapia is characterized by estuarine areas subject to salinity variations between FW and seawater (SW) with tidal frequency. Mozambique tilapia can be acclimated to SW without mortality if the fish stay at least one day in brackish water (up to 18 ppt) through transference (Baldisserotto et al. 1994, Cnaani and Hulata 2011), but direct transfer provokes mortality (Yang et al. 2016). Blue tilapia (*O. aureus*) requires four days and Nile tilapia eight days in brackish water through transference to SW to avoid any mortality. Mozambique tilapia grew more rapidly in brackish water, but fish kept in brackish water appeared to depend on food-related calcium for growth as branchial Ca^{2+} uptake provides no more than 20% of growth related Ca^{2+}-accumulation (Vonck et al. 1998). However, some populations of Nile tilapia do not survive in salinities above 22 ppt. The red tilapia (*O. mossambicus* × *O. hornorum*) juveniles tolerate up to 19 ppt and adults up to 29 ppt. Usually hybridization with Mozambique tilapia increases salinity tolerance compared to the other species used (Cnaani and Hulata 2011, Zhu et al. 2018). The salinity tolerance of wild tilapia from the Nile river can be ranked as redbelly tilapia > mango tilapia (*Sarotherodon galilaeus*) > blue tilapia > Nile tilapia. Maximum salinity levels for growth of these species are: 29 ppt for redbelly tilapia, 15-20 ppt for mango tilapia, 10-15 ppt for blue tilapia and 5-10 ppt for Nile tilapia (Payne and Collinson 1983).

SW-adapted Mozambique tilapia presents higher standard metabolic rate and plasma osmolality than FW-adapted ones (Zicos et al. 2014, Pavlosky et al. 2019) and type IV ionocytes in the gills (Furukawa et al. 2014). Some authors consider that there are only three types of ionocytes; therefore, to them only type III (with NKCC) would be present in SW tilapia (Pavlosky et al. 2019). Salts were secreted by the basolateral NKCC and NKA, as well as a chloride channel (*cftr*) for Cl⁻, NOMK for K⁺ and the "leaky" tight junction for Na⁺. The NHE3 antiporter is also present in the apical membrane (Furukawa et al. 2014). SW-adapted Mozambique tilapia showed higher expression of *nkcc1a*, *nkaα1b* and *cftr* than FW-adapted ones (Figure 3), while the last presented higher expression of *ncc*, *nkaα1a* and *aqp3* (aquaporin 3 gene).

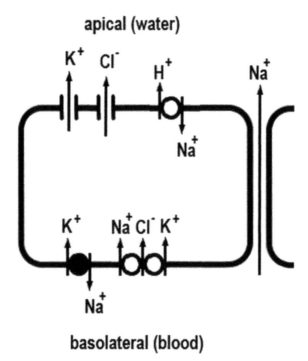

Fig. 3. Schematic model of gill ionocytes type IV in seawater-adapted Mozambique tilapia. Na⁺ transport occurs through a paracellular pathway. Full circles: transporters that need energy to execute their function. Empty circles: transporters that do not need energy directly to execute their function.

Branchial gene expression of transporters of ion transport in Mozambique tilapia reared in tidal regimen of salinity changing between FW and SW every 6 h was similar to those reared in steady-state SW (Pavlosky et al. 2019). Transference to FW induces transformation of type IV ionocytes in type I within 3 h (Lin et al. 2004). Mozambique tilapia might maintain internal water and ion balance via switching of the NKA α-isoforms (resulting in different affinities) in the kidney rather than regulating renal NKA activity between FW and SW (Yang et al. 2016). Additionally, the expression of one variant of *nkaα1* was higher in the anterior and posterior intestines of both Mozambique and Nile tilapia SW-adapted (one month) than FW-adapted specimens. The expression of *nkcc2* and *atp6v* (H^+-ATPase, v-type) were also higher, while *ncc* was lower in the intestine of SW-adapted fish (Ronkin et al. 2015). Acclimation to SW increased 16-fold the secretion of bicarbonate in the intestine, as well as the expression of *slc26a3* (Cl^-/HCO_3^- co-transporter) in the medium intestine (Ruíz-Jarabo et al. 2017).

Nile tilapia transferred from FW to SW at 15 and 20 ppt showed gill ionocytes with more mitochondria and a more developed tubular system arising from the basolateral membrane and a pre-acclimation with diet supplemented with NaCl (8%) during three weeks in FW did not cause changes in plasma osmolality, Cl^- concentration, plasma level of cortisol and gill NKA 24 h after transfer to SW (Fontaínhas-Fernandes et al. 2001). Another study showed FW tilapias fed high-salt diets (8-10% NaCl) for two weeks or more showed lower ion plasma changes and higher survival after transference to high salinities (Cnaani and Hulata 2011).

The acclimation of Mozambique tilapia to SW involves the crisis period and the stabilization period (Hwang et al. 1989). Immediately post-transfer to SW, the critical problem faced by teleosts is dehydration, caused by osmotic removal of water in gill and gut epithelial (Bath and Eddy 1979). This crisis period in tilapia appears to be within 6 to 12 h (Hwang et al. 1989, Fontaínhas-Fernandes et al. 2001). The subsequent 12–24 h mark the beginning of the stabilization period, in which the plasma osmolality and Cl^- concentration start to decrease. Zhu et al. (2018) verified that during the critical time of acclimation, there were significant increases in the NKA α1 expression in the gills of the four tilapia species in the first 24 h after transfer, with a reduction afterward to 48 h but still higher than those in freshwater.

Wendelaar-Bonga and Van der Meij (1989) found that ionocytes are more frequently observed at 3–5 days after the transfer of Mozambique tilapia from FW to SW. They usually occurred in groups of two or more cells, i.e. as accessory (increased in size and number after transition), mature and degenerated ionocytes. Lee et al. (1996) observed that reversible changes of ionocytes occurred within 24 h after tilapia were transferred to various hypotonic milieus. In addition, oxygen consumption decreased with salinity, indicating a reduction in activity level at high salinity in hybrid Mozambique tilapia (*Oreochromis mossambicus* × *O. urolepishornorum*) (Sardella et al. 2004). Therefore, living in SW, from a physiological point of view, takes more energy than living in the less saline environments.

Osmoregulation in the face of changing environmental salinity is largely mediated through the neuroendocrine system and involves the activation of ion uptake and extrusion mechanisms in osmoregulatory tissues (McCormick and Bradshaw 2006). Consistent with its role in maintaining ion balance, PRL reduces water permeability and increases ion uptake, at least in part, by upregulating NCC mRNA expression in the gills (Breves et al. 2010). In addition, circulating levels of PRL are higher in FW tilapia than SW tilapia and when tilapias are moved from FW to SW, plasma PRL levels decrease (Seale et al. 2006). Two PRL receptors (PRLRs) have been identified in tilapia: PRLR1 and PRLR2 (Fiol et al. 2009). Branchial PRLR1 is up-regulated when tilapias are transferred from SW to FW and down-regulated after transfer from FW to SW (Breves et al. 2011). The function of PRLR2, however, is not clear (Fiol et al. 2009). Unlike PRLR1 mRNA, PRLR2 mRNA expression has been shown to increase in response to an increase in external salinity in Mozambique tilapia, which confers an increased ability to adapt to a hyper-osmotic environment (Fiol et al. 2009, Seale et al. 2012a).

GH exerts acute osmoregulatory actions and enhances SW adaptation and stimulates the differentiation of ionocytes toward SW adaptation in Nile tilapia (Xu et al. 1997). GH can affect both the ultrastructural features of ionocytes and the ability of gills to extrude Na^+ after fish are transferred to SW (Prunet et al. 1994). Mozambique tilapia reared in a tidally-changing salinity can compensate for large changes in external osmolality while maintaining osmoregulatory parameters within a narrow range closer to that observed in SW-acclimated fish (Moorman et al. 2014). Additionally, Nile tilapia showed a markedly improved SW survival and reduced Na^+, Ca^{2+} and Mg^{2+} plasma levels when directly transferred from FW to 62.5% SW 24 h after a recombinant GH (reGH) injection (0.25 or 2.5 $\mu g\ g^{-1}$) (Xu et al. 1997).

Differences in the abundance of the water channel, aquaporin 3, and the stretch activated Ca^{2+} channel, transient receptor potential vanilloid 4 in PRL cells of FW and SW fish may explain their differing osmosensitivity and osmoreceptive output in differing acclimation salinities (Seale et al. 2012b). The differential genes unions from FW to SW in Nile tilapia were classified into three categories. In the constant change category (1), steroid biosynthesis, fat digestion and absorption, complement and coagulation cascades. In the change-then-stable category (2), ribosomes, oxidative phosphorylation, signaling pathways for peroxisome proliferator activated receptors, and fat digestion and absorption. In the stable-then-change category (3), protein export, protein processing in endoplasmic reticulum, tight junction, thyroid hormone synthesis, antigen processing and presentation, glycolysis/gluconeogenesis and glycosaminoglycan biosynthesis-keratan sulfate (Xu et al. 2015).

5. High ammonia levels

Sakala and Musuka (2014) and Rebouças et al. (2015) observed that high concentrations of ammonia affected both the growth and survival of redbreast tilapia (*Coptodon* (*Tilapia*) *rendalli*) and Nile tilapia, respectively. Cortisol level increased by about two folds in Nile tilapia exposed to high ammonia levels (Zeitoun et al. 2016). Under chronic ammonia exposure, hyper-secretion of mucus is another phenomenon that may occur in tilapias. This is considered to be a common protective phenomenon in ammonia exposed gills that prevents ammonia fractions from their further entrance in the gills (El-Shebly et al. 2011). Soufy et al. (2007) stated that hyperplasia may indeed have a protective function but it may also inhibit the respiration and osmoregulatory functions of the gills.

Another consequence of high ammonia levels is the reduction of GH levels and gill hyperplasia in Nile tilapia (El-Shebly et al. 2011). Growth hormone is deeply involved in osmoregulatory functions that include the regulation of gill ionocytes (Sakamoto and McCormick 2006). Zhao et al. (2018) verified that MicroRNA-26a may be involved in the remission of physiological disturbances resulting from ammonia stress. MicroRNA-26a is abundantly expressed in the brain and gill tissues of tilapia and ammonia stress leads to a remarkable decrease in miR-26a level (Zhao et al. 2018).

Acknowledgments

B. Baldisserotto was the recipient of a research fellowship by Conselho Nacional de Desenvolvimento Tecnológico (CNPq, Brazil).

References cited

Aride, P.H.R., R. Roubach and A.L. Val. 2007. Tolerance response of tambaqui *Colossoma macropomum* (Cuvier) to water pH. Aquacult. Res. 38: 588–594.

Baldisserotto, B. 2003. Osmoregulatory adaptations of freshwater teleosts. pp. 179–201. *In*: A.L. Val and B.J. Kapoor [eds.]. Fish Adaptations. Science Publishers, Inc. Enfield, USA.

Baldisserotto, B., C. Kamunde, A. Matsuo and C.M. Wood. 2004. A protective effect of dietary calcium against acute waterborne cadmium uptake in rainbow trout. Aquat. Toxicol. 67: 57–73.

Baldisserotto, B., O.M. Mimura and L.C. Salomão. 1994. Urophyseal control of plasma ionic concentration in *Oreochromis mossambicus* (Pisces) exposed to osmotic stress. Ciênc. Nat. 6: 39–50.

Bart, A.N., B. Prasad and D.P. Thakur. 2013. Effects of incubation water hardness and salinity on egg hatch and fry survival of Nile tilapia *Oreochromis niloticus* (Linnaeus). Aquacult. Res. 44: 1085–1092.

Bath, R.N. and F.B. Eddy. 1979. Salt and water balance in rainbow trout *Salmo gairdneri* rapidly transferred from fresh water to seawater. J. Exp. Biol. 83: 193–202.

Bhatnagar, A. and P. Devi. 2013. Water quality guidelines for the management of pondfish culture. Int. J. Environ. Sci. 3: 1980–1993.

Breves, J.P., S. Watanabe, T. Kaneko, T. Hirano and E.G. Grau. 2010. Prolactin restores branchial mitochondrion-rich cells expressing $Na(^+)/Cl(^-)$ cotransporter in hypophysectomized Mozambique tilapia. Am. J. Physiol. 299: R702–R710.

Breves, J.P., A.P. Seale, R.E. Helms, C.K. Tipsmark, T. Hirano and E.G. Grau. 2011. Dynamic gene expression of GH/PRL-family hormone receptors in gill and kidney during freshwater-acclimation of Mozambique tilapia. Comp. Biochem. Physiol. 158: 194–200.

Chasiotis, H., D. Kolosov, P. Bui and S.P. Kelly. 2012. Tight junctions, tight junction proteins and paracellular permeability across the gill epithelium of fishes: A review. Respir. Physiol. Neurobiol. 184: 269–281.

Cnaani, A. and G. Hulata. 2011. Improving salinity tolerance in tilapias: Past experience and future prospects. Isr. J. Aquacult-Bamid. 63: 1–21.

Copatti, C.E., B. Baldisserotto, C.F. Souza and L.O. Garcia. 2019. Protective effect of high hardness in pacu juveniles (*Piaractus mesopotamicus*) under acidic or alkaline pH: Biochemical and haematological variables. Aquaculture 50: 250–257.

Copatti, C.E., K.C.S. Bolner, F.L. Rosso, V.L. Loro and B. Baldisserotto. 2015. Tolerance of piava juveniles to different ammonia concentrations. Semina. Ciênc. Agrár. 36: 3991–4002.

Copatti, C.E., L.O. Garcia, M.A. Cunha, D. Kochhann and B. Baldisserotto. 2011. Inter-action of water hardness and pH on growth of silver catfish, *Rhamdia quelen*, juveniles. J. World Aquacult. Soc. 42:580–585.

Dymowska, A.K., P.P. Hwang and G.G. Goss. 2012. Structure and function of ionocytes in the freshwater fish gill. Resp. Physiol. Neurob. 184: 282–292.

Eddy, F.B., O.S. Bamford and G.M.O. Maloiy. 1981. Na⁺ and Cl⁻ effluxes and ionic regulation in *Tilapia grahami*, a fish living in conditions of extreme alkalinity. J. Fish Biol. 91: 249–353.

El-Shafey, A.A.M. 1998. Effect of ammonia on respiratory functions of blood of *Tilapia zilli*. Comp. Biochem. Physiol. A 121: 305–313.

El-Shebly, A.A. and H.A.M. Gad. 2011. Effect of chronic ammonia exposure on growth performance, serum growth hormone (GH) levels and gill histology of Nile tilapia (*Oreochromis niloticus*). J. Microbiol. Biotech. Res. 1: 183–197.

Fiol, D.F., E. Sanmarti, R. Sacchi and D. Kultz. 2009. A novel tilapia prolactin receptor is functionally distinct from its paralog. J. Exp. Biol. 212: 2006–2014.

Flik, G., J.C. Fenwick, Z. Kolar, N. Mayer-Gostan and S.E. Wendelaar Bonga. 1986a. Effects of low ambient calcium levels on whole-body Ca²⁺ flux rates and internal calcium pools in the freshwater cichlid teleost, *Oreochromis mossambicus*. J Exp. Biol. 120: 249–264.

Flik, G., J.C. Fenwick, Z. Kolar, N. Mayer-Gostan and S.E. Wendelaar Bonga. 1986b. Effects of ovine prolactin on calcium uptake and distribution in *Oreochromis mossambicus*. Am. J. Physiol. 250: R161–R166.

Flik, G., J.A. Van der Velden, H.C.M. Seegers, Z. Kolar and S.E. Wendelaar Bonga. 1989. Prolactin cell activity and sodium fluxes in tilapia (*Oreochromis mossambicus*) after long-term acclimation to acid water. Gen. Comp. Endocrinol. 75: 39–45.

Flik, G., W. Atsma, J.C. Fenwick, F. Rentier-Delrue, J. Smal and S.E. Wendelaar Bonga. 1993. Homologous recombinant growth hormone and calcium metabolism in the tilapia, *Oreochromis mossambicus*, adapted to fresh water. J. Exp. Biol. 185: 107–119.

Flik, G., P.M. Verbost and S.E. Wendelaar Bonga. 1995. Calcium transport process in fishes. pp. 317-342. *In*: C.M. Wood and T.J. Shuttleworth [eds.]. Cellular and Molecular Approaches to Fish Ionic Regulation, Fish Physiology, vol. 14. Academic Press, San Diego.

Fontaínhas-Fernandes, A., F. Russell-Pinto, E. Gomes, A. Reis-Henriques and J. Coimbra. 2011. The effect of dietary sodium chloride on some osmoregulatory parameters of the teleost, *Oreochromis niloticus*, after transfer from freshwater to seawater. Fish Physiol. Biochem. 23: 307–316.

Furukawa, F., S. Watanabe, M. Inokuchi and T. Kaneko. 2011. Responses of gill mitochondria-rich cells in Mozambique tilapia exposed to acidic environments (pH 4.0) in combination with different salinities. Comp. Biochem. Physiol. A 158: 468–476.

Furukawa, F., S. Watanabe, K. Kakumura, J. Hiroi and T. Kaneko. 2014. Gene expression and cellular localization of ROMKs in the gills and kidney of Mozambique tilapia acclimated to fresh water with high potassium concentration. Am. J. Physiol. Regul. Integr. Comp. Physiol. 307: R1303–1312.

Furukawa, F., S. Watanabe, T. Kanekoand and K. Uchida. 2010. Changes in gene expression levels of somatolactin in the pituitary and morphology of gill mitochondria-rich cells in Mozambique tilapia after transfer to acidic freshwater (pH 3.5). Gen. Comp. Endocrinol. 166: 549–555.

Ginneken, V.V., R.V. Eersel, P. Balm, M. Nieveen and G. Van den Thillart. 1997. Tilapia are able to withstand long-term exposure to low environmental pH, judged by their energy status, ionic balance and plasma cortisol. J. Fish Biol. 51: 795–806.

Gonzalez, R.J., C.M. Wood, R.W. Wilson, M.L. Patrick, H. Bergman, A. Narahara, et al. 1998. Effects of water pH and calcium concentration on ion balance in fish of the Rio Negro, Amazon. Physiol. Zool. 71: 15–22.

Hwang, P.P., C.M. Sun and S.M. Wu. 1989. Changes of plasma osmolarity, chloride concentration and gill Na-K-ATPase activity in tilapia *Oreochromis mossambicus* during seawater acclimation. Mar. Biol. 100: 295–299.

Hyde, G.N., A.P. Seale, E. Gordon Grau and R.J. Borski. 2004. Cortisol rapidly suppresses intracellular calcium and voltage-gated calcium channel activity in prolactin cells of the tilapia (*Oreochromis mossambicus*). Am. J. Physiol. Endocrinol. Metab. 286: E626–E633.

Inokuchi, M., J. Hiroi, S. Watanabe, P.P. Hwang and T. Kaneko. 2009. Morphological and functional classification of ion-absorbing mitochondria-rich cells in the gills of Mozambique tilapia. J. Exp. Biol. 212: 1003–1010.

Klaren, P.H.M., S.E. Wendelaar Bonga and G. Flik. 1997. Evidence for P-2-purinoceptor-mediated uptake of Ca^{2+} across a fish (*Oreochromis mossambicus*) intestinal brush border membrane. Biochem. J. 322: 129–134.

Kumai, Y., D. Nesan, M.M. Vijayan and S.F. Perry. 2012. Cortisol regulates Na^+ uptake in zebrafish (*Daniorerio*), larvae via the glucocorticoid receptor. Mol. Cell. Endocrinol. 364: 113–125.

Kwong, R.W.M. and S.F. Perry. 2013. Cortisol regulates epithelial permeability and sodium losses in zebrafish exposed to acidic water. J. Endocrinol. 217: 253–264.

Kwong, R.W.M., Y. Kumai and S.F. Perry. 2014. The physiology of fish at low pH: The zebrafish as a model system. J. Exp. Biol. 217: 651–662.

Lee, T.H., P.P. Hwang, H.C. Lin and F.L. Huang. 1996. Mitochondria-rich cells in the branchial epithelium of the teleost, *Oreochromis mossambicus*, acclimated to various hypotonic environments. Fish Physiol. Biochem. 15: 513–523.

Lemos, C.H.daP., C.V.diM. Ribeiro, C.P.B. de Oliveira, R.D. Couto and C.E. Copatti. 2018. Effects of interaction between pH and stocking density on the growth, haematological and biochemical responses of Nile tilapia juveniles. Aquaculture 495: 62–67.

Li, Z., E.Y. Lui, J.M. Wilson, Y.K. Ip, Q. Lin, T.J. Lam, et al. 2014. Expression of key ion transporters in the gill and esophageal-gastrointestinal tract of euryhaline Mozambique tilapia *Oreochromis mossambicus* acclimated to fresh water, seawater and hypersaline water. PLoS One 9: e87591.

Lin, C.H., C.L. Huang, C.H. Yang, T.H. Lee and P.P. Hwang. 2004. Time-course changes in the expression of Na, K-ATPase and the morphometry of mitochondrion-rich cells in gills of euryhaline tilapia (*Oreochromis mossambicus*) during freshwater acclimation. J. Exp. Zool. A 301: 85–96.

Lin, C.-H., T.-H. Shih, S.-T. Liu, H.-H. Hsu and P.-P. Hwang. 2015. Cortisol regulates acid secretion of H^+-ATPase-rich ionocytes in zebrafish (*Danio rerio*) embryos. Front. Physiol. 6: 328.

Lin, C.-H., W.-C. Kuan, B.-K. Liao, A.-N. Deng, D.-Y. Tseng and P.-P. Hwang. 2016. Environmental and cortisol-mediated control of Ca^{2+} uptake in tilapia (*Oreochromis mossambicus*). J. Comp. Physiol. B 186: 323–332.

Loretz, C.A., C. Pollina, S. Hyodo, Y. Takei, W. Chang and D. Shoback. 2004. cDNA cloning and functional expression of a Ca^{2+}-sensing receptor with truncated C-terminal tail from the Mozambique tilapia (*Oreochromis mossambicus*). J. Biol. Chem. 279: 53288–53297.

Loretz, C.A., C. Pollina, S. Hyodo and Y. Takei. 2009. Extracellular calcium-sensing receptor distribution in osmoregulatory and endocrine tissues of the tilapia. Gen. Comp. Endocrinol. 161: 216–228.

Loretz, C.A., C. Pollina, A.L. Herberger, S. Hyodo and Y.Takei. 2012. Skeletal tissues in Mozambique tilapia (*Oreochromis mossambicus*) express the extracellular calcium-sensing receptor. Comp. Biochem. Physiol. A 163: 311–318.

McCormick, S.D. 2001. Endocrine control of osmoregulation in teleost fish. Am. Zool. 41: 781–794.

McCormick, S.D. and D. Bradshaw. 2006. Hormonal control of salt and water balance in vertebrates. Gen. Comp. Endocrinol. 147: 3–8.

Mai, W.-J., J.-L. Yan, L. Wang, L. Zheng, Y. Xen and W.-N. Wang. 2010. Acute acidic exposure induces p53-mediated oxidative stress and DNA damage in tilapia (*Oreochromis niloticus*) blood cells. Aquat. Toxicol. 100: 271–281.

Mommsen, T.P., M.M. Vijayan and T.W. Moon. 1999. Cortisol in teleosts: Dynamics, mechanisms of action, and metabolic regulation. Rev. Fish Biol. Fisher. 9: 211–268.

Moorman, B.P., M. Inokuchi, Y. Yamaguchi, D.T. Lerner, E.G. Grau and A.P. Seale. 2014. The osmoregulatory effects of rearing Mozambique tilapia in a tidally changing salinity. Gen. Comp. Endocrinol. 207: 94–102.

Pavlosky, K.K., Y. Yamaguchi, D.T. Lerner and A.P. Seale. 2019. The effects of transfer from steady-state to tidally-changing salinities on plasma and branchial osmoregulatory variables in adult Mozambique tilapia. Comp. Biochem. Physiol. A 227: 134–145.

Payne, A.I. and R.I. Collinson. 1983. A comparison of the biological characteristics of *Sarotherodon niloticus* (L) with those of *S. aureus* (Steindachner) and other tilapia of the Delta and Lower Nile. Aquaculture 30: 335–351.

Prunet, P., M. Pisamm, J.P. Claireaux, G. Boeuf and A. Rambourg. 1994. Effects of growth hormone on gill chloride cells in juvenile Atlantic salmon (*Salmo salar*). Am. Physiol. Soc. 266: R850–R857.

Rebouças, V.T., F.R.D.S. Lima and D.D.H. Cavalcante. 2015. Tolerance of Nile tilapia juveniles to highly acidic rearing water. Acta Sci. Anim. Sci. 37: 227–233.

Ronkin, D., E. Seroussi, T. Nitzan, A. Doron-Faigenboim and A. Cnaani. 2015. Intestinal transcriptome analysis revealed differential salinity adaptation between two tilapiine species. Comp. Biochem. Physiol. D 13: 35–43.

Ruíz-Jarabo, I., S.F. Gregorio, P. Gaetano, F. Trischitta and J. Fuentes. 2017. High rates of intestinal bicarbonate secretion in seawater tilapia (*Oreochromis mossambicus*). Comp. Biochem. Physiol. A 207: 57–64.

Sakala, M.E. and C.G. Musuka. 2014. The effect of ammonia on growth and survival rate of *Tilapia rendalli* in quail manured tanks. Int. J. Aquacult. 4: 1–6.

Sakamoto, T. and S.D. McCormick. 2006. Prolactin and growth hormone in fish osmoregulation. Gen. Comp. Endocrinol. 147: 24–30.

Sardella, B.A., V. Matey, J. Cooper, R.J. Gonzalez and C.J. Brauner. 2004. Physiological, biochemical and morphological indicators of osmoregulatory stress in 'California' Mozambique tilapia (*Oreochromis mossambicus* × *O. urolepishornorum*) exposed to hypersaline water. J. Exp. Biol. 207: 1399–1413.

Seale, A.P., J.C. Fiess, T. Hirano, I.M. Cooke and E.G. Grau. 2006. Disparate release of prolactin and growth hormone from the tilapia pituitary in response to osmotic stimulation. Gen. Comp. Endocrinol. 145: 222–231.

Seale, A.P., B.P. Moorman, J.J. Stagg, J.P. Breves, D.T. Lerner and E.G. Grau. 2012a. Prolactin177, prolactin188 and prolactin receptor 2 in the pituitary of the euryhaline tilapia, *Oreochromis mossambicus*, are differentially osmosensitive. J. Endocrinol. 21: 89–98.

Seale, A.P., S. Watanabe and E.G. Grau. 2012b. Osmoreception: Perspectives on signal transduction and environmental modulation. Gen. Comp. Endocrinol. 176: 354–360.

Soufy, H., M.K. Soliman, E.M. El-manakhly and A.Y. Gaafar. 2007. Some biochemical and pathological investigations on monosex tilapia following chronic exposure to carbofuran pesticide. Global Vet. 1: 45–52.

Takei, Y. and C.A. Loretz. 2006. Chapter 8: Endocrinology. pp. 271–318. *In*: D.H. Evans and J.B. Claiborne [eds.]. The Physiology of Fishes, 3th ed. CRC Press, Boca Raton, USA.

Van der Salm, A., F. Spanings, R. Gresnigt, S. Wendelaar Bonga and G. Flik. 2005. Background adaptation and water acidification affect pigmentation and stress physiology of tilapia, *Oreochromis mossambicus*. Gen. Comp. Endocrinol. 144: 51–59.

Vonck, A.P.M.A., S.E. Wendelaar Bonga and G. Flik. 1998. Sodium and calcium balance in Mozambique tilapia, *Oreochromis mossambicus*, raised at different salinities. Comp. Biochem. Physiol. A 119: 441–449.

Wang, W.N., J. Zhou, P. Wang, T.T. Tian, Y. Zheng, Y. Liu, et al. 2009. Oxidative stress, DNA damage and antioxidant enzyme gene expression in the Pacific white shrimp, *Litopenaeus vannamei* when exposed to acute pH stress. Comp. Biochem. Physiol. C 150: 428–435.

Watanabe, S., M. Niida, T. Maruyama and T. Kaneko. 2008. Na^+/H^+ exchanger isoform 3 expressed in apical membrane of gill mitochondrion-rich cells in Mozambique tilapia *Oreochromis mossambicus*. Fish. Sci. 74: 813–821.

Wendelaar Bonga, S.E., C.J. Löwik and J.C.A. Van der Meij. 1983. Effects of external Mg^{2+} and Ca^{2+} on branchial osmotic water permeability and prolactin secretion in the teleost fish *Sarotherodon mossambicun*. Gen. Comp. Endocrinol. 52: 222–231.

Wendelaar Bonga, S.E., J.C.A. Van der Meij, W.A. Van Der Krabben and G. Flik. 1984. The effect of water acidification on prolactin cells and pars intermedia PAS-positive cells in the teleost fish *Oreochromis* (formerly *Sarotherodon*) *mossambicus* and *Carassius auratus*. Cell. Tissue Res. 238: 601–609.

Wendelaar-Bonga, S.E. and J.C.A. Van der Meij. 1989. Degeneration and death, by apoptosis and necrosis, of the pavement and chloride cells in the gills of the teleost *Oreochromis mossambicus*. Cell Tiss. Res. 255: 235–243.

Wendelaar-Bonga, S.E., G. Flik, P.H.M. Balm and J.C.A. Van der Meij. 1990. The ultrastructure of chloride cells in the gills of the teleost *Oreochromis mossambicus* during exposure to acidified water. Cell. Tissue Res. 259: 575–585.

Wilkie, M.P. and C.M. Wood. 1996. The adaptations of fish to extremely alkaline environments. Comp. Biochem. Physiol. B. 113: 665–673.

Wilkie, M.P., P. Laurent and C.M. Wood. 1999. The physiological basis for altered Na^+ and Cl^- movements across the gills of rainbow trout (*Oncorhynchus mykiss*) in alkaline (pH 9.5). Physiol. Biochem. Zool. 72: 360–368.

Wood, C.M. 1989. The physiological problems of fish in acid waters. pp. 125–152. *In*: R. Morris, E. Taylor, D. Brown and J. Brown [eds.]. Acid Toxicity and Aquatic Animals. Cambridge University, Cambridge.

Wood, C.M. 2001. Toxic responses of the gill. pp. 1–89. *In*: D. Schlenk, W.H. Benson and V.I. Organs [eds.]. Target Organ Toxicity in Marine and Freshwater Teleosts. Taylor & Francis, London.

Wright, P.A. and C.M. Wood. 2009. A new paradigm for ammonia excretion in aquatic animals: Role of Rhesus (Rh) glycoproteins. J. Exp. Biol. 212: 2303–2312.

Wright, P.A. and C.M. Wood. 2012. Seven things fish know about ammonia and we do not. Respir. Physiol. Neurobiol. 184: 231–240.

Xu, B., H. Miao, P. Zhang and D. Li. 1997. Osmoregulatory actions of growth hormone in juvenile tilapia (*Oreochromis niloticus*). Fish Physiol. Biochem. 17: 295–301.

Xu, Z., L. Gan, T. Li, C. Xu, K. Chen, X. Wang, et al. 2015. Transcriptome profiling and molecular pathway analysis of genes in association with salinity adaptation in Nile tilapia *Oreochromis niloticus*. PLoS One 10: e0136506.

Yang, W.-K., C.-H. Chung, H.C. Cheng, C.-H. Tang and T.-H. Lee. 2016. Different expression patterns of renal Na^+/K^+-ATPase α-isoform-like proteins between tilapia and milkfish following salinity challenges. Comp. Biochem. Physiol. B 202: 23–30.

Zahangir, M.M., F. Haque, G.M. Mostakim and M.S. Islam. 2015. Secondary stress responses of zebrafish to different pH: Evaluation in a seasonal manner. Aquacult. Rep. 2: 91–96.

Zakaria, Y., A. Rahmat, A.H.L. Pihie, N.R. Abdullah and P.J. Houghton. 2009. Eurycomanone induce apoptosis in HepG2 cells via up-regulation of p53. Cancer Cell. Int. 9: 16.

Zeitoun, M.M., K.E.-D.M. El-Azrak, M.A. Zaki, B.R. Nemat-Allah and E.-S.E. Mehana. 2016. Effects of ammonia toxicity on growth performance, cortisol, glucose and hematological response of Nile tilapia (*Oreochromis niloticus*). Aceh J. Anim. Sci. 1: 21–28.

Zhao, Y., H. Zhou, C.L. Ayisi, Y. Wang, J. Wang, X. Chen, et al. 2018. Suppression of miR-26a attenuates physiological disturbances arising from exposure of Nile tilapia (*Oreochromis niloticus*) to ammonia. Biol. Open 7: bio029082.

Zhu, H., Z. Liu, F. Gao, M. Lu, Y. Liu, H. Su, et al. 2018. Characterization and expression of Na^+/K^+-ATPase in gills and kidneys of the Teleost fish *Oreochromis mossambicus*, *Oreochromis urolepishornorum* and their hybrids in response to salinity challenge. Comp. Biochem. Physiol. A 224: 1–10.

Zikos, A., A.P. Seale, D.T. Lerner, E.G. Grau and K.E. Korsmeyer. 2014. Effects of salinity on metabolic rate and branchial expression of genes involved in ion transport and metabolism in Mozambique tilapia (*Oreochromis mossambicus*). Comp. Biochem. Physiol. A. 178: 121–131.

Pathologies in Farmed Tilapia and the Use of Immunostimulants and Vaccines to Prevent or Treat Diseases: An Overview

Alberto Cuesta* and María Ángeles Esteban*

Immunobiology for Aquaculture Group, Department of Cell Biology and Histology, Faculty of Biology, University of Murcia, 30100 Murcia, Spain

1. Introduction

The success of tilapia production in more than 100 countries around the world was due to a sum of several positive factors at the very beginning which changed to negative factors with time. The fish species exhibit suitable and rapid growth, have ease of propagation and rapid reproduction, high tolerance and good adaptation ability to new environmental conditions, easy acceptance of natural foods and also, high resistance to diseases (Pech et al. 2017). Due to these inherent characteristics, tilapia species were perceived as a resilient fish during the initial years of their culture, and it was generally assumed that these fish were capable of tolerating very adverse water quality conditions and many other stressors, better than most commercial farmed species. Due to this erroneous supposition, tilapias were labelled as very "disease-resistant". However, this concept changed during the 1990s when cultured tilapia started to suffer different and severe diseases. World production of tilapia increased rapidly and unfortunately so has the disease incidences (Barkham et al. 2019).

As usual, the course in tilapia disease was the direct result of the introduction of novel tilapia pathogens (via infected eggs and fish), and of the intensification (growing fish at high production densities and in recirculating systems) of fish culture methods globally (Chitmanat et al. 2016). Furthermore, infected tilapia, as also occurs with many other fish species, often do not show any clinical signs of disease despite being infected and they are capable of transmitting a pathogen to other specimens very easily (Plumb 1997). Once a pathogen is introduced into a recirculating system, it is difficult to eradicate. Fortunately, it is possible to considerably reduce the risk

*Corresponding authors: alcuesta@um.es; aesteban@um.es

of introducing pathogens to one farm by implementing simple management and biosecurity methods.

In addition, it is mandatory to improve the knowledge of the fish immunity and pathogen biology. Fish represents the first group of vertebrates in which a complete immune system is present, comprising both innate and acquired immune responses. Fish primary lymphoid organs are thymus and head-kidney (HK, equivalent to mammalian bone marrow), while the main secondary lymphoid tissues are spleen and mucosa-associated lymph tissues (MALT), dispersed in the skin, gut or gills. In tilapia, the description of the structure and cell-types present in the thymus, HK and spleen tissues were described at the same time than the generation of immunoglobulins (Ig) after immunization (Sailendri and Muthukkaruppan 1975a, b). Afterwards, leucocyte characterization has been parallel to other fish species and not great differences are denoted among teleost fish species, in spite of being the greater group of vertebrates.

Immune response can be divided into humoral and cellular factors, according to the soluble or cellular nature of the mediators and players, respectively, or into innate/non-specific and acquired/specific depending on the first or successive encounters with the antigen. Humoral immunity is played by different molecules secreted by either leucocytes or other cells in the body, with innate properties [complement system, lysozyme, antimicrobial peptides (AMPs), C-reactive protein, interferon (IFN)] or specific to antigens (Igs) (Magnadottir et al. 2005, Ye et al. 2013, Stosik et al. 2018). By the other side, different leucocyte subpopulations are responsible for the innate and specific cellular immune responses including phagocytosis, respiratory burst, cellular-mediated cytotoxicity, antibody production or antigen presentation among others (Evans et al. 2001, Nakanishi et al. 2011, Scapigliati 2013, Esteban et al. 2015). Further characterization of the tilapia immune system organization, functioning and regulation is under characterization, which would help in the design of preventive and therapeutic measures for important pathogens affecting tilapia culture.

In the present review, we will first focus only on the significant pathogens of tilapia at present providing a short overview of the field, and this paper will end with the main prophylactic measures available to prevent the incidence of diseases, including the available vaccines.

2. Tilapia diseases

Tilapia could be a very important culture fish species and has an important role in aquaculture and fisheries economy in the world. However, some important infectious diseases can be considered as one of the most important challenges in aquaculture development. The clinically significant tilapia pathogens fall into the broad categories of bacteria, viruses, and protozoa. Mycotic (fungal) diseases as well as those diseases caused by metazoan ecto- and endoparasites are usually not-significant for tilapia industry (Walakira et al. 2014). Usually, the main diseases affecting tilapia culture present a rate of morbidity and mortality which is related to age, water temperature and other culture conditions.

2.1. Bacterial diseases

Different bacteria including *Streptococcus iniae, S. agalactiae, Vibrio anguillarum, V. harveyi, Photobacterium damselae, Flavobacterium columnare, Edwardsiella tarda, Aeromonas* sp., *Vibrio* sp. and *Francisella* sp. cause mortality in farmed tilapia (Wang et al. 2000).

Streptococcus spp. From the different pathogens found in farmed tilapia, the most common, widespread and severe tilapia disease is caused by two species of bacteria in the genus *Streptococcus* (Perera et al. 1994). *Streptococcus* spp. are Gram-positive, non-acid fast, non-motile, oxidase-positive, catalase-negative *cocci*. *S. agalactiae* (including the previously described *S. difficilis/difficile*) is the major cause of streptococcosis in farmed tilapia while *S. iniae* also causes mortality but to a lesser extent. No obvious differences in the clinical signs of the diseases caused by these bacteria are evident. Streptococcosis can theoretically affect all fish sizes (Shoemaker et al. 2000, 2001, De-Hai et al. 2007). However, fish from 100 g to market size are usually more susceptible to the disease (causing higher economic losses) than smaller fish (Osman et al. 2017). These bacteria are the most important threats of the mid and last stages of tilapia culture because streptococcal disease could kill big fish from 150-300 g onwards.

The main entry site of *S. agalactiae* is the gastrointestinal epithelium (Iregui et al. 2016). The most consistent pathological hallmarks in infected tilapia are a marked congestion of internal organs and septicaemia, mainly in the liver, spleen and kidney (Zamri-Saad et al. 2010, Vásquez-Machado et al. 2019). Furthermore, the bacteria have tropism for the central nervous system (it provokes meningoencephalitis) and infected fish exhibit abnormal behaviour like swirling, loss of appetite, disorientation, emaciation, lethargy and bent bodies. Sick fish also have eye lesions, such as endo- or exophthalmia, haemorrhages and unilateral or bilateral opacification of the eye. Infected fish usually have diffused bleeding on skin and 2- to 3-mm abscesses symmetrically positioned on the inferior jaw. In general, these abscesses quickly burst and become haemorrhagic ulcers, which do not heal. Larger abscesses of approximately 5 mm which contain purulent material can also be observed at the base of the pectoral fins. The base of the tail is a common site for large abscesses (10-20 cm). When fish are able to survive this infection, these abscesses usually remain visible.

Regarding the internal signs of sick fish, septicaemia is frequent because bacteria swiftly reach the blood system and are disseminated to all internal organs. Major clinical signs associated with this condition are haemorrhages and inflammation in main organs, as was previously indicated. Furthermore, empty stomach and gut, as well as a big gall bladder, are commonly observed, which are typical signs of the absence of digestive activity. In severely infected fish, adhesions of the internal organs together and with the peritoneal cavity walls are common as well as the presence of fibrinous material in the peritoneal cavity (Ortega et al. 2018). Impression smears of internal organs (brain, liver, spleen and kidney) can be stained with Gram stain and they are being used to make on-farm diagnosis (Gram-positive cocci) in true cases of streptococcosis.

Outbreaks usually occur when fish have been exposed to stress (e.g. increase in water temperature, low dissolved oxygen and poor water quality or overcrowding for a long period of time). The disease is transmitted horizontally from fish to fish (via cannibalism, skin injuries, etc.), and also from the environment to the fish. Regarding the control and treatment, a decrease in feeding and/or stocking density can help to control or reduce the mortality rate. One of the hypotheses to explain this phenomenon is that the bacteria are present in the water and their uptake is facilitated by feeding. Decrease in stocking density helps to keep lower both the stress level and the pathogen load within the population. Antibiotics are only effective in treating a *Streptococcus* outbreak if treatment is applied during the early stages of the course of the disease. However, in most cases, oral antibiotics are ineffective as infected fish have a reduced appetite. Unfortunately, at present, there are no effective commercial vaccines to prevent *Streptococcus* outbreaks in tilapia.

Flavobacterium columnare. Previously called *Flexibacter columnaris, Cytophaga columnare* or *Myxobacterium columnare* is a common pathogen in early stages (especially at the fry and fingerling stages) and it is one of the most common diseases in tilapia culture. Once established, the disease is highly contagious and may spread horizontally from fish to fish. *F. columnare* is a Gram-negative, rod-shaped bacterium forming typical "hay stacks" or "columns" in wet-mount preparations (hence the name).

Infected fish generally show lethargy, anorexia, weak swimming, and mortality. Infected fish often display external lesions such as skin and gill erosion, and necrosis. In acute cases, these lesions may spread quickly and lead to high mortality within a matter of hours. The initial lesions are seen only as a paler area that lacks the normal shiny appearance. These sores are usually surrounded by a zone with a distinct reddish shade. Early signs of infection also include fin erosion. Lesions on the back often extend down the sides, giving the appearance of a 'saddle', typical of columnaris disease. The mouth can be severely affected. The gill lesions are typically necrotic and the filaments disintegrate as the bacteria invade them with the infected fish breathing rapidly and 'gasping' at the surface due to lack of oxygen. *F. columnare* utilizes fish mucus as a nutrient source but it is not known if growth in mucus or mucin results in differential protein expression and/or increased virulence of this bacteria towards fish (Shoemaker and LaFrentz 2015).

Less commonly, the infection is observed internally but, sometimes, the bacteria can reach the blood system resulting in a systemic infection. Again, the best possibility to eliminate the occurrence of columnaris is to alleviate the stress in the farmed fish population. This disease is frequently associated with high temperature fluctuations, trauma, and poor water quality, stressful handling and overcrowding. Proper diet, maintaining good water quality and avoidance of excessive handling could help to avoid the fish from being stressed.

Aeromonas. Another important bacterial disease with high impact at some farms is the disease named aeromonas septicaemia ("*Aeromonas*"). The disease is caused by *Aeromonas hydrophila,* which is the causal agent of haemorrhagic septicaemia syndrome or red pest on skin (Austin and Austin 1999). The disease results in the clinical signs of generalized external reddening and haemorrhagic septicaemia with symptoms such as lethargy, weakness, loss of appetite, red discoloration at the anus and the base of the fins, haemorrhagic eyes, gills, internal organs, and muscle, blood tinged abdominal fluid, and swollen kidney, spleen, and liver. *Aeromonas* is generally associated with poor water quality (e.g. high organic load and low dissolved oxygen) or over-crowding or after stressful fish handling. Usually the bacteria temporarily respond to antibiotic therapy for short times (Yardimci and Aydin 2011). The intracellular survival of *A. hydrophila* in host macrophages and the biochemical mechanisms involved in this phagocytosis process that affect infection has recently been studied as a valid model for the therapeutic and functional study of this bacterial infection (Fernandes et al. 2019).

2.2. Viral diseases

In our days, there is a continuous emergence of new viral diseases in aquaculture. They may be motivated by virus factors, animal host factors, environmental factors, and/or anthropogenic factors (Kibenge 2019). The most important viral threats in tilapia culture and industry in the world are iridovirus, herpes-like virus and viral nervous necrosis.

Tilapia iridovirus or tilapia lake virus (TiLV, *Tilapia tilapinevirus*). In the last years, some viruses are emerging in farmed fish (Kibenge 2019). TiLV is one of them that affects tilapia and this iridovirus was the only documented viral disease in tilapia till some years ago. TiLV is a new fish Orthomyxovirus and considered the most important threat for the culture of tilapia worldwide (FAO 2017). It was discovered as the etiological cause of massive losses of tilapia in 2009 (Bacharach et al. 2016). TiLV caused fish mortality rates of 10–90% in farmed tilapia and also in the wild population in at least 12 countries across three continents (Asia, Africa, South America) (Jansen et al. 2018). This wide dispersion of the virus is explained due to the movement of infected fish, which provoked also the dissemination of this pathological agent (Dong et al. 2017, FAO 2017). In some occasions, mortalities are associated to TiLV though no external clinical signs of infection were noted (Mugimba et al. 2018). Fortunately, TiLV is inactivated in tilapia fillets stored at −20°C for 90-120 days (Thammatorn et al. 2019) reducing the risk for human consumers.

It was unknown whether different tilapia strains are equally susceptible to TiLV infection. Recently, it has been demonstrated that the susceptibility and post challenge mortality levels of grey (*Oreochromis niloticus* × *O. aureus*) and red tilapia (*Oreochromis* spp.) to experimental TiLV infection are very similar. Furthermore, equal virus concentration at the time of death in target organs and proinflammatory cytokine responses in target and lymphoid organs were observed (Mugimba et al. 2019). It should be necessary to continue the search for a less susceptible tilapia strains to this virus, in order to improve tilapia culture.

Other viruses in tilapia. Although the incidence of other viruses has been also documented in tilapia species, their importance and magnitude in the aquaculture is not so high, as of now compared to TiLV. In this sense, infectious pancreatic necrosis virus (IPNV), nodavirus (NNV), tilapia larvae encephalitis virus (TLEV) or the infectious spleen and kidney necrosis virus (ISKNV) have also been identified in cultured tilapia specimens (Bigarré et al. 2009, Shlapobersky et al. 2010, Sinyakov et al. 2011, Suebsing et al. 2016, Subramaniam et al. 2016, Mulei et al. 2018). Further attention needs to be paid to the incidence and distribution of these viruses in the culture of tilapia around the world.

2.3. Parasites

After introducing tilapia fries into culture ponds, there is a high probability of infection to some bacteria but also to different parasites (Monogenean and Protozoan) (Soler-Jiménez et al. 2017).

    Trichodina, or "Trich", is a protozoan parasite that has severely impacted tilapia production. Trich can provoke extremely high mortality rates, mainly in young fish. The parasites heavily infect the gill (which become less efficient than usual) and body surfaces. Infected fish display rapid breathing, weakness, and uncoordinated swimming. *Trhichodina* populations can be temporarily controlled with copper sulphate and salt or formalin. However, treated fish remain carriers even after treatment and it is nearly impossible to eliminate trich from a system once it has been introduced (Hassan 1999).

    Other parasites affecting tilapia culture are isopods (clinical signs are swimming in pain and jumping over the water surface), monogeneans (the fish swim near the surface and eat less, become more pigmented and present swelling of the gills) and fish louses (the fish have bleeding ulcers, swimming in pain and rubbing their body against the pond's walls in an attempt to scrape off the parasites) (Shoemaker et al. 2006).

3. Benefit of prophylactics in tilapia aquaculture

Understanding tilapia farming life cycle is essential to analyze biosecurity risks and come up with biosecurity solutions suitable at farm, local, national and international levels. Unfortunately, control measures practiced by fish farmers in many countries are not very effective and well understood, largely due to insufficient information that can guide researchers, policy makers and farmers to develop control or preventive strategies against potential aquatic diseases (Akoll and Mwanja 2012). Protecting the culture environment from deteriorating and cultured animals from falling sick is the best possible option for production of healthy aquatic animals. However, at the moment, quarantine facilities are non-existent in some countries and limited biosecurity measures have been implemented to supervise new introductions and/or occurrence of diseases in farmed fish. Furthermore, inadequate measures to prevent escapes of farmed fish to the wild could also imply a great risk to the wild stocks (Hickley et al. 2008). In fact, the management and handling of tilapias in many emergent countries could be considered inadequate, since most of the actions carried out in the different production stages do not act in accordance with the good practices established by the Food and Agriculture Organization (FAO) and the World Health Organization (WHO), through the Codex Alimentarius Commission. These practices focus on water management, food handling, usage of chemicals and drugs, and product safety during harvesting. When these safety measures are applied, the risks of biological, physical and chemical contamination are drastically reduced, as well as the possible losses caused by opportunistic bacteria. It is well known that the occurrence of diseases in tilapia farms is mainly due to several factors that act individually or jointly in an aquaculture crop. For these reasons, the implementation of strategies is essential for its optimal management, with the objective of achieving a sustainable production under safety and good management practices programs, as well as an alternative to effective vaccination as a preventive and corrective treatment (Pech et al. 2017).

To be able to maintain the good health of fish during the culture process is one of the key points for the economic sustainability of this industry because losses (mainly due to diseases and environmental stressors) are one of the major impediments. The two major approaches for effective fish health management are prophylactics and therapeutics. Sometimes therapeutics cannot be avoided but prophylactic approaches can be considered even more important than therapeutics because they could reduce the use of disinfectants and/or antimicrobial drugs, which are harmful to the host and the environment (Opiyo et al. 2018).

There are many prophylactic measures followed in aquaculture operations to prevent non-desirable diseases. Security is one of the most important. Security has to be established at different levels, very high in the breeding centre, then in multiplier hatcheries and also in each one of the production systems. In each farm, the first step in disease prevention is to buy fingerlings from a highly regarded source. Afterwards, each producer can further reduce the risk by implementing in his/her routine work the following simple methods such as maintaining good personal hygiene, hand-washing with antibacterial soaps, disinfectant foot baths, live-haul truck disinfection, using well or municipal water wherever possible and limiting visitors. Regarding animals, good practices include limiting fish introductions, avoiding over-crowding and always maintaining a good fish nutrition (Ojwala et al. 2018).

Better management practices for optimum culture environments include, for example, avoiding the accumulation of metabolites and uneaten feed in the pond bottom, causing degradation of the culture environment and stress in fish in the long run. Furthermore, the enormous power of microbes as bioremediation agents can be exploited for improving the health and production of cultured aquatic organisms (Ponsano et al. 2019) including the use of probiotics. Furthermore, microbes have been extensively used to clean up environmental toxic pollutants.

Selective breeding for genetic resistance can also be considered but there are some negative counterparts, for example, genetic selection programs are time-consuming and usually involve huge costs. Nonetheless, the selection for disease resistance is not easy because the possibility of increasing susceptibility to a non-target pathogen has to be considered. The result can be the loss of production traits, as it was recorded in dairy and beef cattle and poultry (Agha et al. 2018, Chen et al.2019a). These ambitious selection programs have to be considered as a multi-disciplinary approach in which many professionals (from biological and veterinary scientists including immunologists, epidemiologists, virologists, pathologists, and environmental experts and specialists in production systems management) have to be involved.

4. Vaccines

Development of vaccines is considered the most significant medical achievement in human history after antibiotics, which paved the way for the prophylactic in health care systems. As per the latest report (4th October 2016) USDA has approved 141 vaccines and 74 bacterins for use in veterinary practice, while there were only nine such products approved for use in aquatic animals as on 30th August 2016. This suggests the need for developing and licensing vaccines for aquatic animals. Lack of vaccines for aquatic animals could be attributed to the lack of effective delivery systems. Recently, several organizations have developed mineral oil and nanoparticle-based delivery methods for use in aquatic animals. As the vaccine delivery through oral, injectable and immersion means is being standardized, it should be possible to develop vaccines for different species of finfish and shellfishes in the near future.

There are different laboratory studies evaluating the protection conferred by vaccines in tilapia. Among the pathogens tested, vaccines for *A. hydrophila*, *S. agalactiae*, *S. iniae*, *F. columnare*, *Francisella asiatica*, *F. noatunensis*, *Edwardsiella tarda*, *E. ictaluri*, *Vibrio vulnificus* bacteria, *Ichthyophthirius multifiliis* and *Cryptocaryon irritans* parasites have been developed (Table 1) (Bercovier et al. 1997, Håstein et al. 2005, Brudeseth et al. 2013, Sommerset et al. 2005, Liu et al. 2016, Munang'andu et al. 2016). In general, vaccines conferred good protection and induced the production of specific antibodies. Although the vaccine types are very varied, most of the studies focus on *S. agalactiae* and results indicate that, among the different vaccine types, inactivated *S. agalactiae* vaccines gave rise to better protection efficiency when compared with other strategies (e.g. live attenuated, recombinant, DNA vaccines). The most effective immunoprotection method is injection and Freund's incomplete adjuvant seemed to be suitable for tilapia vaccines. Many factors, such as immunization duration and number, fish size and challenge dose, may have some influences on vaccine efficacy (Liu et al. 2016, Munang'andu et al. 2016). Result of this is the commercialization by several companies of inactivated *S. agalactiae* vaccines for tilapia farming (Merck, PharmaQ). Further characterization of the antigenic properties of the pathogens, the best antigens, the immune response elicited as well as how to prepare and administrate the vaccines needs to be improved before vaccines are finally formulated and commercialized.

5. Immunostimulants

Immunostimulants are substances able to promote the immune response or any of its components in order to better fight against pathogens. They represent good prophylactic measures, mainly through activation of the innate immunity, against a broad range of pathogens. However, most of the immunostimulants also play other functions in the body due to pleiotropic actions and multiple properties. Then, this term in englobed in a more general classification within nutraceuticals. Nutraceuticals, a food (or a part of food) that provides medical or health benefits, may have a very important role in the prevention and/or treatment of a disease. These include herbal and other natural products, dietary supplements and functional foods (Van Doan et al. 2019a), whose functions go beyond the animal or cell nutrition.

Table 1. Overview of the most important pathogens for tilapia species and the use of prophilactic vaccines or immunostimulants

Pathogen	Vaccines[a]		Immunostimulants[b]			
	Types	Effectiveness	Probiotics	Prebiotics	Plants	Others
Streptococcus agalactiae	Inactivated Attenuated Recombinant DNA	Medium-High	Yes	Yes	Yes	Sodium alginate b-glucans
Streptococcus iniae	Inactivated Attenuated recombinant	Medium-High	Yes	Yes	Yes	N-acetylglucosamine
Flavobacterium columnare	Inactivated attenuated	Medium-High	Yes			
Aeromonas hydrophila	Inactivated Attenuated recombinant	Medium-High	Yes			Spirulin
Aeromonas veronii				Yes	Yes	
Francisella asiatica	Attenuated	High				
Edwardsiella tarda	Inactivated Attenuated recombinant	Medium-High	Yes		Yes	Several polysaccharides
Vibrio vulnificus	Inactivated	Medium-High				
Pleiomonas shigelloides					Yes	
Ichthyophthirius multifiliis	Live, inactivated	Medium				

Type of vaccines[a] and immunostimulants[b] used for the prevention of diseases in tilapia and their efficacy upon challenge. Medium-High, 50-100 of protection; High, >80% protection; Yes, protection.

Nutraceuticals gained importance in the last decade due to the increasing cost and side effects of therapeutic pharmaceutical agents. Among them are prebiotics, probiotics, polyunsaturated fatty acids and natural antioxidants, which can also improve the fish immune status (Durigon et al. 2019). In aquaculture, these ingredients, or fed additives, have many positive demonstrated effects on farmed fish such as improved growth, immunity, water quality, nutrient digestibility, stress tolerance and reproductive performance and they are eco-friendly. Furthermore, some of them can also have a direct inhibitory action on pathogens (Ellison et al. 2018). In sharp contrast to vaccines, these substances are usually cheaper, have no side-effects and provide good protection against most of the adverse situations during the farming schedule such as disease or stress. In fact, they increase the general immune status, and the innate immune responses in particular, which gives good protection for any pathogen and is not limited to the one vaccinated.

Probiotics are probably the most studied immunostimulants in tilapia. Probiotics are live microorganisms that when administered in proper dosages improve digestion, nutrition, growth and welfare. In this sense, the use of different probiotics, namely lactic acid bacteria, have proved to be very promising in tilapia, including *Lactobacillus acidophilus, L. brevis, L. plantarum, Bacillus subtilis, B. licheniformis, B. pumilus,* and *B. amyloliquefaciens* (Hai 2015). But other probiotics have also demonstrated good effects on tilapia such as *Aspergillus oryzae, Bifidobacterium bifidum, Enterococcus faecium, Micrococcus luteus, Pediococcus acidilactici, P. fluorescens, Streptococcus salivarius* subsp. *thermophilus, Psychrobacterm aritimus, Rummeliibacillus stabekisii, Paenibacillus ehimensis* and *Saccharomyces cerevisiae* (Hai 2015, Chen et al. 2019b, Makled et al. 2019, Tan et al. 2019). Additionally, some pathogenic bacteria for some fish species are also beneficial for tilapia such as *Citrobacter freundii* and *P. fluorescens*. The use of these probiotics produces immunostimulation and increase the disease resistance in tilapia (Table 1). Probiotics were able to increase the number of leucocytes, the humoral immune activities in either the serum or mucus including lysozyme, peroxidase and complement activities, as well as the activities of the macrophages and neutrophils such as phagocytosis, respiratory burst, production of nitrogen reactive species, alkaline phosphatase and myeloperoxidase activities (Hai 2015, Gobi et al. 2018, Chen et al. 2019b, Tan et al. 2019, Van Doan et al. 2019b). At gene level, probiotics were also able to stimulate the expression of the cytokine genes interleukin-1β (IL-1β) in tilapia, tumour necrosis factor-α (TNFα), transforming growth factor-β (TGFβ), or heat shock protein 70 (HSP70) among others. Besides this, dietary administration of probiotics, single or combined, resulted in decreased numbers of bacteria in tilapia specimens, signs of infection and even partial to complete protection against experimental infections with *S. agalactiae* or *A. hydrophila*. Interestingly, commercial formulations containing probiotics have been used for tilapia culture and demonstrated improved nutrition, growth, immunity, microbiota and reproduction (Mehrim et al. 2015, Standen et al. 2016, Abarike et al. 2018).

Administered alone, or in combination with probiotics, the application of prebiotics to tilapia has been scarcely investigated. Prebiotics are defined as non-digestible food ingredients that beneficially affect the host by selectively stimulating the growth and/or activity of one or a limited number of bacterial species already resident in the colon, and thus attempt to improve host health (Gibson and Roberfroid 1995). Therefore, their action is indirect and depends on the presence of intestinal bacteria, raising the importance in the combination of proper pre and probiotics balance (called then symbiotic). In general, prebiotics are fermented by *Bifidobacterium* spp. and *Lactobacillus* spp. leading to the formation of short-chain fatty acids (SCFAs), mainly acetic, propionic and butyric acids, which cause a pH drop in the gut, lead to local and systemic health effects and might be absorbed by the host and used as energy sources (Gibson and Roberfroid 1995). The benefits of some prebiotics including mannanooligosaccharides (MOS), fructooligosaccharides (FOS, or oligofructose), short-chain fructooligosaccharides (scFOS), galactooligosaccharides (GOS), xylooligosaccharides (XOS) and inulin, in their laboratory or commercial formulations, have been demonstrated to promote fish growth, immunity and disease resistance in fish (Guerreiro et al. 2018). Very few studies have been addressed in this sense in tilapia. Thus, the use of commercial formulations, alginate, b-glucans or prebiotics isolated from plants administered by feeding, alone or in combination with probiotics, produced increases in the growth in tilapia, feed conversion, innate immunity (lysozyme, phagocytosis, respiratory burst, or complement activities) and disease resistance against *A. hydrophila, A. veronii* or *S. agalactiae* challenge (Table 1) (Zhou et al. 2011, Widanarni and Tanbiyaskur 2015, Van Doan et al. 2016, 2017, Pilarski et al. 2017, Sewaka et al. 2019).

At present, one of the most promising nutraceuticals for fish farming is the use of plants since they could be used to replace fish meal from the fish diets, are readily available and affordable to the industry. In this sense, most of the studies focus on the use of herbal/medicinal plants, and their extracts, and vegetable oils; however, we will only focus on those studies aiming to investigate their immunostimulant potential, and not restricted to nutrition. Regarding tilapia species immunity, a vast number of medicinal plants and herbs, as a whole or by extracts, have been administered and their role in immunity documented: increased leucocyte numbers, lysozyme, MPO, phagocytosis or respiratory burst activities, Ig production and disease resistance upon bacterial challenge (Table 1) (Gabriel 2019). Some examples of this are *Allium sativum, Moringa oleifera, Citrus* spp., *Astragalus membranaceus, Aloe barbadensis, Rosmarinus officinalis, Thymus vulgaris, Trigonella foenum graecum, Euphorbia hirta, Cuminum cyminum, Pimenta dioica* or *Echinacea purpurea* amongst many others. The composition and nature of the active compounds, named phytochemicals, greatly varies with the plant species, part (roots, leaves, flowers, fruits, etc.), the ripeness stage, the extraction procedures, and many other factors. Overall, the active compounds consist of polysaccharides, polyphenols, carotenoids, terpenoids and alkaloids among others. These substances show potent biological activities such as antimicrobial (bacteria, fungi and virus), anti-cancer, antioxidant, analgesic, antiparasitic, anti-allergenic, antispasmodic, antihyperglycemic, anti-inflammatory and immunomodulatory, which makes them very good candidates for the prevention and therapy of several diseases and infections in fish. For example, a transcriptomic analysis of genetically improved farmed tilapia (GIFT) fed with resveratrol, a well-known stilbene polyphenol, produced a significant alteration in the expression of genes and pathways related to innate immunity and biological processes leading to improved immunity (Zheng et al. 2019). At practical level, phytochemicals are suggested as therapeutic and preventive agents for aquaculture since these natural products might have direct antibacterial killing activities. Thus, extracts of *A. sativum* and *Syzygium aromaticum* significantly reduced the mortality of tilapia specimens infected with antibiotic-resistant *Enterococcus faecalis*, which demonstrates the goodness of phytochemicals to replace or reduce antibiotics in aquaculture (Zheng et al. 2019). Further implementation of phytochemicals isolation, characterization and application to fish aquaculture is nowadays mandatory for the sector.

There are many other studies conducted in tilapia species demonstrating the use and goodness of nutraceuticals or immunostimulants. Vitamins are necessary micronutrients for animals. Vitamins A, C or E are potent antioxidant molecules, which when administered to tilapia resulted in improved innate and acquired immunity both as well as reversed the negative impact of toxicants or oxidative stress (Michael et al. 1998, Guha and Khuda-Bukhsh 2002, Hung et al. 2007, Harabawy and Mosleh 2014, Guzmán-Guillén et al. 2015, Abo-Al-Ela et al. 2017, Abdel Rahman et al. 2018, Alkaladi 2019). Micronutrients such as Se, Zn, Cu, Mn, or Fe minerals are also used as immunostimulants for tilapia(Hung et al. 2007). Interestingly, tilapia feeding with vitamins or micronutrient minerals resulted in increased nutritive value of the fillets destined for human consume, which might be used to improve the human nutrition (Farzad et al. 2019). Further studies about the description of immunostimulants in tilapia would help to implement preventive strategies for the successful culture of this fish species.

6. Future remarks

Several aspects are necessary in the farming of tilapia species to be both biologically and economically sustainable. One of this is the necessity to raise awareness in the producers for fish and farm management protocols to improve welfare and biosecurity. This would imply, for example, the control of fish and fish by-products trade-off between fish farms and countries, in which policy makers also need to implement strategies to help with this objective.

Other of the fields to investigate is in the knowledge of the pathogens of tilapia and in the tilapia-pathogen interactions. In this sense, further studies in the pathogen infection processes, target tissues, replication strategies, and spreading/evading mechanisms would greatly help to know and control them. For example, generation of 'omic' tools would help to identify and propose the generation of proper vaccines, as the recent complete genome sequence of a TiLV isolate obtained from Nile tilapia points to (Al-Hussinee et al. 2018). *S. galactiae* is widely spread in Southeast Asia, which causes disease and important economic losses in aquaculture, but has been also recorded as a food borne disease causing bacteremia in humans. However, new studies are needed to know if the transmission is from aquaculture to humans, or vice versa. Furthermore, there are no data if some reservoir (still non-identified) exists. Cross-border collaborations in human and animal health are needed to complete the epidemiological picture of this disease (Barkham et al. 2019) as in the other tilapia diseases.

In addition, but greatly connected to this, the understanding of the tilapia immune system and its functioning against pathogens is mandatory. However, from the applied point of view, there is no great necessity to go further in the study of tilapia immune factors, which are similar to other fish species, but it is greatly necessary to i) identify the antigens in the pathogens for vaccine development, and ii) to search and identify more and new prophylactics, preferably of natural origin, to improve tilapia immunity and reduce the use of antibiotics. Goals obtained in the research laboratories related to these aspects would ensure the future existence of effective commercial vaccines and treatments to be applied in the tilapia aquaculture around the world.

Finally, genome of several tilapia species is being studied by the Broad Institute of MIT and Harvard (https://www.broadinstitute.org/tilapia/tilapia-genome-project). The new available data as results of the applications of the 'omics techniques to all these species of interest in aquaculture will open new challenges and opportunities for our understanding and improvement of the culture of tilapia. Furthermore, recent developments in bioinformatics could cover the way for new generation selection programs in aquatic animal breeding programs.

Acknowledgments

This work was supported by the *Fundación Seneca de la Región de Murcia (Grupo de Excelencia* grant no. 19883/GERM/15).

References cited

Abarike, E.D., J. Cai, Y. Lu, H. Yu, L. Chen, J. Jian, et al. 2018. Effects of a commercial probiotic BS containing *Bacillus subtilis* and *Bacillus licheniformis* on growth, immune response and disease resistance in Nile tilapia, *Oreochromis niloticus*. Fish Shellfish Immunol. 82: 229–238.

Abdel Rahman, A.N., A.A. Khalil, H.M. Abdallah and M. ElHady. 2018. The effects of the dietary supplementation of *Echinacea purpurea* extract and/or vitamin C on the intestinal histomorphology, phagocytic activity, and gene expression of the Nile tilapia. Fish Shellfish Immunol. 82: 312–318.

Abo-Al-Ela, H.G., A.F. El-Nahas, S. Mahmoud and E.M. Ibrahim. 2017. Vitamin C modulates the immunotoxic effect of 17α-methyltestosterone in Nile tilapia. Biochemistry 56: 2042–2050.

Agha, S., W. Mekkawy, N. Ibanez-Escriche, C.E. Lind, J. Kumar, A. Mandal, et al. 2018. Breeding for robustness: Investigating the genotype-by-environment interaction and micro-environmental sensitivity of genetically improved farmed tilapia (*Oreochromis niloticus*). Anim. Genet. 49: 421–427.

Akoll, P. and W.W. Mwanja. 2012. Fish health status, research and management in East Africa: Past and present. African J. Aquatic Sci. 37: 117–129.

Al-Hussinee, L., K. Subramaniam, M.S. Ahasan, B. Keleher and T.B. Waltzek. 2018. Complete genome sequence of a tilapia lake virus isolate obtained from Nile tilapia (*Oreochromis niloticus*). Genome Announc. 6: e00580-18.

Alkaladi, A. 2019. Vitamins E and C ameliorate the oxidative stresses induced by zinc oxide nanoparticles on liver and gills of *Oreochromis niloticus*. Saudi J. Biol. Sci. 26: 357–362.

Austin, B. and D.A. Austin. 1999. Bacterial Fish Pathogens Diseased of Farmed and Wild Fish. Praxis Publishing Ltd.

Bacharach, E., N. Mishra, T. Briese, M.C. Zody, J.E. Kembou Tsofack, R. Zamostiano, et al. 2016. Characterization of a novel Orthomyxo-like virus causing mass die-offs of tilapia. MBio, 7: e00431-00416.

Barkham, T., R.N. Zadoks, M.N.A. Azmai, S. Baker, V.T.N. Bich, V. Chalker, et al. 2019. One hypervirulent clone, sequence type 283, accounts for a large proportion of invasive *Streptococcus agalactiae* isolated from humans and diseased tilapia in Southeast Asia. PLoS Negl. Trop. Dis. 13: e0007421.

Bercovier, H., C. Ghittino and A. Eldar. 1997. Immunization with bacterial antigens: Infections with streptococci and related organisms. Dev. Biol. Stand. 90: 153–160.

Bigarré, L., J. Cabon, M. Baud, M. Heimann, A. Body, F. Lieffrig, et al. 2009. Outbreak of betanodavirus infection in tilapia, *Oreochromis niloticus* (L.), in fresh water. J. Fish Dis. 32: 667–673.

Bowser, P.R., G.A. Wooster, R.G. Getchell and M.B. Timmons. 1998. *Streptococcus iniae* infection of tilapia *Oreochromis niloticus* in a recirculation production facility. J. World Aquac. Soc. 29: 335–339.

Brudeseth, B.E., R. Wiulsrød, B.N. Fredriksen, K. Lindmo, K.-E. Løkling, M. Bordevik, et al. 2013. Status and future perspectives of vaccines for industrialised fin-fish farming. Fish Shellfish Immunol. 35: 1759–1768.

Chen, C.H., B.J. Li, X.H. Gu, H.R. Lin and J.H. Xia. 2019a. Marker-assisted selection of YY supermales from a genetically improved farmed tilapia-derived strain. Zool. Res. 40: 108-112.

Chen, S.-W., C.-H. Liu and S.-Y. Hu 2019b. Dietary administration of probiotic *Paenibacillus ehimensis* NPUST1 with bacteriocin-like activity improves growth performance and immunity against *Aeromonas hydrophila* and *Streptococcus iniae* in Nile tilapia (*Oreochromis niloticus*). Fish Shellfish Immunol. 84: 695–703.

Chitmanat, C., P. Lebel, N. Whangchai, J. Promya and L. Lebel. 2016. Tilapia diseases and management in river-based cage aquaculture in northern Thailand. J. Applied Aquacult. 28: 9-16.

De-Hai, X., C.A. Shoemaker and P.H. Klesius. 2007. Evaluation of the link between gyrodactylosis and streptococcosis of Nile tilapia, *Oreochromis niloticus* (L.). J. Fish Dis. 30: 233-238.

Dong, H.T., P. Ataguba, T. Khunrae, T. Rattanarojpong and S. Serapin. 2017. Evidence of TiLV infection in tilapia hatcheries from 2012 to 2017 reveals probable global spread of the disease. Aquaculture 479: 579–583.

Durigon, E.G., D.F. Kunz, N.C. Peixoto, J. Uczay and R. Lazzari. 2019. Diet selenium improves the antioxidant defense system of juveniles Nile tilapia (*Oreochromis niloticus* L.). Braz. J. Biol. 79: 527–532.

Ellison, A.R., T.M. Uren Webster, O. Rey, C. Garcia de Leaniz, S. Consuegra, P. Orozco-ter Wengel, et al. 2018. Transcriptomic response to parasite infection in Nile tilapia (*Oreochromis niloticus*) depends on rearing density. BMC Genomics 19: 723.

Esteban, M.Á., A. Cuesta, E. Chaves-Pozo and J. Meseguer. 2015. Phagocytosis in teleosts: Implications of the new cells involved. Biology (Basel). 4: 907–922.

Evans, D.L., J.H. Leary and L. Jaso-Friedmann. 2001. Nonspecific cytotoxic cells and innate immunity: Regulation by programmed cell death. Dev. Comp. Immunol. 25: 791–805.

Farzad, R., D.D. Kuhn, S.A. Smith, S.F. O'Keefe, N.V.C. Ralston, A.P. Neilson, et al. 2019. Trace minerals in tilapia fillets: Status in the United States marketplace and selenium supplementation strategy for improving consumer's health. PLoS One 14: e0217043.

Fernandes, D.C., S.F. Eto, A.C. Moraes, E.J.R. Prado, A.S.R. Medeiros, M.A.A. Belo, et al. 2019. Phagolysosomal activity of macrophages in Nile tilapia (*Oreochromis niloticus*) infected *in vitro* by *Aeromonas hydrophila*: Infection and immunotherapy. Fish Shellfish Immunol. 87: 51–61.

Food and Agriculture Organization of the United Nations (FAO): FAO Fisheries and Aquaculture Circular No. 1135/5 2017. 2017. Available at: http://www.fao.org/3/a-i6875e.pdf.

Gabriel, N.N. 2019. Review on the progress in the role of herbal extracts in tilapia culture. Cogent Food Agric. 5.

Gibson, G.R. and M.B. Roberfroid. 1995. Dietary modulation of the human colonic microbiota: Introducing the concept of prebiotics. J. Nutr. 125: 1401–1412.

Gobi, N., B. Vaseeharan, J.-C. Chen, R. Rekha, S. Vijayakumar, M. Anjugam, et al. 2018. Dietary supplementation of probiotic *Bacillus licheniformis* Dahb1 improves growth performance, mucus and serum immune parameters, antioxidant enzyme activity as well as resistance against *Aeromonas hydrophila* in tilapia *Oreochromis mossambicus*. Fish Shellfish Immunol. 74: 501–508.

Guerreiro, I., A. Oliva-Teles and P. Enes. 2018. Prebiotics as functional ingredients: Focus on Mediterranean fish aquaculture. Rev. Aquac. 10: 800–832.

Guha, B. and A.R. Khuda-Bukhsh. 2002. Efficacy of vitamin-C (L-ascorbic acid) in reducing genotoxicity in fish (*Oreochromis mossambicus*) induced by ethyl methane sulphonate. Chemosphere 47: 49–56.

Guzmán-Guillén, R., A.I. Prieto Ortega, A. Martín-Caméan and A.M. Cameán. 2015. Beneficial effects of vitamin E supplementation against the oxidative stress on Cylindrospermopsin-exposed tilapia (*Oreochromis niloticus*). Toxicon 104: 34–42.

Hai, N.V. 2015. Research findings from the use of probiotics in tilapia aquaculture: A review. Fish Shellfish Immunol. 45: 592–597.

Harabawy, A.S.A. and Y.Y.I. Mosleh. 2014. The role of vitamins A, C, E and selenium as antioxidants against genotoxicity and cytotoxicity of cadmium, copper, lead and zinc on erythrocytes of Nile tilapia, *Oreochromis niloticus*. Ecotoxicol. Environ. Saf. 104: 28–35.

Hassan, A.M.A.E.H. 1999. Saudi Arabia Trichodiniasis in farmed freshwater tilapia in Eastern Saudi Arabia. Mar. Sci. 10: 157–168.

Håstein, T., R. Gudding and O. Evensen. 2005. Bacterial vaccines for fish – An update of the current situation worldwide. Dev. Biol. (Basel). 121: 55–74.

Hickley, P., M. Muchiri, R. Britton and R. Boar. 2008. Economic gain versus ecological damage from the introduction of non-native freshwater fish: Case studies from Kenya. Open Fish. Sci. J. 1: 36–46.

Hung, S.-W., C.-Y. Tu and W.-S. Wang. 2007. *In vivo* effects of adding singular or combined anti-oxidative vitamins and/or minerals to diets on the immune system of tilapia (*Oreochromis hybrids*) peripheral blood monocyte-derived, anterior kidney-derived, and spleen-derived macrophages. Vet. Immunol. Immunopathol. 115: 87–99.

Iregui, C.A., J. Comas, G.M. Vasquez and N. Verjan. 2016. Experimental early pathogenesis of *Streptococcus agalactiae* infection in red tilapia *Oreochromis* spp. J. Fish Dis. 39: 205–215.

Jansen, M.D., H.T. Dong and C.V. Mohan. 2018. Tilapia lake virus: A threat to the global tilapia industry. Rev. Aquacult. 2018: 1-15.

Kibenge, F.S.B. 2019. Emerging viruses in aquaculture. Curr. Opin. Virol. 34: 97–103.

Liu, G., J. Zhu, K. Chen, T. Gao, H. Yao, Y. Liu, et al. 2016. Development of *Streptococcus agalactiae* vaccines for tilapia. Dis. Aquat. Organ. 122: 163–170.

Magnadottir, B., S. Lange, S. Gudmundsdottir, J. Bogwald and R. Dalmo. 2005. Ontogeny of humoral immune parameters in fish. Fish Shellfish Immunol. 19: 429–439.

Makled, S.O., A.M. Hamdan and A.-F.M. El-Sayed. 2019. Growth promotion and immune stimulation in Nile tilapia, *Oreochromis niloticus*, fingerlings following dietary administration of a novel marine probiotic, *Psychrobacter maritimus* S. Probiotics Antimicrob. Proteins. doi.org/10.1007/s12602-019-09575-0

Mehrim, A.I., F.F. Khalil and M.E. Hassan. 2015. Hydroyeast Aquaculture® as a reproductive enhancer agent for the adult Nile tilapia (*Oreochromis niloticus* Linnaeus, 1758). Fish Physiol. Biochem. 41: 371–381.

Michael, R.D., C.G. Quinn and S. Venkatalakshmi. 1998. Modulation of humoral immune response by ascorbic acid in *Oreochromis mossambicus* (Peters). Indian J. Exp. Biol. 36: 1038–1040.

Mugimba, K.K., A.A. Chengula, S. Wamala, E.D. Mwega, C.J. Kasanga, D.K. Byarugaba, et al. 2018. Detection of tilapia lake virus (TiLV) infection by PCR in farmed and wild Nile tilapia (*Oreochromis niloticus*) from Lake Victoria. J. Fish Dis. 41: 1181-1189.

Mugimba, K.K., S. Tal, S. Dubey, S. Mutoloki, A. Dishon, Ø. Evensen, et al. 2019. Gray (*Oreochromis niloticus* × *O. aureus*) and Red (*Oreochromis* spp.) tilapia show equal susceptibility and proinflammatory cytokine responses to experimental tilapia lake virus infection. Viruses 11: E893.

Mulei, I.R., P.N. Nyaga, P.G. Mbuthia, R.M. Waruiru, L.W. Njagi, E.W. Mwihia, et al. 2018. Infectious pancreatic necrosis virus isolated from farmed rainbow trout and tilapia in Kenya is identical to European isolates. J. Fish Dis. 41: 1191–1200.

Munang'andu, H.M., J. Paul and Ø. Evensen. 2016. An overview of vaccination strategies and antigen delivery systems for *Streptococcus agalactiae* vaccines in Nile tilapia (*Oreochromis niloticus*). Vaccines 4: E48.

Nakanishi, T., H. Toda, Y. Shibasaki and T. Somamoto. 2011. Cytotoxic T cells in teleost fish. Dev. Comp. Immunol. 35: 1317–1323.

Ojwala, R.A., E.O. Otachi and N.K. Kitaka. 2018. Effect of water quality on the parasite assemblages infecting Nile tilapia in selected fish farms in Nakuru county, Kenya. Parasitol. Res. 117: 3459-3471.

Opiyo, M.A., E. Marijani, P. Muendo, R. Odede, W. Leschen and H. Charo-Karisa. 2018. A review of aquaculture production and health management practices of farmed fish in Kenya. Int. J. Vet. Sci. Med. 6: 141-148.

Ortega, C., I. García, R. Irgang, R. Fajardo, D. Tapia-Cammas, J. Acosta, et al. 2018. First identification and characterization of *Streptococcus iniae* obtained from tilapia (*Oreochromis aureus*) farmed in Mexico. J. Fish Dis. 41: 773–782.

Osman, K.M., K.S. Al-Maary, A.S. Mubarak, T.M. Dawoud, I.M.I. Moussa, M.D.S. Ibrahim, et al. 2017 Characterization and susceptibility of streptococci and enterococci isolated from Nile tilapia (*Oreochromis niloticus*) showing septicaemia in aquaculture and wild sites in Egypt. BMC Vet. Res. 13: 357.

Pech, Z.G.H., M.R. Castaneda-Chavez and F. Lango-Reynoso. 2017. Pathogenic bacteria in *Oreochromis niloticus* var. stirling tilapia culture. Fish Aqua. J8: 197.

Perera, R.P., S.K. Johnson, M.D. Collins and D.H. Lewis. 1994. *Streptococcus iniae* associated mortality of *Tilapia nilotica* × *T. aurea* hybrids. J. Aquat. Anim. Health 6: 335–340.

Pilarski, F., C.A. Ferreira de Oliveira, F.P.B. Darpossolo de Souza and F.S. Zanuzzo. 2017. Different β-glucans improve the growth performance and bacterial resistance in Nile tilapia. Fish Shellfish Immunol. 70: 25–29.

Plumb, J.A. 1997. Infectious diseases of tilapia. pp. 212–228, *In*: B.A. Costa-Pierce and J.E. Rakocy (eds.), Tilapia Aquaculture in the Americas. World Aquaculture Society, Bataon Rouge, Louisiana.

Ponsano, E.H.G., T.L.M. Grassi, E.F.E. Santo, L.K.F. de Lima and R.C. Pereira. 2019. Production and use of microbial biomass helping sustainability in tilapia production chain. Biotech. 9: 325.

Sailendri, K. and V. Muthukkaruppan. 1975a. The immune response of the teleost, *Tilapia mossambica* to soluble and cellular antigens. J. Exp. Zool. 191: 371–381.

Sailendri, K. and V. Muthukkaruppan. 1975b. Morphology of lymphoid organs in a cichlid teleost, *Tilapia mossambica* (Peters). J. Morphol. 147: 109–121.

Scapigliati, G. 2013. Functional aspects of fish lymphocytes. Dev. Comp. Immunol. 41: 200–208.

Sewaka, M., C. Trullas, A. Chotiko, C. Rodkhum, N. Chansue, S. Boonanuntanasarn, et al. 2019. Efficacy of synbiotic Jerusalem artichoke and *Lactobacillus rhamnosus* GG-supplemented diets on growth performance, serum biochemical parameters, intestinal morphology, immune parameters and protection against *Aeromonas veronii* in juvenile red tilapia (*Oreochromis* spp.). Fish Shellfish Immunol. 86: 260–268.

Shlapobersky, M., M.S. Sinyakov, M. Katzenellenbogen, R. Sarid, J. Don and R.R. Avtalion. 2010. Viral encephalitis of tilapia larvae: Primary characterization of a novel herpes-like virus. Virology 399: 239–247.

Shoemaker, C.A., J.J. Evans and P.H. Klesius. 2000. Density and dose: Factors affecting mortality of *Streptococcus iniae* infected tilapia (*Oreochromis niloticus*). Aquaculture 188: 229–235.

Shoemaker, C.A., P.H. Klesius and J.J. Evans. 2001. Prevalence of *Streptococcus iniae* in tilapia, hybrid striped bass and channel catfish from fish farms in the United States. Amer. J. Vet. Res. 62: 174–177.

Shoemaker, C.A., D.-H. Xu, J.J. Evans and P.H. Klesius. 2006. Parasites and diseases. pp. 561–582. *In*: C. Lim and C.D. Webster (eds.), Tilapia: Biology, Culture and Nutrition. The Haworth Press, Inc., Binghamton, NY.

Shoemaker, C.A. and B.R.LaFrentz. 2015. Growth and survival of the fish pathogenic bacterium, *Flavobacterium columnare*, in tilapia mucus and porcine gastric mucin. FEMS Microbiol. Lett. 362.

Sinyakov, M.S., S. Belotsky, M. Shlapobersky and R.R. Avtalion. 2011. Vertical and horizontal transmission of tilapia larvae encephalitis virus: The bad and the ugly. Virology 410: 228–233.

Soler-Jiménez, L.C., A.I. Paredes-Trujillo and V.M. Vidal-Martínez. 2017. Helminth parasites of finfish commercial aquaculture in Latin America. J. Helminthol. 91: 110–136.

Sommerset, I., B. Krossøy, E. Biering and P. Frost. 2005. Vaccines for fish in aquaculture. Expert Rev. Vaccines 4: 89–101.

Standen, B.T., D.L. Peggs, M.D. Rawling, A. Foey, S.J. Davies, G.A. Santos, et al. 2016. Dietary administration of a commercial mixed-species probiotic improves growth performance and modulates the intestinal immunity of tilapia, *Oreochromis niloticus*. Fish Shellfish Immunol. 49: 427–435.

Stosik, M.P., B. Tokarz-Deptuła and W. Deptuła. 2018. Specific humoral immunity in *Osteichthyes*. Cent. Eur. J. Immunol. 43: 335–340.

Subramaniam, K., M. Gotesman, C. Smith, N. Steckler, K. Kelley, J. Groff, et al. 2016. Megalocytivirus infection in cultured Nile tilapia *Oreochromis niloticus*. Dis. Aquat. Organ. 119: 253–258.

Suebsing, R., P.J. Pradeep, S. Jitrakorn, S. Sirithammajak, J. Kampeera, W.A. Turner, et al. 2016. Detection of natural infection of infectious spleen and kidney necrosis virus in farmed tilapia by hydroxynapthol blue-loop-mediated isothermal amplification assay. J. Appl. Microbiol. 121: 55–67.

Tan, H.Y., S.-W. Chen and S.-Y. Hu. 2019. Improvements in the growth performance, immunity, disease resistance, and gut microbiota by the probiotic *Rummeliibacillus stabekisii* in Nile tilapia (*Oreochromis niloticus*). Fish Shellfish Immunol. 92: 265–275.

Thammatorn, W., P. Rawiwan and W. Surachetpong. 2019. Minimal risk of tilapia lake virus transmission via frozen tilapia fillets. J. Fish Dis. 42: 3–9.

Tilapia Genome Project. Broad Institute. https://www.broadinstitute.org/tilapia/tilapia-genome-project.

Van Doan, H., W. Tapingkae, T. Moonmanee and A. Seepai. 2016. Effects of low molecular weight sodium alginate on growth performance, immunity, and disease resistance of tilapia, *Oreochromis niloticus*. Fish Shellfish Immunol. 55: 186–194.

Van Doan, H., S.H. Hoseinifar, W. Tapingkae and P. Khamtavee. 2017. The effects of dietary kefir and low molecular weight sodium alginate on serum immune parameters, resistance against *Streptococcus agalactiae* and growth performance in Nile tilapia (*Oreochromis niloticus*). Fish Shellfish Immunol. 62: 139–146.

Van Doan, H., S.H. Hoseinifar, K. Sringarm, S. Jaturasitha, B. Yuangsoi, M.A.O. Dawood, et al. 2019a. Effects of Assam tea extract on growth, skin mucus, serum immunity and disease resistance of Nile tilapia (*Oreochromis niloticus*) against *Streptococcus agalactiae*. Fish Shellfish Immunol. 93: 428–435.

Van Doan, H., S.H. Hoseinifar, W. Tapingkae, M. Seel-audom, S. Jaturasitha, M.A.O. Dawood, et al. 2019b. Boosted growth performance, mucosal and serum immunity, and disease resistance Nile tilapia (*Oreochromis niloticus*) fingerlings using corncob-derived xylooligosaccharide and *Lactobacillus plantarum* CR1T5. Probiotics Antimicrob. Proteins. doi.org/10.1007/s12602-019-09554-5

Vásquez-Machado, G., P. Barato-Gómez and C. Iregui-Castro. 2019 Morphological characterization of the adherence and invasion of *Streptococcus agalactiae* to the intestinal mucosa of tilapia *Oreochromis* sp.: An *in vitro* model. J. Fish Dis. 2019 42: 1223–1231.

Walakira, J., P. Akoll, M. Engole, M. Serwadd, M. Nkambo, V. Namulawa, et al. 2014. Common fish diseases and parasites affecting wild and farmed Tilapia and catfish in Central and Western Uganda. Uganda J. Agri. Sci. 15: 113–125.

Widanarni, T. 2015. Application of probiotic, prebiotic and synbiotic for the control of Streptococcosis in Tilapia *Oreochromis niloticus*. Pakistan J. Biol. Sci. 18: 59–66.

Yardimci, B. and Y. Aydin. 2011. Pathological findings of experimental Aeromonas hydrophila infection in Nile tilapia (*Oreochromis niloticus*). Ankara Univ. Vet. Fak. 58: 47–54.

Ye, J., I.M. Kaattari, C. Ma and S. Kaattari. 2013. The teleost humoral immune response. Fish Shellfish Immunol. 35: 1719–1728.

Zamri-Saad, M., M.N. Amal and A. Siti-Zahrah. 2010. Pathological changes in red tilapias (*Oreochromis* spp.) naturally infected by *Streptococcus agalactiae*. J. Comp. Pathol. 143: 227–229.

Zheng, Y., G. Hu, W. Wu, Z. Zhao, S. Meng, L. Fan, et al. 2019. Transcriptome analysis of juvenile genetically improved farmed tilapia (*Oreochromis niloticus*) livers by dietary resveratrol supplementation. Comp. Biochem. Physiol. Part C Toxicol. Pharmacol. 223: 1–8.

Zhou, Z., S. He, Y. Liu, Y. Cao, K. Meng, B. Yao, et al. 2011. Gut microbial status induced by antibiotic growth promoter alters the prebiotic effects of dietary DVAQUA® on *Aeromonas hydrophila*-infected tilapia: Production, intestinal bacterial community and non-specific immunity. Vet. Microbiol. 149: 399–405.

Sex Determination and Differentiation of Tilapia

Gonzalo de Alba*, Francisco Javier Sánchez-Vázquez and
José Fernando López-Olmeda

Department of Physiology, Faculty of Biology, University of Murcia, 30100 Murcia, Spain

1. Introduction

In aquaculture, tilapia has become one of the most popular genera of commercial fish species. The most cultivated tilapia species is *Oreochromis niloticus,* which accounts for 85% of the world's tilapia production because of its good availability, and its numerous productive and reproductive advantages. Tilapia's productive interest lies in its rapid growth and good disease resistance (Jalabert 1989, Lapeyre 2008). In addition to these productive advantages, several other reproductive aspects make tilapia a great species for aquaculture. Firstly, females present high fecundity with successive breeding cycles every 2- to 4-week intervals all year round (Tacon et al. 2000, Lapeyre 2008). Secondly, tilapia exhibits parental care, which means that after fertilization, females hold eggs inside their oral cavities (mouthbrooding) to safeguard offspring's survival against predators and other external hazards (Rana 1988). However, there are other reproductive aspects that compromise fish farming protocols and production. On the one hand, tilapia exhibits precocious sexual maturation, which may occur at around three months of age depending on both environmental conditions and genetic differences (De Silva and Radampola 1990, Suresh and Bhujel 2013). Hence, especially in females, high breeding frequency involves using most energy reserves to fulfil reproduction needs, which limits muscle growth and productive yield (Beardmore et al. 2001). On the other hand, due to the asynchronicity of ovarian development, the female gonad involves different development stages that make it difficult to control not only her reproductive cycle but also, therefore, synchronization of breeding and the capacity of obtaining fertilised eggs at will. Moreover, in Nile tilapia, fertility rates depend on both the weight (body size) of females and environmental conditions.

Due to high energy demands for reproduction in females, tilapia aquaculture has focused on the production of male-monosex populations (Omasakai et al. 2016),

*Corresponding author: gonzalode.alba@um.es

which provides a series of relevant advantages. For instance, monosex production allows the control of the reproduction process and both genetic and phenotypic variability. Furthermore, the use of monosex populations allows productive parameters to be intensified (males grow faster and bigger than females and they reach the market size earlier), which optimises the resources used in the tilapia industry (i.e. facilities and technical staff) (Beardmore et al. 2001, Pinto et al. 2018). The main techniques to obtain monosex populations are based on using thermal, hormonal and genetic protocols to perform the sexual reversion of female to male during sensitive periods for sexual determinations (Baroiller and Jalabert 1989).

So as regards tilapia reproduction, knowing in-depth the interaction between the genetic and environmental components that establish sex determination is most important. In addition, it is essential to know the stages of sexual differentiation and the endocrine mechanisms which control the transformation of an undifferentiated gonadal tissue into ovaries or testicles, thus establishing the phenotype of the adult fish.

2. Sex determination

Excellent fish flexibility is also reflected in the sexual determination mechanisms found among species of the same genus, and within populations of the same species (Trewavas 1983). Sex determination is a process that involves a series of mechanisms that establish an individual's gender (male/female). In fish, this process is influenced by genetic and/or environmental factors, which give rise to genotypic sex determination (GSD) and to environmental sex determination (ESD), respectively. The inheritance of sex is determined by sexual factors (related to GSD) and environmental effects (related to ESD), which are described in the next sections (Ospina-Alvarez and Piferrer 2008).

2.1. Genetic influence on sex determination

Genotypic sex determination is the most frequent mechanism in fish species. GSD is determined by the major and secondary sexual factors contained in sex chromosomes by establishing the main axis of chromosomal inheritance (Guinguen et al. 2019). This type of chromosomal inheritance occurs in fish species whose sex chromosomes are heteromorphic to the rest (they morphologically differ from other chromosomes). Chromosomal inheritance in fish comprises monofactorial systems as XX/XY or WZ/ZZ (formed by a single pair, or multiple pairs, of sex chromosomes) and multifactorial systems. The monofactorial system is the commonest in fish in which heterogametic sex (sex individual with a different type of sex chromosomes) is only one of the sexes, either the male (in XX/XY system) or female (WZ/ZZ system). However, in multifactorial systems, both sexes are heterogametic and sex is determined by three main factors or more (Devlin and Nagahama 2002).

The Nile tilapia is probably one of the best-documented sex determination systems in fish (Baroiller et al. 2009). Decades ago, the main genetic sex determination mechanisms in Nile tilapia were determined thanks to the use of chromosomal manipulation techniques (gynogenesis, androgenesis or triploidization) and progeny tests associated with employing hormonal therapies for sex reversion. It was then that it was established that *O. niloticus* presented a monofactorial XX/XY system with some particularities (Pruginin et al. 1975). In *O. niloticus*, as in *O. mossambicus*, heterogametic determination mechanisms are present in the male sex (XX-XY system). However, in other species of the same genus such as *O. aureus*, it is the female sex that exhibits heterogametic chromosomes (WZ-ZZ system) (Carrasco et al. 1999, Campos-Ramos et al. 2001, Cnaani and Kocher 2008, McAndrew et al. 2016).

In recent years, research efforts have been made to identify the genes and region (locus) of the chromosomes that determine sex. To date, several genes have been described as master sex-determining genes in different teleost species: *doublesex-and mab-3-related transcription factor 1* (*dmrt1*), *SRY-box transcription factor 3* (*sox3*), *gonadal somatic cell-derived factor* (*gsdf*), *anti-müllerian hormone* (*amh*), *sexually dimorphic on the Y chromosome* (*sdY*) and *interferon regulatory factor 9* (*irf9*) (Guinguen et al. 2019). Although major sex-determining genes have not yet been found in tilapia, the use of whole-genome sequencing indicated strong evidence for *amh* as a major gene that controls sex differentiation in Nile tilapia (Caceres et al. 2019). Several studies have characterized 13 genes to be sex-linked markers (regions) in six different tilapia species (Lee et al. 2003, Shirak et al. 2007, Cnaai and Kocher 2008, Rahther et al. 2019). The identification of the candidate genes for sex determination was performed in most of these studies by using both linkage and physical mappings in the regions with different expression profiles from the first early development stages in order to associate the phenotypic sex (gender) with specific linkage groups (LGs). Thus in the male heterogametic system of Nile tilapia (XY), the major sex-determining regions are described in LG1 and LG23 (Lee at al. 2003, Eshel et al. 2012, Li et al. 2015, Conte et al. 2017). For blue tilapia, which presents a female heterogametic system (ZW), two sex-determining regions have been identified in LG1 and LG3 as sex-specific markers (Campos-Ramos et al. 2001, Lee et al. 2004, Bian et al. 2019). However, further studies in this field are needed to increase our knowledge so as to identify genomic regions to track sex-linked loci in breeding programmes to control the sex of the population's individuals.

As very high diversity in sex determination mechanisms and multiple sex-determining regions have been described for different tilapia species, they are a unique model to study not only the genes that determine sex in fish, but also the evolutionary origin of sex determination in vertebrates.

2.2. Environmental influence on sex determination

Regarding sex determination, some fish possess certain phenotypic plasticity (from the same genotype) in initial development stages depending on temperature. Nile tilapia's sex is determined by a system composed of three components. Two of them are, as described in the previous section, related to genetics and are key determining factors of tilapia sex. These include a major and determining locus and secondary genetic factors. In Nile tilapia, environmental variables can influence sex determination (Baroiller et al. 2009). The main environmental variables to impact the sex determination of tilapia are temperature and pH (Baroiller and D'Cotta 2001), while other environmental factors, such as photoperiod and salinity, may also have an influence, but have been less investigated (Abucay et al. 1999).

Temperature is the main environmental factor to influence sex determination in most teleost species. Out in the wild, warmer (27°C) or colder (23°C) water temperatures in different geographical locations induce a male or female sex ratio in Southern flounder (Honeycutt et al. 2019). In the sex determination of tilapia, the impact of temperature has been studied in several tilapia species (Baroiller et al. 1995a, b, Wang and Tsai 2000). Nile tilapia development involves several thermosensitive periods, which coincide with the periods during which gonadal tissue is more sensitive to hormonal stimuli (Rougeot et al. 2008, Baroiller and D'Cotta 2001, Nakamura 1975, Baroiller et al. 1999). Some research works have aimed to determine this thermosensitivity window. Most studies generally agree that thermal treatments in the early larval development stages have the strongest effects on tilapia sex determination (Kwon et al. 2000, Ijiri et al. 2008, Rougeot et al. 2008, Baroiller et al. 2009). Exposure to high temperature (36°C) in the critical sex determination stage (9-15 dpf, days post fertilization) for about 10-28 days in Nile tilapia induces sex reversal of genotypic XX-females to XX-pseudomales (D'Cotta et al. 2001a, b, Kwon et al. 2000, Ijiri et al. 2008, Baroiller et al. 2009). However, exposure to lower or higher temperatures prior to that sex determination stage has no effect on sex determination (Baroiller et al. 1995a, Tessema et al. 2006, Sun et al. 2018, Zhao et al. 2019). Apart from following higher- or lower-temperature protocols, gonad development and sex ratio are also influenced by daily temperature fluctuations that fish experience in nature (higher temperatures in the daytime, cooler temperatures at night), which has been described in several fish species such as zebrafish (*Danio rerio*) (Villamizar et al. 2012) and Senegalese Sole (*Solea senegalensis*) (Blanco-Vives et al. 2011). In these research works, the fish that undergo daily thermocycles present higher percentages of females than when facing constant average temperature and reverse thermocycles (cooler temperatures in the daytime, higher temperatures at night). These results suggest that the thermal sensitivity of sex determination mechanisms varied throughout the day, with higher and lower masculinization levels after exposures to high temperature at night and in the daytime, respectively. Continuing to investigate the chronosensitivity of sex determination genes to heat shock could improve the efficiency of masculinization protocols and minimize the negative effects derived from high temperatures such as cellular stress, mortality, developmental malformations and incomplete masculinization process.

Although the effect of different temperature regimes on the sex ratio has been well studied, very few studies have focused on whether that sensitivity varies between breeding pairs (Baroiller and D'Cotta 2001, Tessema et al. 2006). Indeed these studies show that maternal and paternal mating patterns strongly influence male proportions in temperature-treated progenies, which suggests some heritability in thermal sensitivity in Nile tilapia, as well as a significant parental effect on TSD.

Recently, some authors investigated how high-temperature treatments induce sex reversal in Nile tilapia females (Tao et al. 2020, Zhao et al. 2020). They also showed the effect of high temperature on the transcriptional changes of *dmrt1*, *gsdf* and other sex-determination mechanisms. In addition, high temperature effects on sex reversal can also be observed molecularly. The main molecular changes were found at the *vasa*, *cyp19a1* and *dmrt1* expression levels. Therefore, the sex reversal process has been described in morphological terms. Although very few morphological changes were observed in early gonad differentiation stages, the gonad of XX females acquired the morphological patterns of testis (sperm germ cells in different development stages) from 99 dpf.

GSD is the main sex determination process in Nile tilapia. Hence, the phenotypic character of Nile tilapia (sex) is determined by a combination of genomic-environmental processes, although temperature can strongly impact the phenotypic sex ratio (i.e. temperature-influenced GSD). Temperature treatments can be helpful for inducing masculinization, which is actually the preferred method in tilapia aquaculture. Moreover, several hormonal agents and techniques have been applied to obtain monosex populations (Beardmore 2001). These topics are described in detail in Chapter 9 of the present book.

3. Gonadal development and sexual differentiation

In tilapia, and also in all vertebrates, gonads are formed from the development of primordial germ cells (PGCs) and mesodermal somatic tissue (Bhatta et al. 2012). On 3-4 dph (days post hatching), primordial germinative cells, which migrate to the developing gonads and lie around the posterior intestine, surround the outer lateral plate mesoderm layer and are located in this place after celomic cavity formation (Kobayashi et al. 2000, 2002, 2008). These primordial cells start to proliferate in females from 9 to 14 dph, whereas their numbers remain constant until day 14 dph in males (Ijiri et al. 2008). In this sexually undifferentiated stage, poor blood vessel development in the gonadal stroma takes place, along with a few steroid-producing cells (Nakamura et al. 1998). After early embryonic development, the migration of PGCs leads to gonia formation through successive mitotic divisions. During the last of these divisions, a differentiation process through meiosis begins, during which gonia become oocytes or spermatocytes (described in detail in the next sections). Therefore, the sexual differentiation of tilapia gonads starts in relatively early larval stages (between approximately 15 and 30 dph), although complete sexual maturity is reached at 3-4 months of age (Yoshikawa and Oguri 1978, De Silva and Radampola 1990, Strüssmann and Nakamura 2003, Suresh and Bhujel 2013). Several stages are involved in the differentiation process, starting with undifferentiated tissues and moving to matured reproductive systems for males (testis) or females (ovary). Regarding external characters, Nile tilapia presents a clear sexual dimorphism in relation to their reproductive structures (Figure. 1). Thus, males have only two ventral orifices (anus and genitourinary orifice), while females have three (anus, genital pore, urinary pore) (Rana 1988).

3.1. Testicular morphoanatomy and development

Testicles are responsible for producing sperm and sex hormones in males. In tilapia, testicles are elongated in shape, positioned in the dorsolateral situation and joined through the dorsal wall. Babiker and Ibrahim (1979) offered an extremely detailed description of male testes in different differentiation and maturation stages.

a   b

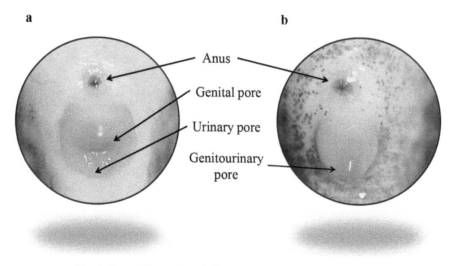

Anus

Genital pore

Urinary pore

Genitourinary
pore

Fig. 1. Sexual dimorphism in female (a) and male (b) Nile tilapia.

In their studies, the colour and shape of testicles varied depending on the gonadal differentiation stage. In initial or immature stages, male tilapia present flesh colours. However, when males approach the maturity stages, testes displayed creamy colours and occupied the total visceral cavity length. In addition, the spermiduct leaves each testicle to reach its joint with the urogenital papilla. Studies performed in *O. mossambicus* (Bhatta et al. 2012) and *O. niloticus* (Babiker and Ibrahim 1979, Nakamura and Nagahama 1989, Melamed et al. 1997, de Graaf and Huisman 1999) showed that testicular development stages increase at the same time as body size and the gonadosomatic index (GSI) do. Tilapia has a lobular testicular structure-type composed of a set of lobes with several cysts. A cyst constitutes the functional unit of the testicle, and it is the place where spermatogenesis occurs. In this spermatic cyst, a few Sertoli cells and primary spermatogonia are found. After successive mitotic and meiotic divisions, these cells give rise to spermatocytes, spermatids and spermatozoa. Therefore, in the testicular lobe, spermatocysts loaded with spermatozoa ready to be released to sperm can be found in the final stage (Grier and Fishelson 1995, Fishelson 2003).

Testicular development starts around 15 dph. At this point, the gonadal tissue that will give rise to the testicular apparatus is distributed along the enteric mesentery (Nakamura et al. 1998). Laxed-undifferentiated tissue is composed of PGCs and somatic tissue, which will be essential for the formation of the male reproductive system. Thus, PGCs will be involved in the synthesis and differentiation of spermatozoa. Meanwhile, somatic tissue will be needed to bring about the structural and functional basis of germ cells by also forming the seminiferous tubes and connective tissue of the testicular apparatus. Between both germ and somatic tissues, specialised somatic tissue cells can be found: Sertoli and Leydig cells. While Leydig cells will be responsible for the production of sex steroids, Sertoli cells will be in charge of structural and nutritional functions, and of phagocytization of cytoplasmic residues and residual germ cells. Sertoli cells are involved in the production of the hormones and factors needed for the differentiation, development and survival of germ cells.

The histological changes that take place during testicular development are described in detail in *O. niloticus* (Babiker and Ibrahim 1979, Nakamura and Nagahama 1989, Fishelson 2003) and *O. mossambicus* (Bhatta et al. 2012). Most previous research works have divided the Nile tilapia testicular development process into seven phases: differentiation, renovation, mitotic proliferation, meiotic proliferation, spermiogenesis, spermiation, and sperm maturation. In the first phase (15 dph), the differentiation of PGCs to spermatogonial germ cells or spermatogonias A occurs (21-23 dph) (Melo et al. 2019). From 25 dph, a space begins to form in the cleft of testicular stroma that will form the efferent duct, accompanied by a proliferation of epithelial cells (23-26 dph) (Iriji et al. 2008). After differentiation, the renewal and multiplication of spermatogonia A occurs. These spermatogonia A will proliferate rapidly by successive mitosis divisions, which result in different types of spermatogonias B (30-50 dph). In order to characterise sexual differentiation in male tilapia, the presence of spermatogonias (before meiotic proliferation) and somatic cells has been identified as a marker sign (Nakamura and Nagahama 1989, Vilela et al. 2003). Following the beginning of meiotic divisions, the transformation of spermatogonias into spermatocytes (first meiotic division) and spermatids (second meiotic division) occurs. Spermatids undergo morpho-functional maturation during spermatogenesis, which gives rise to spermatozoa (Nakamura and Nagahama 1989). During spermatogonial proliferation, the increase in Leydig cells is slow. However, Leydig cell numbers start increasing rapidly after 70 dph and appear between spermatogonia cysts in interstitial tissue (Gier and Fishelson 1995, Nakamura and Nagahama 1989). The breakdown of Sertoli cells causes cyst rupture and spermatozoa are released into the lobular lumen and spermiducts. On its way through spermiducts, sperm undergoes a capacitating process in which it acquires the ability to move and fertilise. From this stage (70 dph), smaller sized males can reach sexual maturity (14-20 cm). The maturity stage is characterised by the marked presence of sperm germ cells in different development stages that line sperm ducts as well as Leydig cells groups in interstitial tissue to mark the beginning of active spermatogenesis, the production of sex steroids in the testicle and, hence, the onset of male fertility.

3.2. Ovarian morphoanatomy and development

The ovary is the last effector organ of the female reproductive axis and is responsible for regulating the production of viable eggs and sex hormones needed for the ovarian development and differentiation process. The ovary of tilapia presents similar morphological characteristics to the cystic ovaries of other teleosts (Nagahama 1983). A detailed visual inspection (appearance, size, shape, colour) during the maturity stages of tilapia ovaries has been made by several authors (Babiker and Ibrahim 1979, Shoko et al. 2015). In their description, ovaries are small in size with a flesh creamy colour in the first immature stages. As maturation progresses, ovaries become yellowish in colour, are oval-shaped, and occupy about one third (depending on both size and individuals) of this animal's coelom cavity (Babiker and Ibrahim 1979).

Ovarian development constitutes a highly complex process which is timely regulated by both environmental and endocrine pathways. On the first days of development, the peritoneum extension leads to the formation of the germinative epithelium, from which ovarian follicles form. The tunica albuginea can be found below the germinal epithelium, which is characterized as a dense connective tissue composed of muscle fibres and blood vessels. The tunica albuginea presents several folds (ovigerous lamellae) that are the structural and functional support of follicular development. Germ and follicular cells are located in ovarian lamellae, which is where oogenesis occurs. After oocytes form in ovarian stroma, they are released to the lumen of the ovary (Nakamura et al. 1998).

Histological ovarian morphological changes have been studied in tilapines (Ibrahim and Babiker 1979, Nakamura et al. 1998, Coward and Bromage 2000, 2002). The interest in ovarian structure, morphology and development is reflected in the different studies carried out in mouthbreeders, such as *O. niloticus* (Babiker and Ibrahhim 1979, Alves et al. 1983, Tacon et al. 2000), *O. mossambicus* (Dadzie 1974, Bhatta et al. 2012) and *O. aureus* (Garcia and Philip 1986). In most of these studies, the authors describe recrudescence in seven development stages depending on the histological morphology characteristics (nucleus, cytoplasm and follicular layer) and biochemical properties: oogonic proliferation, oogeneis, folliculogenesis, alveolar cortical formation, vitellogenesis, final maturation, and ovulation (Fig. 2). Around the first day of ovary development, which occurs between 8-15 dph, proliferative germ cells give rise to oogonias. The first sign of morphological differentiation in females is the ovarian cavity formation (22-26 dph) (Iriji et al. 2008). Oogonias are divided through mitotic phases and give rise to primary oocytes, which are surrounded by granulosa cells. These cells are, in turn, surrounded by a cellular connective tissue monolayer, also known as the Teca cell layer. Hence, the granulosa and Teca layers, together with the oocyte, form the functional complex of the ovary, namely, the ovarian follicle.

After mitotic divisions, two series of meiotic cell divisions occur in the ovarian cycle (around 30 dph). In the prophase of the first meiotic division, the oocyte nucleus is subjected to five successive phases (leptotene, zigotene, pachytene, diplotene, and diacinesis) with meiotic arrest. In the first three phases, primary follicle growth takes place. After this primary growth, follicle growth is arrested in the diplotene stage, in which the follicle undergoes secondary growth. This secondary growth is composed of two phases: previtellogenic (early, middle, late) and vitellogenic. Previtellogenic stages take place in the first secondary growth phase. In these stages, the oocyte increases in size and accumulates mRNAs, nutritional reserves and other necessary components for subsequent vitellogenesis, and also for the embryonic and larval fertilization and development processes (Nakamura et al. 1998). In early previtellogenic stages, the growing oocyte undergoes a series of transformations in its nucleus, in which numerous producing ribosomomas nucleoli appear in its periphery (perinuclear phase). In this phase, the oocyte begins to synthesise ribosomal RNAs, which are transported to the ooplasm to encode the lipids, proteins and enzymes involved in vitellogenesis. Following middle and late previtellogenesis,

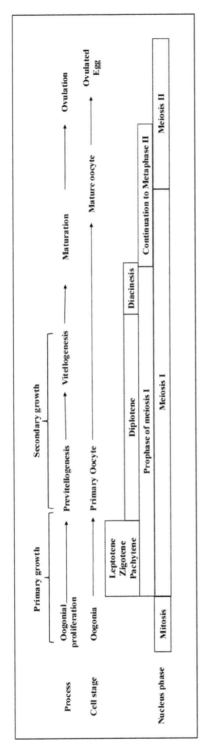

Fig. 2. Diagram of the oogenesis process correlated with nucleus mitosis and meiosis in fish. From left to right: from oogonia to primary oocytes, oocytes in the previtellogenic and vitellogenic stages, mature oocytes and ovulated eggs. See the text for a detailed description.

vitello particles assemble the cortical alveoli in the periphery, which incorporate glycoproteins as follicular development progresses (cortical alveolar stage). Previtellogenesis generally constitutes the first secondary growth stage, when all the necessary material is prepared for the second secondary growth stage: vitellogenesis. In this stage, the incorporation of plasma proteins of vitello precursor (vitellogenin) and lipoproteins occurs. After vitellogenesis, hormonal changes take place that trigger the resumption of meiotic arrest in the prophase of the cell cycle to continue to final oocyte maturation. It is in this phase when the oocyte nucleus moves towards the animal pole of the egg and arrests in the metaphase of meiosis II. In this stage, the hydration and proteolysis of yolk proteins stimulate oocyte growth. Then a hormonal change (mainly peaks in the luteinizing and maturation-inducing hormones) allows the cycle to resume by extruding the oocyte away from the follicle complex to the lumen of the ovary. The oocyte development cycle finishes with fertilization (Nakamura et al. 1998, Devlin and Nagahama 2002).

When ovarian development ends, the presence of oocytes in all development stages marks the beginning of the sexual maturity stage, in which females are able to perform the reproduction process and reach first maturity with a small size, and age depends on tilapia strains and environmental conditions (Babiker and Ibrahim 1979, Trewavas 1983). With Nile tilapia, females are sexually mature at a size of 20-30 cm and a weight of 150-250 g under natural conditions. However, under intensive aquaculture conditions, females mature more quickly and reach first maturity when they are lighter (30-50 g) (De Silva and Radampola 1990, de Graaf and Huisman 1999). During the reproductive period, tilapia ovaries go through an ovarian recrudescence period during which the ovary increases in size in the reproductive cycle. Hence, the ovarian cycle begins vitellogenesis from previtellogenic stages. Eight days after the ovarian cycle starts, the ovary is mostly occupied (60-70%) by oocytes in the last development stage (late vitellogenesis/maturing oocytes) (Coward and Bromage 1998). In Nile tilapia, the increase in mature oocytes correlates with a higher gonadomatic index (GSI) (Babiker and Ibrahim 1979, de Graaf and Huisman 1999, Melamed et al. 2000). The GSI reaches its highest levels (5.5%) on day 14 after spawning, which indicates its availability to begin new ovulation and, therefore, another spawning event (see the next chapter). After ovulation, persistence of ovarian follicles (POF) has been described in some tilapia species, although more studies are needed to describe their possible steroidogenic functionality regarding mainly progestins (Coward and Bromage 2000). The reabsorption of both the oocyte and ovarian follicle (atresia) is a frequent phenomenon in tilapia species and an essential one to maintain ovarian homeostasis. Although the morphological changes of oocyte atresia have been studied in tilapia, the endocrine mechanisms involved in regulating the degeneration and reabsorption of ovarian follicles have been less investigated (Srisakultiew 1993).

3.3. Influence of temperature on testicular and ovarian development

The gonadal cycle is influenced by different factors like temperature, salinity (Viera et al. 2019) and feeding (Sales et al. 2020). Of these, water temperature is the variable that most affects gonadal development and sex determination (Baroiller and D'Cotta 2001, 2009). The temperature effect may differ depending on the species, strains or age of tilapia. For instance, high temperatures (37°C) for long periods (45-60 days) lead to the permanent loss of germ cells in ovaries and infertility stages in *O. niloticus* (Pandit et al. 2015). In addition to germ cell thermosensitivity, the effect of temperature has also been observed on vitellogenesis. *In vitro* studies into *O. mossambicus* hepatocytes have shown greater vitellogenin synthesis at 28°C compared to 23°C and 33°C (Kim and Takemura 2003). The effect of temperature on the testicular function and spermatogenesis has been investigated by several research works. The use of low temperature (20°C) in male Nile tilapia seems to positively affect the primary spermatogonia generation by increasing Leydig and Sertoli cell proliferation, and conserving the reservoir and renovation of germ cells (Alvarenga and França 2009, Melo et al. 2016). Other studies have shown that applying high temperatures (30-35°C) causes faster germ cell differentiation and spermatogenesis (Vilela et al. 2003, Alvarenga and França 2009, Lacerda et al. 2018). However, as observed in tilapia females, the application of very high temperatures (36-37°C) provokes deleterious effects on testicular germ and somatic cells, characterised by loss of spermatogenic cells in testes (Jin et al. 2019a).

3.4. Endocrine control of sexual differentiation

In the last decade, research into Nile tilapia reproduction has focused on identifying the factors involved in sex determination and differentiation. Gonadal transcriptome and microRNAs analyses have indicated that the main sexual steroids and expression patterns of the genes that encode stereidogenic enzymes play an important role in processes such as gonadal development and sexual differentiation (Fig. 3) (Yoshiura et al. 2003, Ijiri et al. 2008, Tao et al. 2013, 2018, Eshel et al. 2014, Wang et al. 2016).

Regardless of an individual's genotype, the involvement of androgens and oestrogens in the sexual differentiation process results in the individual's masculinization or feminization, respectively (Yamamoto 1969). The proportion of androgens and oestrogens during stereidogenesis is determined by the key enzyme aromatase (Cyp19), which catalyses the conversion of androgens into 17β-Estradiol. Two genes (gonadal aromatase, *cyp19a* and brain aromatase, *cyp19b*), which encode this enzyme, have been described in many fish species, including Nile tilapia (Chang et al. 2005, Piferrer and Guinguen 2008). In ovaries, *cyp19a* is expressed from the first days of life (5 dph), and its expression continues to exponentially increase during ovarian differentiation (9-19 dpf) (Tao et al. 2013, 2018). Moreover, an increase in *cyp19a* levels has been linked with fish feminization (D'Cotta et al. 2001a, b). Blocking this enzyme during sensitive periods to temperature or aromatase inhibitors leads to the inhibition of oestrogen synthesis, which results in masculinization (male sex reversal) (Baroiller et al. 1995b, 2009, Kwon et al. 2000, D'Cotta et al. 2001a, Tessema et al. 2006, Sun et al. 2014, 2018). In addition to Cyp19, other transcription factors have been identified to play a role in tilapia ovarian differentiation. To date, the latest studies have also considered Foxl2 (forkhead transcriptional factor L2) to be a marker of female sexual differentiation in undifferentiated stages of fish (Kobayashi et al. 2004, Wang et al. 2007). As observed for *cyp19a*, *foxl2* expression

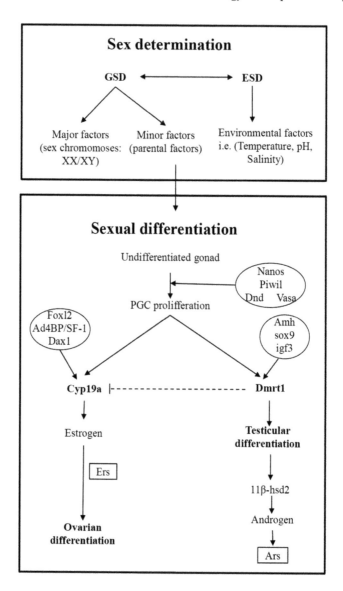

Fig. 3. Diagram of the factors involved in the sex determination and differentiation of Nile tilapia. Sex determination causes the differentiation of an undifferentiated gonad. Following PGCs proliferation, certain transcription factors and stereidogenic enzymes participate in gonadal differentiation. In tilapia, the main factors of ovarian and testicular differentiation are Cyp19a and Dmrt1, respectively. The presence of oestrogens in females triggers differentiation in the ovary. In contrast, the presence of androgens is the result of testicular differentiation. The solid lines indicate stimulation, while the dashed lines indicate inhibition or suppression. The transcription factors that participate in the differentiation pathways are surrounded by circles. GSD, Genotipic Sex Determination; ESD, Environmental Sex Determination; PGC, Primordial Germ Cell; Ers, Oestrogen receptors; Ars, Androgen receptors. Modified from Piferrer and Guinguen (2008), and adapted to Nile tilapia.

is higher in females than in males on 9 dpf. After this day, *foxl2* levels begin to sharply rise in XX females, which correlates with *cyp19a* levels (Wang et al. 2007, Iriji et al. 2008). Foxl2 seems to induce *cyp19a* expression. This might explain why the levels of both genes increase in parallel. In addition to *foxl2*, the role of other transcription factors like Dax1 (dose-sensitive sex reversal, adrenal hypoplasia, a critical region in the X chromosome) and Ad4BP/SF-1 has been studied. Dax1 has been suggested to suppress testicular differentiation which would, in turn, stimulate ovarian differentiation (Wang et al. 2002, Baroiller at al. 2009). Role of Ad4BP/SF-1 in the regulation of stereidogenic enzymes and the stimulation of aromatase transcription has been suggested, but more research is necessary (Yoshiura et al. 2003).

Oestrogens play a leading role as inductors of ovarian differentiation. However, androgens are a product of testicular differentiation (Tao et al. 2013). Thus, the increase in androgen levels (11-ketotestoterone, 11-KT) and P450c 11β-hydroxylase enzyme (responsible for synthesizing the 11-KT precursor) are the result, and not the cause, of testicular differentiation (Baroiller and D'Cotta 2001, Sudhakumari et al. 2005, Ijiri et al. 2008, Zhang et al. 2010). Several transcription factors have been suggested to be involved in the pathway of differentiation to tilapia testis. Dmrt1 (Doublesex and mab-3-related transcription factor 1) may be the main factor involved in testicular differentiation as it seems to suppress *cyp19a* expression and oestrogen synthesis in *O. niloticus* (Shirak et al. 2006, Ijiri et al. 2008, Kobayashi and Nakamura 2009, Wang et al. 2010, Tao et al. 2013, 2018, Rather et al. 2019). Other factors, such as Amh (antimüllerian hormone), Sox9 (the HMG-box protein 9 gene related to Sry) and Igf3 (insulin-like growth factor 3) are involved in Dmrt1 regulation and may also play an important role in testicular differentiation. For instance, Dmrt1 stimulates the expression of Sox family members such as *sox30* and *sox9,* which are involved in both the development of testis efferent ducts and the spermiogenesis process (Tang et al. 2019). In relation to Amh, it also plays a central role in gonad development and spermatogenesis. Several studies also indicate a role of *sox9* in the transcription of *amh* in tilapia as both expression levels increase in parallel for 10-15 dpf (days post-fertilization), with the highest levels in XY males on 20-25 dpf (Da Cotta et al. 2001, Iriji et al. 2008, Kobayashi et al. 2008). Igf3 has also been reported as a decisive factor in tilapia late spermatogenesis as it is regulated by androgens and intervenes in regulating the spermatocyte to spermatid transition (Li et al. 2020).

Finally, sex steroid receptors may also play an important role in the signalling pathway of sexual differentiation. Oestrogen (Esr1, Esr2a, and Esr2b), androgen (Ar1 and Ar2) and progestin receptors (Pgr) have even been detected in first development stages in tilapia (Gale 1996, Chang et al. 1999, Sudhakumari et al. 2005, Wang et al. 2005, Tao et al. 2013). Recently, other markers and genes involved in developmental processes, such as *nanos, piwil, dnd, vasa* and *pum,* appear to be implicated in the specification and maintenance of PGCs during Nile tilapia's ontogenic development (Kobayahashi et al. 2002, Jin et al. 2019b).

The previous factors involved in controlling steroidogenesis and gonadal function presented seasonal and daily changes in the gene expression synchronized to light: dark cycles, as reported in several teleost species like zebrafish (Di Rosa et al.

2016, Paredes et al. 2019), Senegalese Sole (Oliveira et al. 2009) and Nile tilapia (De Alba et al. 2019). In addition to light cycles, studies performed in zebrafish revealed the influence of daily and seasonal temperature changes on the stereidogenesis pathway and the reproduction process (Villamizar et al. 2012). These recent findings highlight the rhythmic nature of the mechanisms that intervene in the sexual differentiation of fish, as well as the influence of light and temperature cycles on their expression patterns.

4. Concluding remarks

Sex-determination genes are less conserved in fish compared to other vertebrates like mammals. Molecular markers and whole sequencing genomes are effective in identifying and isolating the loci related to GSD. Thanks to the genetic conservation of sex determination in tilapia, this species has been characterized as a relevant model for identifying the genes involved in the morphological, molecular and biochemical aspects of gonadal development and differentiation. Transcriptomic and microarrays analyses are very useful tools for studying the impact of hormonal and temperature protocols on the reproductive physiology of tilapia.

The expansion of aquaculture and the introduction of new species highlight the need to continue to strive to know the mechanisms that control the sex determination and differentiation of fish species. Tilapia aquaculture also requires better knowledge about techniques based on controlling the genotype and sex ratio of populations. Selecting productive characteristics of both sexual phenotypes will guarantee improvements in the tilapia industry's productive and reproductive efficiency.

Acknowledgments

The present research was partially funded by Projects CHRONOLIPOFISH (RTI2018-100678-A-I00), BLUESOLE (AGL2017-82582-C3-3-R) and CRONOFISH (RED2018-102487-T), granted by the Spanish Ministry of Science, Innovation and Universities, and co-funded with FEDER, to JFLO and FJSV. JFLO was also funded by a "*Ramón y Cajal*" research fellowship (RYC-2016-20959).

References cited

Abucay, J.S., G.C. Mair, D.O.F. Skibinski and J.A. Beardmore. 1999. Environmental sex determination: The effect of temperature and salinity on sex ratio in *O. niloticus* L. Aquaculture 173: 219–234.

Alvarenga, É.R.D. and L.R.D. França. 2009. Effects of different temperatures on testis structure and function, with emphasis on somatic cells, in sexually mature Nile tilapias (*Oreochromis niloticus*). Biol. Reprod. 80: 537–544.

Alves, M.M., H.D.S. Leme, R.A. Lopes, S.O. Petenusci and C. Haiyashi. 1983. Rhythm of development in the oocyte of the tilapia *Oreochromis niloticus* L. (Pisces: Cichlidae); a morphometric and histochemical study. Gegenbaurs Morphologisches Jahrbuch 129: 575–592.

Babiker, M. and H. Ibrahim. 1979. Studies on the biology of reproduction in the cichlid *Tilapia nilotica* (L.): Gonadal maturation and fecundity. J. Fish Biol. 14: 437–448.

Baroiller, J.F. and B. Jalabert. 1989. Contribution of research in reproductive physiology to the culture of tilapias. Aquatic Living Resour. 2: 105–116.

Baroiller, J.F., F. Clota and E. Geraz. 1995a. Temperature sex determination in two Tilapia, *Oreochromis niloticus* and the red Tilapia (red Florida strain): Effect of high or low temperature. pp. 158–160. *In*: F. Goetz and P. Thomas [eds.]. The Reproductive Physiology of Fish. University of Texas. Austin. USA.

Baroiller, J.F., D. Chourrout, A. Fostier and B. Jalabert. 1995b. Temperature and sex chromosomes govern sex ratios of the mouthbrooding cichlid fish *Oreochromis niloticus*. J. Exp. Zool. 273: 216–223.

Baroiller, J.F., Y. Guiguen and A. Fostier. 1999. Endocrine and environmental aspects of sex differentiation in fish. Cell. Mol. Life Sci. 55: 910–931.

Baroiller, J.F. and H. D'cotta. 2001. Environment and sex determination in farmed fish. Comp. Biochem. Physiol. C 130: 399–409.

Baroiller, J.F., H. D'Cotta and E. Saillant. 2009. Environmental effects on fish sex determination and differentiation. Sex. Dev. 3: 118–135.

Beardmore, J.A., G.C. Mair and R.I. Lewis. 2001. Monosex male production in finfish as exemplified by tilapia: Applications, problems, and prospects. *In*: Reproductive Biotechnology in Finfish Aquaculture 1: 283–301.

Bhandari, R.K., M. Nakamura, T. Kobayashi and Y. Nagahama. 2006. Suppression of steroidogenic enzyme expression during androgen-induced sex reversal in Nile tilapia (*Oreochromis niloticus*). Gen. Comp. Endocrinol. 145: 20–24.

Bhatta, S., T. Iwai, T. Miura, M. Higuchi, G. Maugars and C. Miura. 2012. Differences between male and female growth and sexual maturation in tilapia (*Oreochromis mossambicus*). Engi. Tech. 8: 57–65.

Bian, C., J. Li, X. Lin, X. Chen, Y. Yi, X. You, et al. 2019. Whole genome sequencing of the blue tilapia (*Oreochromis aureus*) provides a valuable genetic resource for biomedical research on tilapias. Mar. Drugs 17: 386.

Blanco-Vives, B., L.M. Vera, J. Ramos, M.J. Bayarri, E. Mañanós and F.J. Sánchez-Vázquez. 2011. Exposure of larvae to daily thermocycles affects gonad development, sex ratio, and sexual steroids in *Solea senegalensis*, Kaup. J. Exp. Zool. A: Ecol. Gen. Physiol. 315: 162–169.

Cáceres, G., M.E. López, M.I. Cádiz, G.M. Yoshida, A. Jedlicki, R. Palma-Véjares, et al. 2019. Fine mapping using whole-genome sequencing confirms anti-Müllerian hormone as a major gene for sex determination in farmed Nile tilapia (*Oreochromis niloticus* L.). G3-Genes, Genom. Genet. 9: 3213–3223.

Campos-Ramos, R., S.C. Harvey, J.S. Masabanda, L.A. Carrasco, D.K. Griffin, B.J. McAndrew, et al. 2001. Identification of putative sex chromosomes in the blue tilapia, *Oreochromis aureus*, through synaptonemal complex and FISH analysis. Genetica 111: 143–153.

Carrasco, L.A., D.J. Penman and N. Bromage. 1999. Evidence for the presence of sex chromosomes in the Nile tilapia (*Oreochromis niloticus)* from synaptonemal complex analysis of XX, XY and YY genotypes. Aquaculture 173: 207–218.

Chang, X., T. Kobayashi, T. Todo, T. Ikeuchi, M. Yoshiura, H. Kajiura-Kobayashi, et al. 1999. Molecular cloning of estrogen receptors α and β in the ovary of a teleost fish, the tilapia (*Oreochromis niloticus*). Zool. Sci. 16: 653–659.

Chang, X., T. Kobayashi, B. Senthilkumaran, H. Kobayashi-Kajura, C.C. Sudhakumari and Y. Nagahama. 2005. Two types of aromatase with different encoding genes, tissue distribution and developmental expression in Nile tilapia (*Oreochromis niloticus*). Gen. Comp. Endocrinol. 141: 101–115.

Cnaani, A. and T.D. Kocher. 2008. Sex-linked markers and microsatellite locus duplication in the cichlid species *Oreochromis tanganicae*. Biol. Letters 4: 700–703.

Conte, M.A., W.J. Gammerdinger, K.L. Bartie, D.J. Penman and T.D. Kocher. 2017. A high quality assembly of the Nile tilapia (*Oreochromis niloticus*) genome reveals the structure of two sex determination regions. BMC Genomics 18: 341.

Coward, K. and N.R. Bromage. 1998. Histological classification of oocyte growth and the dynamics of ovarian recrudescence in *Tilapia zillii*. J. Fish. Biol. 2: 285–302.

Coward, K. and N.R. Bromage. 2000. Reproductive physiology of female tilapia broodstock. Rev. Fish Biol. Fish. 10: 1–25.

Coward, K. and N.R. Bromage. 2002. Stereological point-counting: An accurate method for assessing ovarian function in tilapia. Aquaculture 212: 383–401.

D'Cotta, H., A. Fostier, Y. Guiguen, M. Govoroun and J.F. Baroiller. 2001a. Search for genes involved in the temperature induced gonadal sex differentiation in the tilapia, *Oreochromis niloticus*. J. Exp. Zool. 290: 574–585.

D'Cotta, H., A. Fostier, Y. Guiguen, M. Govoroun and J.F. Baroiller. 2001b. Aromatase plays a key role during normal and temperature induced sex differentiation of tilapia *Oreochromis niloticus*. Mol. Reprod. Dev. 59: 265–276.

Dadzie, S. 1974. Oogenesis and the stages of maturation in the female cichlid fish, *Tilapia mossambica*. (G. J. S.) Ghana J. Scie. 1: 1–5.

De Graaf, G.J. and E.A. Huisman. 1999. Reproductive biology of pond reared Nile tilapia, *Oreochromis niloticus* L. Aquaculture Res. 30: 25–33.

de Alba, G., N.M.N. Mourad, J.F. Paredes, F.J. Sánchez-Vázquez and J.F. López-Olmeda. 2019. Daily rhythms in the reproductive axis of Nile tilapia (*Oreochromis niloticus*): Plasma steroids and gene expression in brain, pituitary, gonad and egg. Aquaculture 507: 313–321.

De Silva, S.S. and K. Radampola. 1990. Effect of dietary protein level on the reproductive performance of *Oreochromis niloticus*. Asian Fish. Soc. Manila. 1: 559–563.

Devlin, R.H. and Y. Nagahama. 2002. Sex determination and sex differentiation in fish: An overview of genetic, physiological, and environmental influences. Aquaculture 208: 191–364.

Di Rosa, V., J.F. López-Olmeda, A. Burguillo, E. Frigato, C. Bertolucci, F. Piferrer, et al. 2016. Daily rhythms of the expression of key genes involved in steroidogenesis and gonadal function in zebrafish. PLoS ONE 11: e0157716.

Eshel, O., A. Shirak, J.I. Weller, G. Hulata and M. Ron. 2012. Linkage and physical mapping of sex region on LG23 of Nile tilapia (*Oreochromis niloticus*). G3-Genes, Genom. Genet. 2: 35–42.

Eshel, O., A. Shirak, L. Dor, M. Band, T. Zak, M. Markovich-Gordon and G. Hulata. 2014. Identification of male-specific *amh* duplication, sexually differentially expressed genes and microRNAs at early embryonic development of Nile tilapia (*Oreochromis niloticus*). BMC Genomics 15: 774.

Fishelson, L. 2003. Comparison of testes structure, spermatogenesis, and spermatocyto genesis in young, aging, and hybrid cichlid fish (Cichlidae, Teleostei). J. Morphol. 256: 285–300.

Gale, W.L. 1996. Sexual differentiation and steroid-induced sex inversion in Nile tilapia (*Oreochromis niloticus*): 1. Characterization of a gonadal androgen receptor, 2. Masculinization by immersion in methyldihydrotestosterone. Master Thesis, Oregon State University, Oregon, USA.

Garcia, T. and P. Phillip. 1986. Oocyte development in *Oreochromis aureus*. Rev. Invest. Mar. 7: 63–70.

Grier, H.J. and L. Fishelson. 1995. Colloidal sperm-packaging in mouthbrooding tilapiine fishes. Copeia 4: 966–970.

Guiguen, Y., A. Fostier and A. Herpin. 2019. Sex determination and differentiation in fish: Genetic, genomic, and endocrine aspects. Part 1. pp. 35–63. *In*: Wang, H.P, F. Pifferer and S. Chen. [eds.]. Sex Control in Aquaculture. John Wiley & Sons. Chichester, UK.

Honeycutt, J.L., C.A. Deck, S.C. Miller, M.E. Severance, E.B. Atkins, J.A. Luckenbach, et al. 2019. Warmer waters masculinize wild populations of a fish with temperature-dependent sex determination. Scientific Reports 91: 1–13.

Ijiri, S., H. Kaneko, T. Kobayashi, D.S. Wang, F. Sakai, B. Paul-Prasanth, et al. 2008. Sexual dimorphic expression of genes in gonads during early differentiation of a teleost fish, the Nile tilapia *Oreochromis niloticus*. Biol. Reprod. 78: 333–341.

Jalabert, B. 1989. Contribution of research in reproductive physiology to the culture of tilapias. Aquatic Living Resour. 2: 105–116.

Jin, Y.H., A. Davie and H. Migaud. 2019a. Temperature-induced testicular germ cell loss and recovery in Nile tilapia *Oreochromis niloticus*. Gen. Comp. Endocrinol. 283: 113227.

Jin, Y.H., A. Davie and H. Migaud. 2019b. Expression pattern of *nanos, piwil, dnd, vasa* and *pum* genes during ontogenic development in Nile tilapia *Oreochromis niloticus*. Gene 688: 62–70.

Kim, B.H. and A. Takemura. 2003. Culture conditions affect induction of vitellogenin synthesis by estradiol-17β in primary cultures of tilapia hepatocytes. Comp. Biochem. Physiol. B 135: 231–239.

Kobayashi, Tohru, H. Kajiura-Kobayashi and Y. Nagahama. 2000. Differential expression of *vasa* homologue gene in the germ cells during oogenesis and spermatogenesis in a teleost fish, tilapia, *Oreochromis niloticus*. Mech. Develop. 99: 139–142.

Kobayashi, T., H. Kajiura-Kobayashi and Y. Nagahama. 2002. Two isoforms of *vasa* homologs in a teleost fish: Their differential expression during germ cell differentiation. Mech. Develop. 111: 167–171.

Kobayashi, T., L. Zhou and Y. Nagahama. 2004. Molecular cloning and gene expression of Foxl2 in the Nile tilapia, *Oreochromis niloticus*. Biochem. Bioph. Res. Co. 320: 83–89.

Kobayashi, T., H. Kajiura-Kobayashi, G. Guan and Y. Nagahama. 2008. Sexual dimorphic expression of DMRT1 and Sox9a during gonadal differentiation and hormone induced sex reversal in the teleost fish Nile tilapia (*Oreochromis niloticus*). Dev. Dynam. 237: 297–306.

Kwon, J.Y., V. Haghpanah, L.M. Kogson-Hurtado, B.J. McAndrew and D.J. Penman. 2000. Masculinization of genetic female Nile tilapia (*Oreochromis niloticus*) by dietary administration of an aromatase inhibitor during sexual differentiation. J. Exp. Zool. 287: 46–53.

Lacerda, S.M.S.N., S.R. Batlouni, S.B.G. Silva, C.S.P. Homem and L.R. França. 2018. Germ cells transplantation in fish: The Nile tilapia model. Anim. Reprod. 3: 146–159.

Lapeyre, B.A. 2008. Control of reproduction in Nile tilapia (*Oreochromis niloticus*) by manipulation of environmental factors. Ph.D. Thesis, University of Göttingen, Germany.

Lee, B.Y., D.J. Penman and T.D. Kocher. 2003. Identification of a sex-determining region in Nile tilapia (*Oreochromis niloticus*) using bulked segregant analysis. Anim. Genet. 34: 379–383.

Lee, B.Y., G. Hulata and T.D. Kocher. 2004. Two unlinked loci controlling the sex of blue tilapia (*Oreochromis aureus*). Heredity 92: 543–549.

Li, M., Y. Sun, J. Zhao, H. Shi, S. Zeng, K. Ye, et al. 2015. A tandem duplicate of anti-Müllerian hormone with a missense SNP on the Y chromosome is essential for male sex determination in Nile tilapia, *Oreochromis niloticus*. PLoS Genet. 11: 11.

Li, M., X. Liu, S. Dai, H. Xiao, S. Qi, Y. Li, et al. 2020. Regulation of spermatogenesis and reproductive capacity by Igf3 in tilapia. Cell. Mol. Life Sci. 1: 1–18.

McAndrew, B.J., D.J. Penman, M. Bekaert and S. Wehner. 2016. Tilapia genomic studies. pp.

105–129. *In*: S. MacKenzie and S. Jentoft [eds.]. Genomics in Aquaculture. Academic Press, Cambridge, Massachussets, USA.

Melamed, P., G. Gur, H. Rosenfeld, A. Elizur and Z. Yaron. 1997. The mRNA levels of GtH Iβ, GtH IIβ and GH in relation to testicular development and testosterone treatment in pituitary cells of male tilapia. Fish Physiol. Biochem. 17: 93–98.

Melamed, P., G. Gur, H. Rosenfeld, A. Elizur, R.W. Schulz and Z. Yaron. 2000. Reproductive development of male and female tilapia hybrids (*Oreochromis niloticus* × *O. aureus*) and changes in mRNA levels of gonadotropin (GtH) Iβ and IIβ subunits. J. Exp. Zool. 286: 64–75.

Melo, R.M.C., Y.M. Ribeiro, R.K. Luz and N. Bazzoli. 2016. Influence of low temperature on structure and dynamics of spermatogenesis during culture of *Oreochromis niloticus*. Anim. Reprod. Sci. 172: 148–156.

Melo, L.H., R.M. Melo, R.K. Luz, N. Bazzoli and E. Rizzo. 2019. Expression of *Vasa, Nanos2* and *Sox9* during initial testicular development in Nile tilapia (*Oreochromis niloticus*) submitted to sex reversal. Reprod. Fert. Develop. 31: 1637–1646.

Nagahama, Y. 1983. The functional morphology of teleost gonads. Pp. 223–275. *In*: Y. Nagahama [ed.]. Fish Physiology. Academic Press. Cambridge, Massachusetts, USA.

Nakamura, M. 1975. Dosage-dependent changes in the effect of oral administration of methyltestosterone on gonadal sex differentiation in *Tilapia mossambica*. Bulletin of the Faculty of Fisheries Hokkaido University 26: 99–108.

Nakamura, M. and Y. Nagahama. 1989. Differentiation and development of Leydig cells, and changes of testosterone levels during testicular differentiation in tilapia *Oreochromis niloticus*. Fish Physiol. Biochem. 7: 211–219.

Nakamura, M., T. Kobayashi and X. Chang. 1998. Gonadal sex differentiation in teleost fish. J. Exp. Zool. 281: 362–372.

Oliveira, C., L.M. Vera, J.F. López-Olmeda, J.M. Guzmán, E. Mañanós, J. Ramos, et al. 2009. Monthly day/night changes and seasonal daily rhythms of sexual steroids in Senegal sole (*Solea senegalensis*) under natural fluctuating or controlled environmental conditions. Comp. Biochem. Physiol. Part A: Mol. Int. Physiol. 152: 168–175.

Omasaki, S.K., H. Charo-Karisa, A.K. Kahi and H. Komen. 2016. Genotype by environment interaction for harvest weight, growth rate and shape between monosex and mixed sex Nile tilapia (*Oreochromis niloticus*). Aquaculture 458: 75–81.

Ospina-Alvarez, N. and F. Piferrer. 2008. Temperature-dependent sex determination in fish revisited: Prevalence, a single sex ratio response pattern, and possible effects of climate change. PLoS ONE 3:e2837.

Pandit, N.P., R.K. Bhandari, Y. Kobayashi and M. Nakamura. 2015. High temperature-induced sterility in the female Nile tilapia, *Oreochromis niloticus*. Gen. Comp. Endocrinol. 213: 110–117.

Paredes, J.F., M. Cowan,, J.F. López-Olmeda, J.A. Muñoz-Cueto and F.J. Sánchez-Vázquez. 2019. Daily rhythms of expression in reproductive genes along the brain-pituitary-gonad axis and liver of zebrafish. Comp. Biochem. Physiol. A: Mol. Int. Physiol. 231: 158–169.

Piferrer, F. and Y. Guiguen. 2008. Fish gonadogenesis. Part II: Molecular biology and genomics of sex differentiation. Rev. Fish. Sci. 16: 35–55.

Pinto, C.S.M., J.R. Verani, D.M. Antoniutti and H.L. Stempniewski. 2018. Estudo comparado do crescimento de machos de *Oreochromis niloticus* em diferentes periodos de cultivo. Boletim do Instituto de Pesca 16: 19–27.

Pruginin, Y., S. Rothbard, G. Wohlfarth, A. Halevy, R. Moav and G. Hulata. 1975. All-male broods of *Tilapia nilotica* × *T. aurea* hybrids. Aquaculture 6: 11–21.

Rana, K. 1988. Reproductive biology and the hatchery rearing of tilapia eggs and fry. pp. 343–406. *In*: Recent Advances in Aquaculture. Springer, Dordrecht, Holland.

Rather, M.A. and B.C. Dhandare. 2019. Genome-wide identification of double sex and Mab-3-related transcription factor (DMRT) genes in Nile tilapia (*Oreochromis niloticus*). Biotech. Reports 24: e00398.

Rougeot, C., C. Prignon, C.V.N. Kengne and C. Mélard. 2008. Effect of high temperature during embryogenesis on the sex differentiation process in the Nile tilapia, *Oreochromis niloticus*. Aquaculture 276: 205–208.

Sales, C.F., A.P.B. Pinheiro, Y.M. Ribeiro, A.A. Weber, F.O. Paes-Leme, R.K. Luz, et al. 2020. Effects of starvation and refeeding cycles on spermatogenesis and sex steroids in the Nile tilapia *Oreochromis niloticus*. Mol. Cell. Endocrinol. 500: 110643.

Shirak, A., E. Seroussi, A. Cnaani, A.E. Howe, R. Domokhovsky, N. Zilberman, et al. 2006. Amh and Dmrta2 genes map to tilapia (*Oreochromis* spp.) linkage group 23 within quantitative trait locus regions for sex determination. Genetics 174: 1573–1581.

Shirak, A., E. Seroussi, N. Zilberman, R. Domokhovsky, A. Cnaani, T.D. Kocher, et al. 2007. Mapping of candidate genes for sex determination in tilapias. Aquaculture 272: 271–272.

Shoko, A.P., S.M. Limbu, H.D.J. Mrosso and Y.D. Mgaya. 2015. Reproductive biology of female Nile tilapia *Oreochromis niloticus* (Linnaeus) reared in monoculture and polyculture with African sharptooth catfish *Clarias gariepinus* (Burchell). Springer Plus 4: 275.

Srisakultiew, P. 1993. Studies on the reproductive biology of *Oreochromis niloticus* L. Ph.D. Thesis, University of Stirling, Scotland, UK.

Strüssmann, C.A. and M. Nakamura. 2003. Morphology, endocrinology, and environmental modulation of gonadal sex differentiation in teleost fishes. Fish Physiol. Biochem. 26: 13–29.

Sudhakumari, C.C., B. Senthilkumaran, T. Kobayashi, H. Kajiura-Kobayashi, D.S. Wang, M. Yoshikuni, et al. 2005. Ontogenic expression patterns of several nuclear receptors and cytochrome P450 aromatases in brain and gonads of the Nile tilapia *Oreochromis niloticus* suggests their involvement in sex differentiation. Fish Physiol. Biochem. 31: 129.

Sun, L.N., X.L. Jiang, Q.P. Xie, J. Yuan, B.F. Huang, W.J. Tao, et al. 2014. Transdifferentiation of differentiated ovary into functional testis by long-term treatment of aromatase inhibitor in Nile tilapia. Endocrinology 155: 1476–1488.

Sun, L.X., J. Teng, Y. Zhao, N. Li, H. Wang and X.S. Ji. 2018. Gonad transcriptome analysis of high-temperature-treated females and high-temperature-induced sex-reversed neomales in Nile tilapia. Int. J. Mol. Sci. 19: 689.

Suresh, V. and R.C. Bhujel. 2013. Tilapias. pp. 338–364. *In*: J.S. Lucas and P.C. Southgate [eds.]. Aquaculture: Farming Aquatic Animals and Plants. Blackwell Publishing, Oxford, UK.

Tacon, P., J.F. Baroiller, P.Y. Le Bail, P. Prunet and B. Jalabert. 2000. Effect of egg deprivation on sex steroids, gonadotropin, prolactin, and growth hormone profiles during the reproductive cycle of the mouthbrooding cichlid fish *Oreochromis niloticus*. Gen. Comp. Endocrinol. 117: 54–65.

Tang, Y., X. Li, H. Xiao, M. Li, Y. Li, D. Wang, et al. 2019. Transcription of the Sox30 gene is positively regulated by Dmrt1 in Nile tilapia. Int. J. Mol. Sci. 20: 5487.

Tao, W., J. Yuan, L. Zhou, L. Sun, Y. Sun, S. Yang, et al. 2013. Characterization of gonadal transcriptomes from Nile tilapia (*Oreochromis niloticus*) reveals differentially expressed genes. PLoS ONE 8:e63604.

Tao, W., J. Chen, D. Tan, J. Yang, L. Sun, J. Wei, et al. 2018. Transcriptome display during tilapia sex determination and differentiation as revealed by RNA-Seq analysis. BMC Genomics 19: 363.

Teng, J., Y. Zhao, H.J. Chen, H. Wang and X.S. Ji. 2020. Transcriptome profiling and analysis of genes associated with high temperature–induced masculinization in sex-undifferentiated Nile tilapia gonad. Marine Biotechnol 22: 1–13.

Tessema, M., A. Müller-Belecke and G. Hörstgen-Schwark. 2006. Effect of rearing temperatures on the sex ratios of *Oreochromis niloticus* populations. Aquaculture 258: 270–277.

Trewavas, E. 1983. Tilapiine Fishes of the Genera *Sarotherodon, Oreochromis* and *Danakilia*. British Museum Natural History, London, UK.

Vieira, A.B.C., A.A. Weber, Y.M. Ribeiro, R.K. Luz, N. Bazzoli and E. Rizzo. 2019. Influence of salinity on spermatogenesis in adult Nile tilapia (*Oreochromis niloticus*) testis. Theriogenology 131: 1–8.

Villamizar, N., L. Ribas, F. Piferrer, L.M. Vera and F.J. Sánchez-Vázquez. 2012. Impact of daily thermocycles on hatching rhythms, larval performance and sex differentiation of zebrafish. PLoS ONE 7: e52153.

Vilela, D., R.N. Resources and H.P. Godinho. 2003. Spermatogenesis in teleost: Insights from the Nile tilapia (*Oreochromis niloticus*) model. Fish Physiol. Biochem. 28: 187–190.

Wang, D.S., T. Kobayashi, B. Senthilkumaran, F. Sakai, C.C. Sudhakumari, T. Suzuki, et al. 2002. Molecular cloning of DAX1 and SHP cDNAs and their expression patterns in the Nile tilapia, *Oreochromis niloticus*. Biochem. Bioph. Res. Co. 297: 632–640.

Wang, L.H. and C.L. Tsai. 2000. Effects of temperature on the deformity and sex differentiation of tilapia, *Oreochromis mossambicus*. J. Exp. Zool. 286: 534–537.

Wang, D.S., B. Senthilkumaran, C.C. Sudhakumari, F. Sakai, M. Matsuda, T. Kobayashi, et al. 2005. Molecular cloning, gene expression and characterization of the third estrogen receptor of the Nile tilapia, *Oreochromis niloticus*. Fish Physiol. Biochem. 31: 255.

Wang, D.S., T. Kobayashi, L.Y. Zhou, B. Paul-Prasanth, S. Ijiri, F. Sakai, et al. 2007. Foxl2 up-regulates aromatase gene transcription in a female-specific manner by binding to the promoter as well as interacting with ad4 binding protein/steroidogenic factor 1. Mol. Endocrinol. 21: 712–725.

Wang, W., W. Liu, Q. Liu, B. Li, L. An, R. Hao, et al. 2016. Coordinated microRNA and messenger RNA expression profiles for understanding sexual dimorphism of gonads and the potential roles of microRNA in the steroidogenesis pathway in Nile tilapia (*Oreochromis niloticus*). Theriogenology 85: 970–978.

Yamamoto, T.O. 1969. Sex differentiation. pp. 117–175. *In*: Fish Physiology. Academic Press. Cambridge, Massachusetts, USA.

Yoshikawa, H. and M. Oguri. 1978. Sex differentiation in a cichlid, *Tilapia zillii*. Bulletin of the Japanese Society of Scientific Fisheries (Japan).

Yoshiura, Y., B. Senthilkumaran, M. Watanabe, Y. Oba, T. Kobayashi and Y. Nagahama. 2003. Synergistic expression of Ad4BP/SF-1 and cytochrome P-450 aromatase (ovarian type) in the ovary of Nile tilapia, *Oreochromis niloticus*, during vitellogenesis suggests transcriptional interaction. Biol. Reprod. 68: 1545–1553.

Zhang, W., L. Zhou, B. Senthilkumaran and B. Huang. 2010. Molecular cloning of two isoforms of 11β-hydroxylase and their expressions in the Nile tilapia, *Oreochromis niloticus*. Gen. Comp. Endocrinol. 165: 34–41.

Zhao, Y., Y. Mei, H.J. Chen, L.T. Zhang, H. Wang and X.S. Ji. 2019. Profiling expression changes of genes associated with temperature and sex during high temperature-induced masculinization in the Nile tilapia brain. Physiol. Genomics 51: 159–168.

Zhao, Y., H.J. Chen, Y.Y. Wang, Y. Mei, L.B. Huang, H. Wang, et al. 2020. Gonad development examination of high-temperature–treated genetically female Nile tilapia. Aquaculture 515: 734535.

CHAPTER
9

Reproductive Physiology of Tilapia

Gonzalo de Alba*, Francisco Javier Sánchez-Vázquez and
José Fernando López-Olmeda

Department of Physiology, Faculty of Biology, University of Murcia, 30100 Murcia, Spain

1. Introduction

The input of environmental information finely tunes the activation of the brain neuroendocrine machinery that triggers the hormonal cascade, which leads to the reproduction process (Yaron and Levavi-Sivan 2011). Although different neuroendocrine systems have been identified in fish reproduction, the Brain-Pituitary-Gonadal-Liver axis (BPGL axis) is the main mechanism to control fish reproduction (Figure 1). At each level of this system, a series of key factors intervene from gametogenesis to the release of high quality gametes ready for fertilization. Environmental information is collected by a series of chemical receptors located in the olfactory epithelium, which are photoreceptors in the retina and in the pineal organ. These structures receive light signals (day/night cycle and day length) and transduce them into electrochemical and humoral (melatonin) signals. Specifically, the pineal gland acts as an intrinsic main clock that intervenes in the periodic regulation of melatonin secretion (higher production at night), whose function is to modulate and synchronize biological rhythms, mainly in peripheral tissues (Falcon et al. 2007). Thus, environmental stimuli arrive at the hypothalamus through sensory neuron pathways. In this region, stimulus-sensitive neurosecretory cells are responsible for releasing a neuropeptide that plays a primary role in reproduction: the gonadotropin-releasing hormone (Gnrh). This hormone acts primarily on the glandular pituitary gland (adenohypophysis) by stimulating the synthesis and release of gonadotropins into the bloodstream. These gonadotropins stimulate the production of steroid hormones at the gonadal level by acting as last effectors of gamete development and release. The gonadal steroids exert feedback effects in the pituitary and the brain that modulate gonadotropins and neuropeptides secretion (Yaron et al. 2001; Weltzien et al. 2004; Yaron and Levavi-Sivan, 2011) (Figure 1).

Better and integrative knowledge of the environmental, physiological and neuroendocrine factors that regulate the reproductive physiology of tilapia is needed to improve the breeding protocols established in tilapia aquaculture.

*Corresponding author: gonzalode.alba@um.es

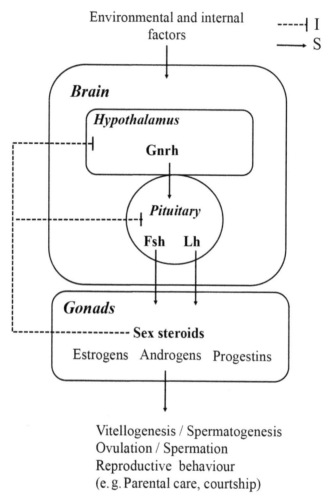

Fig. 1. Schematic representation of the Brain-Pituitary-Gonadal (BPG) axis of fish.
Gnrh – Gonadotropin releasing-hormone; Fsh – follicle-stimulating hormone;
Lh – luteinizing hormone.

2. The brain-pituitary-gonadal axis

2.1. Brain neurohormones

The brain plays its role at the highest level of the reproductive axis in fish (Figure 2). It is responsible for receiving, modulating and transducing environmental stimuli into neurohormonal signals that act at the following levels involved in the reproduction process. Other neuroendocrine systems of the BPGL axis considerably influence fish reproduction, such as the dopaminergic system, kisspeptins and the gonadotropin inhibitory hormone (Gnih) (Cowan et al. 2017). Moreover in recent years, other neuropeptides, such as Neurokinin B, have become important in tilapia reproduction (Biran et al. 2014).

Gnrh is a decapeptide that belongs to a family with high structural conservation on the phylogenetic scale. The influence of this neurohormone on stimulating the secretion and release of gonadotropins in the pituitary gland is well-known (Yaron et al. 2001). Three Gnrh isoforms have been characterized, which present different stimulation levels of the pituitary gland. Gnrh effects are elicited through binding to the specific Gnrh receptors present in the membrane of adenohypophyseal gonadotropic cells (Chen and Fernald 2008). These receptors are distributed throughout the body, but can be found mainly in gonadotropic and somatotropic cells of the pituitary gland and in extrahypophyseal tissues. In addition, the number of receptors is not constant, but shows variations depending on the reproductive cycle phase (Parhar et al. 2002). With tilapia, several studies have characterized the main Gnrh isoforms and their respective receptors (Parhar et al. 1997, Weber et al. 1997, Soga et al. 2005). On the one hand, in the tilapia brain, Gnrh1 neurons (seabream Gnrh, sbGnrh) have been described in the preoptic area of the hypothalamus and in the pituitary (Parhar et al. 1998). On the other hand, Gnrh2 (chicken Gnrh-II) and Gnrh3 (salmon Gnrh-III) neurons have been detected in the midbrain tegumentum and in the terminal nerve, respectively, and both areas are related to the modulation of reproduction, feeding and behavioral processes (Weber et al. 1997). Different studies have revealed the main role of sbGnrh on the regulation of gonadotropin production by describing the direct effect of sbGnrh on LHβ expression in cultured pituitary cells of tilapia hybrids (Melamed et al. 1996). Other functions attributed to Gnrh forms are mating and nesting, and stimulating the secretion of prolactin and somatolactin in *Oreochromis mossambicus* (Weber et al. 1997). The expression of Gnrh-receptors in tilapia has been described for the Gnrh-1, Gnrh-2 and Gnrh-3 neurons (Levavi-Sivan et al. 2004, Soga et al. 2005). In addition, their distribution and expression depend on the sexual maturity state and sex, with a higher expression in mature females than in males (Parhar et al. 2002, Levavi-Sivan et al. 2004, Martínez-Chávez et al. 2008).

Another key factor that plays an important role in the reproduction axis of vertebrates is the Kisspeptin system. Its stimulatory influence on the neuroendocrine regulation of reproduction has been described in various species, whose function and potency depend on the species. The teleost kisspeptin system was first described in Nile tilapia (Parhar et al. 2004). Although there are two kiss isoforms in most teleost fish (Servili et al. 2011), in tilapia Kiss2 is the only isoform to have been identified (Parhar et al. 2004). *Kiss2* expression has been detected in the brain, pituitary and gonad in both tilapia sexes, where it seems to perform an important function in early gonadal maturation (Park et al. 2012). In addition, its expression depends mainly on the ovarian cycle stage of female tilapia, with the highest levels in immature stages compared to mature and post-ovulatory stages. Park et al. (2016) proved its stimulatory effects at all levels of the tilapia reproductive axis, its participation in the synthesis and release of Gnrh, follicle-stimulating hormone (Fsh), Lh and sex steroids.

The stimulatory influence of Neurokinin B (Nkb) on tilapia reproduction was discovered in the last decade (Biran et al. 2012, 2014, Jin et al. 2016). Nkb is located in different neuronal hypothalamus complexes, although the expression of its receptors appears in Lh cells during the Lh surge prior to ovulation, which suggests a role of this neuropeptide in the final maturation step of oocytes (Biran et al. 2014). In addition, the intraperiotoneal Nkb administration has stimulatory effects on pituitary gonadotropins (direct effects) and all the Gnrh variants (indirect effects due to Gnrh neuron activation) in the brain (Biran el al. 2014). Mizrhai et al. (2019) has recently described how receptors of Nkb are co-expressed in GnrH neurons.

In addition to the previous stimulatory systems, there are other neuroendocrine systems, such as dopamine and Gnih, which act as negative effectors of the BPGL axis. The dopaminergic system acts as the major antagonist of the Gnrh system by inhibiting the secretion of adenohypophyseal hormones and blocking reproduction processes (Beaulieu and Gainetdinov 2011). This inhibitory effect has been described in several freshwater fish species, but not in marine species, for which no negative effects have been clearly defined (Prat et al. 2001). With tilapia, dopamine agonists show an inhibitory *in vivo* and *in vitro* effect on the synthesis of Gnrh receptors as well as on Fsh and Lh release (Levavi-Sivan et al. 2006, Biran et al. 2014). The functional role that dopamine exerts on the regulation of gonadotropins in tilapia is explained by not only the anatomical location of its dopamine fibers, but also by the high specificity of dopamine through its D2-like receptors present in the Lh cells of the adenohypophysis (Jiang et al. 2016).

The existence of the dodecapeptide GnIH was discovered for the first time in birds (Tsutsui et al. 2000), but it has also been found in vertebrates and invertebrates, which suggests a high degree of phylogenetic conservation. The wide distribution of GnIH cells and fibers throughout the brain and pituitary, as well as their proximity to GnRH cells, suggest that this peptide plays a key role as a neuromodulator at the brain level (Muñoz-Cueto et al. 2017). In some vertebrates, especially birds and mammals, GnIH seems to present an inhibitory effect on the synthesis and release of hypothalamic GnRH and pituitary gonadotropins (Muñoz-Cueto et al. 2017). However, in fish, as the role of Gnih remains unclear, more studies in this line are necessary. In tilapia, Gnihorthologs have been described in females (Ogawa et al. 2016). It has also been observed that the effects that Gnih has on the pituitary seem to be species-dependent by stimulating or inhibiting the release of gonadotropins. Unlike what happens with other species, intraperitoneal Gnih administration to tilapia stimulates Fsh and Lh production (Biran et al. 2014). Hence, further studies are required to clearly elucidate the role of Gnih in the tilapia reproduction system.

The few above-mentioned studies reflect the role of the brain as a system capable of integrating sensory and neuroendocrine information to trigger stimulatory or inhibitory responses to neurohormones of the following BPGL axis levels that lead to reproduction. Hence, the importance of continuing to investigate other factors involved at the brain level and regulate reproduction in tilapia and, thus, improve its reproductive management, thanks to the development of hormonal protocols that guarantee optimal reproduction control of tilapia in captivity, should be noted.

2.2. Pituitary hormones: Gonadotropins and their receptors

The second BPG axis level involves the neuroendocrine system of the brain, mainly Gnrh, by regulating the stimulations of adenohypophysis cells. These gonadotropic cells are responsible for the synthesis and release of gonadotropins that will participate in different reproduction processes, such as GthI and GthII, which were first described in salmon (Yaron et al. 2003) (Figure 2). Later they were called follicle-stimulating, Fsh, and Luteinizing Hormone, Lh, because they shared their structural and functional characteristics with tetrapod gonadotropins (Levavi-Sivan et al. 2010). These gonadotropins are formed by two subunits: a common subunit, named the glycoprotein hormone α subunit (Gpα), and two β subunits that are specific

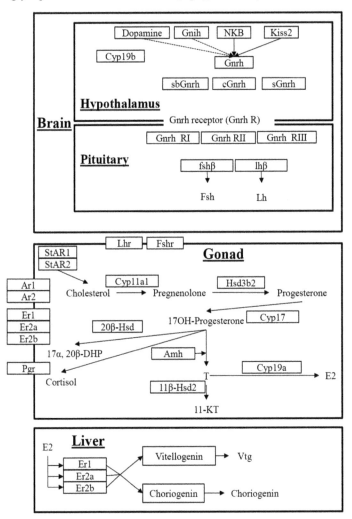

Fig. 2. Factors involved in the Brain-Pituitary-Gonadal-Liver (BPGL) axis of tilapia, including the synthesis pathways of sex steroids in gonads. Continuous and dashed arrows indicate stimulation or inhibition, respectively. Gnrh, gonadotropin releasing-hormone; Gnih, gonadotropin inhibitory hormone; Kiss2, Kisspeptin2; Fsh, follicle-stimulating hormone; Lh, luteinizing hormone; StAR, steroidogenic acute regulatory protein; E2, 17α-estradiol; T, testosterone; 11-KT, 11-ketotestosterone; Amh, Anti-Müllerian Hormone; Ar, androgen receptor; Er, estrogen receptor; Pgr, progestin receptor. For the abbreviations of other enzymes from the sex steroid pathway, please check the text. Scheme adapted to tilapia (see references in the text) from the medaka BPGL axis depicted by Saunders et al. (2015).

to each hormone (Fshβ and Lhβ), which are those with specific biological activity (Levavi-Sivan et al. 2010). The role of Gths and their receptors focuses mainly on stereidogenesis stimulation and gonadal development in testis and ovary (Yan et al. 2012). In fish, the action of gonadotropins is mediated through the binding with their receptors, known as Lhr and Fshr. The expression of both receptors has been observed at the gonadal level, mainly in Sertoli cells. However, only *fshr* expression appears in Leydig cells, and this expression presents a differential pattern depending on the gonadal development phase. Thus, *fshr* expression increases during vitellogenesis and spermatogenesis, whereas *lhr* expression increases during maturation and gamete release (ovulation/spermiation) (Kwok et al. 2005, So et al. 2005).

In Cyprinids, a high degree of conservation in the Gths sequence has been observed. Specifically in tilapia, the existence of *fshβ* (Rosenfeld et al. 1997, 2001, Yaron et al. 2001) and *lhβ* genes (Rosenfeld et al. 1997, Elizur et al. 2000) has been reported. Parhar et al. (2002) and Kasper et al. (2006) have described β subunits of both gonadotropins, which are specially located in the *pars distalisproximalis* and peripheral regions of the *pars intermedia* of the pituitary gland of Nile tilapia, as with most teleosts (Weltzien et al. 2004). These subunits have also been determined in plasma, thanks to the development of specific enzyme-linked immunoabsorbent assays (Aizen et al. 2007a). *In vivo* and *in vitro* studies conducted with tilapia have shown a direct effect of Gnrh on the increased secretion of pituitary Lhβ and Fshβ, and a time-dependent increase in the mRNA levels of these hormones (Levavi-Sivan and Yaron 1993, Melamed et al. 1996, Gur et al. 2000, Aizen et al. 2007b). In Nile tilapia, as in many teleosts, two gonadotropin receptors have been identified, with *fshr* expression found in Sertoli cells and oocyte granulose cells during vitellogenesis, while *lhr* is present in Leyding cells and mature oocytes (Oba et al. 2001). The location of gonadotropins and their receptors indicates the functionality of Fsh and Lh in the reproductive physiology of tilapia. In females, gonadotropins regulate processes related to vitellogenesis, oocyte maturation, and ovulation and steroid secretion, while they regulate spermatogenesis, spermiation and testicular steridogenesis in males (Yaron et al. 2001, 2003) (described in the sections below).

2.3. Gonadal hormones

The gonadal level is the last fish reproductive BPG axis step. At this third level, gonadotropins regulate the expression of the gonadal genes involved in the synthesis of sex steroids (androgens, estrogens, progestins) (Figure 2), and in the production of other growth-related factors (Tokarz et al. 2015). These factors are necessary for the differentiation and cell proliferation of gonads, and they also establish negative feedback by regulating the brain and pituitary levels of the reproductive axis (Yaron and Levavi-Sivan 2011). All steroid hormones derive from a common precursor, cholesterol, which is assimilated by gonadal cells, thanks to the action of the steroidogenic acute regulatory protein (StAR), and is converted into pregnenolone via cytochrome p450 (Cyp11a). In Nile tilapia, two StAR isoforms (StAR1 and StAR 2) have been described to play different roles during gonadal development (Yu et al. 2014). On the steroidogenesis pathway of teleosts, numerous enzymes participate that gives rise to three types of steroids with 18, 19 and 20 carbon atoms, named estrogens (β-estradiol or E2), androgens (11-ketotestosterone or 11-KT and Testosterone) and progestins (17, 20, 21-trihydroxy-4-pregnen-3-one or 20B-S and 17α, 20β -dihydroxy-4-pregnen-3-one or DHP) (Figure 2) (Tokarz et al. 2015). According to Wang et al. (2016), the key stereidogenic enzymes are Hsd3β2 (participates in the conversion of pregnenolone into progesterone) and Cyp17 (converts progesterone into 17OH-Progesterone). In addition, enzymes 20βHsd, 11βHsd and Cyp19 are essential for the synthesis of the main male and female sex steroids, respectively (Figure 2). In females, E2 is the main estrogen, while 11-KT is the chief androgen in males (Ijiri et al. 2008). In addition, both sex steroids (T and E2) can be synthesized from androstenedione, which also exerts its own functions on reproduction. Other hormones like progestins have been related to processes associated with gonadal maturation (MIS, maturation-inducing steroid) and socio-reproductive factors (Chattoraj et al. 2009). Finally, sex steroids bind to their corresponding receptors in the gonad and liver to stimulate the synthesis and release of yolk proteins (Vitellogenin, Vtg, and Egg Yolk-Proteins) required for egg development. In tilapia, sex steroids exert these effects through binding different types of androgen (Ar1 and Ar2), estrogens (Er1, Er2a and Er2b) and progestin receptors (Pgr) (Wang et al. 2005, Tao et al. 2013) (Figure 2).

3. Neuroendocrine regulation of reproductive cycles

The control of both gametogenesis and gamete release processes involves a set of neuroendocrine processes located at four different levels (brain-pituitary-gonad-liver) of the reproductive axis in tilapia. As described in previous sections, the biological activity of hypothalamic systems plays a direct (gonadal) and indirect (pituitary) role in the development of tilapia gametes, and ultimately in spermiation and ovulation.

3.1. Endocrine regulation of spermatogenesis and spermiation

As in most teleost species, the endocrine control of testis development is essential during the spermatogenesis and spermiation processes. Thus, both gonadotropins and sex steroids play a critical role in the stimulation of each testicular development stage: germ cell differentiation to spermatogonia, spermatocytogenesis (spermatogonial renewal and proliferation), meiosis (from spermatogonia to spermatocytes and spermatid), spermiogenesis (from spermatid to spermatozoa), spermiation and sperm maturation. As sperm production in *Oreochromis niloticus* and *Oreochromis aureus* seems to continue throughout the year under natural environmental conditions (Hyder 1972), the research aim has barely focused on studying the endocrine mechanisms that regulate spermatogenesis and spermiation. Thus, studies on male plasma sex steroids and gonadotropins in *O. niloticus* are still scarce compared to other fish species, which means that whether they present variations during the reproductive cycle has not yet been confirmed (Shawky et al. 2018, De Alba et al. 2019). Nevertheless, plasma sex steroids change throughout the reproductive cycle in males of *O. mossambicus* (Cornish 1998). So it is necessary to further investigate the dynamics of the hormonal profiles that regulate spermatogenesis and spermiation to improve our knowledge of male reproduction.

In early development stages, E2 and its receptors appear to enhance the renewal and proliferation of primary spermatogonia in testis through mitotic divisions (Schulz and Miura 2002). After spermatogonial proliferation, the secretion of both gonadotropins Fsh and Lh triggers the onset of meiotic division, which leads to the beginning of spermatogenesis (gonadotropic stimulation). Lack of such gonadotropic stimulation can inhibit spermatogonial divisions and, consequently, spermatogenesis (Nakamura and Nagahama 1989, Kobayashi and Nakamura 2009). In most teleost species, spermatogenesis is regulated mainly by Fsh and androgens (11-KT), while the spermiation and sperm maturation processes are regulated by Lh and progestins (DHP). These premises coincide with a research work performed in tilapia, which showed increased tilapia expression levels for *fshβ*, *lhβ* and their receptors during spermatogenesis and spermiation (Yan et al. 2012). The main role of Fsh in tilapia spermatogenesis has been described as Fsh acts through its receptors (Fshr) on Sertoli cells to produce growth factors (Activin B, Igf-1), and also on the Leydig cells responsible for producing sex steroids (11-KT and DHP) (Oba et al. 2001, Vilela et al. 2003, Kobayashi and Nakamura 2009). The synthesis of growth factors has also been described to be stimulated by the role of 11-KT in Sertolicells (androgenic stimulation). In this way, 11-KT has been characterized from the spermatogonial proliferation period to spermiogenesis (Nakamura and Nagahama 1989, Wang et al. 2016).

On the male gonadal stereidogenic pathway, the participation of Lh in sperm maturation and release should also be considered. Plasma Lh were found at basal levels during early tilapia spermatogenesis, which began to increase during spermiation. In addition, the male *lh* expression in male tilapia pituitary presented parallel fluctuations to *fsh* expression levels (De Alba et al. 2019). Moreover, the stimulatory effect of Lh and 11-KT on Leyding cells has been reported to trigger DHP production (Oba et al. 2001). *In vivo* studies in Nile tilapia show the essential role of DHP in spermatogonial cell proliferation (meiotic divisions) and spermatogenesis by intervening in spermiation, final sperm maturation (morpho-functional changes) and sperm motility enhancement (Oba et al. 2001, Fishelson 2003, Liu et al. 2014). In the pre-spermiation stage, tilapia males present swollen, reddish and prominent genital papilla and a reddish coloration pattern, which indicate their readiness to mate (Rana 1988) (Figure 3A, B, E). Fsh and Lh levels in plasma increase in correlation with testicular growth from early testicular development stages (early spermatogenesis) to spermiation (Melamed et al. 2000). After spermiation, gonadotropins levels drop considerably and rise again as the reproductive cycle progresses toward spermatogenesis and spermiation.

3.2. Endocrine regulation of ovarian development and spawning

In females, neuroendocrine coordination is established mainly by gonadotropins, estrogens and progestins, which regulate the dynamics of different ovarian development stages, which trigger ovulation and the release of eggs. The hormonal profiles of reproductive factors of female tilapia have been reported in detail for each ovarian development stage in *O. niloticus* (Srisakultiew 1993, Rothbard et al. 1991,

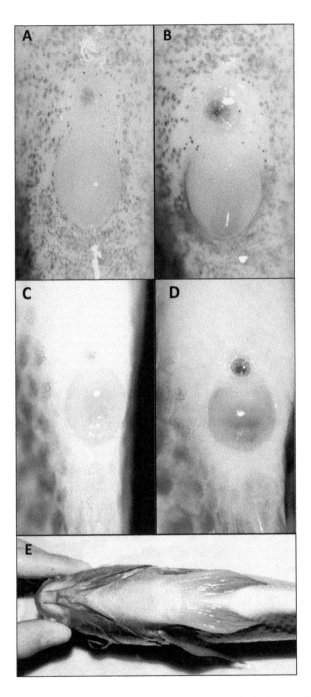

Fig. 3. Macroscopic differences of genital papillas of Nile tilapia during non-breeding and breeding periods. Male (A) and female (C) papillas during non-reproduction periods. Male (B) and female (C) papillas during breeding periods. The reddish coloration pattern can also be observed on most of the body's ventral part length (E).

Tacon et al. 2000, Biswas et al. 2005, Lapeyre 2008, Shawky et al. 2018, Ortiz et al. 2017) and *O. mossambicus* (Cornish 1998, Smith and Haley 1988) (Figure 4). Endocrine factors exert their role in specific ovarian development stages: oogonial mitotic and meiotic divisions, previtellogenesis (early, middle, late), vitellogenesis, oocyte maturation and ovulation.

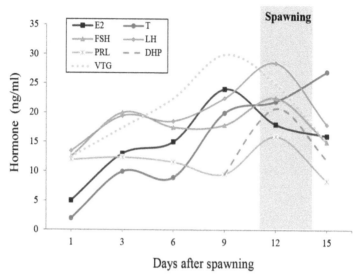

Fig. 4. Representative fluctuations of the plasma levels of the hormones involved in the reproductive cycle of Nile tilapia females in a 12:12LD cycle. Hormonal levels (ng/ml) were corrected by adjusting the Y axis: E2 (x1), T (x2), Fsh (x5), Lh (x3), PRL (x2), DHP (x8) and Vtg (x5). The gray area indicates the natural spawning period of Nile tilapia. Hormone levels are as obtained from several research works performed with Nile tilapia (Tacon et al. 2000, Melamed et al. 2000, Biswas et al. 2005, Aizen et al. 2007a, Ortiz et al. 2017, Shawky et al. 2018).

Although very little is known about the endocrine control of the first previtellogenic phases, E2 and DHP have been described to be the main regulators of the mitotic and meiotic divisions of oogonies. However, androgens (T) acquired an additional role during oocyte development (Miura et al. 2007). A rapid increase in both E2 and T on the first cycle days is associated with an accelerated vitellogenic process. In tilapia, as in most teleost species, vitellogenesis is regulated mainly by estrogens and Fsh. An increase in the plasma levels of E2 induces the hepatic synthesis of Vtg and the oocyte incorporation of vitello protein precursors from plasma (Van Bohemen et al. 1982). Thus, plasmatic E2 and VTG levels run in parallel during previtellogenesis and vitellogenesis. In addition, the stimulatory effect of Gths on E2 synthesis has been suggested as higher levels of *gnrhs* and *gths* mRNA coincide with higher estradiol plasmatic levels in females upon vitellogeneis (Bogomolnaya et al. 1989, Melamed et al. 2000, Levavi-Sivan et al. 2004). In addition, the possible influence of tilapia testosterone as a precursor to E2 (Tacon et al. 2000, Ortiz et al. 2017) and the gonadotropin stimulator (Melamed et al. 1997) on ovarian development has been suggested.

Although the role of Fsh during vitellogenesis in stimulating the follicular production of E2 has been well characterized in tilapia, it would seem that Lh could also play an important role in early oogenesis (primary and secondary vitellogenesis phases) as the parallel fluctuations of both *fshβ* and *lhβ* expressions and plasmatic levels in each sex have recently been described in *O. niloticus* (Melamed et al. 2000, Biswas et al. 2005, Aizen et al. 2007a, b, Ortiz et al. 2017, De Alba et al. 2019). According to Aizen et al. (2007a), the increase in Fsh occurs on the first few days after a new cycle starts, although Lh peaks have also been reported on ovarian cycle days 11-13 (Bogomolnaya et al. 1989, Rothbard et al. 1991). In short, both Lh and Fsh appear to have a stimulatory effect on the oocyte incorporation of plasmatic E2 and Vtg levels (Ortiz et al. 2017). These results suggest that gonadotropins are good indicators of spawning and recovery periods in Nile tilapia.

After vitellogenic growth, Lh and progestins (DHP) have been characterized as being the main regulators of oocyte final maturation and ovulation in tilapia (Tacon et al. 2000, Aizen et al. 2007a). In Nile tilapia, the highest Lh levels in plasma have been observed immediately before ovulation by stimulating the follicle to produce several progestins that induce oocyte final maturation (Levavi-Sivan et al. 2006, Aizen et al. 2007b. The peaks in the plasma levels of DHP and MIH (Maturation-inducing hormone) have been described immediately before spawning, which suggests their influence on the control of oocyte maturation and ovulation (Tacon et al 2000, De Alba et al. 2019). During the pre-spawning period, females present a swollen belly and genital papilla (Figure 3C, D), which indicate imminent spawning, which is a good time for stripping by gently massaging the abdominal cavity (1 h after ovulation) to obtain fertilized eggs. According to several authors, on day 14 of natural reproduction in Nile tilapia, the highest plasma values of sex steroids precede the ovulation period (Tacon 2000, Baroiller and Toguyeni 2004, De Alba et al. 2019). At these time points before ovulation, plasma prolactin (PRL) levels also rise, which suggests an endocrine effect on maternal behaviour (Weber et al. 1997, Tacon et al. 2000). After ovulation, plasmatic E2, Vtg and gonadotropins levels dramatically drop and return to basal levels (Tacon et al. 2000, Poleo et al. 2005, Fujimura and Okada 2007).

The spawning of most genus *Oreochromis* species occurs in the afternoon (Baroiller and Toguyeni 2004). Fertilization takes place directly on female genital papilla, after which females collect the eggs in their oral cavity. If mouth breeding occurs, the female lodges the eggs in her oral cavity to protect them against external agents for approximately the first 12 days after fertilization. While this maternal care lasts, circulating levels of sex steroids remain high, which induces the persistence of postovulatory follicles (POF) and prolongs females' ovarian cycle (Srisakultiew 1993, Tacon et al. 2000, Smith and Haley 1988). There are also reports that low Gnrh1 and Gnrh3 levels during mouthbreeding can be involved in suppressing the ovarian cycle and sexual behaviour (Levavi-Sivan et al. 2006, Das et al. 2018). After spawning and mouthbreeding (if it occurs), the ovary's structural and endocrine system prepares to begin a new ovarian cycle. In most studies on Nile tilapia, an interspawning interval (ISI) lasting 14 days has been described under adequate environmental conditions (12h Light:12h Dark, 12:12 LD, and 27-29°C). However, the tilapia ISI depends on factors like maternal behaviour (egg deprivation), photoperiod, temperature and food composition (De Silva and Radampola 1990, Ridha and Cruz 2000, Tacon et al. 2000, Biswas et al. 2005, Lapeyre 2008) (see the following sections).

Egg characteristics are species-specific. In Nile tilapia, females spawn large batches with up to 2000 eggs that are oval-shaped with yellowish ocher colouration (depending on diet carotenoids). Egg size (1.0-2.0 × 1.5-3.0 mm) and weight (5-9.5 mg) have been correlated with females' size and age (Rana et al. 1988, Baroiller and Toguyeni 2004, Fujimura and Okada 2007). After fertilization, the embryo presents a series of morphogenetic differentiations throughout this period and larval and juvenile stages (Fujimura and Okada 2007) (see Chapter 10 of the present book for more details).

4. Environmental influence on reproduction

In nature, fish are constantly exposed to cyclical changes in environmental factors, such as day/night alternations or seasonal changes in day length or water temperature. Fish can anticipate these predictable changes thanks to biological clocks used to perform important functions under the best environmental conditions. During fish reproduction, the existence of rhythms in the neuroendocrine factors present all along the BPGL axis ensures harmonizing the reproductive system with the most favourable environmental conditions for progeny to be successful (Cowan et al. 2017) (see Chapter 12 of the present book for more details). Actually, external environmental factors strongly influence the reproductive axis by triggering or changing the course of the reproduction timing process. The next sections deal with the influence of light and temperature cycles as the main environmental synchronizers that control the reproductive physiology of tilapia. In addition to these two factors, other external factors have been reported to influence the endocrine axis of tilapia, such as salinity (Baroiller et al. 1997, Abucay et al. 1999), food composition (El-Sayed et al. 2005), stocking density (Ridha and Cruz 1999, Lapeyre 2008) and social aspects (Little et al. 1993, Tacon et al. 2000, Ridha and Cruz 2003).

4.1. Influence of photoperiod on reproduction

The effect of daylight duration (photoperiod) on reproduction has been extensively investigated in fish, and it plays a primary role in species from temperate and arctic latitudes as they are subjected to wide variations during the photoperiod all year long (Migaud et al. 2010). Climate change and foul environmental cues appear to disrupt the fine-tune of the BPG axis and fish reproduction (Servilli et al. 2020). In tilapia, this is a tropical species and, hence, the photoperiod does not seem to strongly impact its reproductive physiology, with tilapia breeding taking place in a wide variety of photoperiods. Nevertheless, reproduction tilapia has been described to have certain seasonal nature (Cornish 1998). Tilapia can apparently reproduce under very different photoperiods, which considerably influence certain reproductive parameters, such as number of spawning, number of eggs, synchronization of spawning, fecundity, survival and larval development (Ridha and Cruz 2000, El-Sayed and Kawanna 2007). Recent studies have shown the molecular mechanisms of transcriptional Nile tilapia ovarian development in different photoperiod regimes (Tang et al. 2019).

Most studies carried out in Nile tilapia indicate a general trend of higher egg production, gonadal development and spawning frequency (ISI) associated with prolonged photoperiod length. Indeed the studies carried out using long (18:6 light:dark, LD) or equinox (12:12 LD) photoperiods reveal a significant increase in number of spawning and fertility, and also better synchronization and positive stimulation of gonadotropins (Biswas et al. 2005, Ridha and Cruz 2000, Campos-Mendoza et al. 2004, Lapeyre 2008, El-Sayed and Kawanna 2007). On the contrary, in experiments performed during short photoperiods (e.g. 6:18 LD), reproductive behavior is inhibited after 3-4 spawning cycles, along with inhibited gonadotropic Fsh and Lh (Biswas et al. 2005). These results agree with El-Sayed and Kawanna (2007) and Lapeyre (2008), who have reported that a 6:18 LD photoperiod negatively affects gonadal development compared to longer photoperiods. In addition to the photoperiod, the effect of light intensity on the BPG axis has also been observed. Ridha and Cruz (2000) used photoperiod combinations of different durations and light intensities to find that most reproductive parameters improved in tilapia with an 18:6 LD photoperiod and 2500 lux. There is also evidence for light wavelength (colour) influencing Nile tilapia reproduction (Volpato et al. 2004). Blue light enhances reproduction by leading to a higher proportion of reproducing fish with active nest constructions.

To summarize, light manipulation (photoperiod, light intensity, colour) has proven useful for controlling tilapia reproduction. Therefore, appropriate lighting protocols should be used to improve tilapia's reproductive efficiency in aquaculture.

4.2. Influence of temperature on reproduction

The plasticity of endocrine mechanisms is a characteristic of poikilothermic animals such as fish, and is a highly influential technique for their aquaculture. In recent years, different temperature protocols have been established to control tilapia's reproductive physiology. This species is capable of surviving within a wide thermal range. However, temperature is also a limiting factor for its natural reproduction. Thus, Nile tilapia needs warm water temperatures to breed in. Some authors set the limit at 22-25°C, with tilapia breeding the whole year if water temperature remains above these temperatures (Stickney 2000). Other studies have suggested an optimal temperature of 25-30°C for spawning, below which spawning frequency decreases and stops below 20°C (Bhujel 2000). In addition to lower temperature limits, very high temperatures can negatively affect tilapia breeding. For instance, the use of high temperatures (37°C) over long periods (45-60 days) means the permanent loss of germ cells in ovaries by inducing infertility in tilapia females (Pandit et al. 2015).

The intensification of reproductive activity related to seasonal changes, such as rainy seasons, has also been described (Baroiller et al. 1997). One of the most widely used techniques to stimulate tilapia reproduction implies cooling down water (22°C) (Srisakultiew and Wee 1988, Lapeyre et al. 2009). In these studies, females were exposed to cool water baths over long or short periods, and returned to an ambient temperature for four weeks to evaluate how cold temperatures induce effects on spawning. The results did not show any differences in using long cold exposure periods compared to the control group (maintained at ambient temperature:

29-30°C). However, when females were exposed to short cool temperature periods (6-24 h), a significant increase in spawning synchronicity was described.

5. Hormonal treatments for breeding stimulation

As described above, temperature and photoperiod are limiting factors for tilapia reproduction. Hence, the manipulation of these environmental factors can be used to induce or inhibit tilapia breeding. In addition, the use of hormonal therapies has been described to have considerable effects on the physiological mechanisms of tilapia reproduction, and protocols involving this technique can be applied. These protocols intend to improve the use of resources (personal, infrastructure, economic) of aquaculture production and to minimize the negative aspects of tilapia's reproductive biology.

Environmental conditions can be used to improve the control of tilapia reproduction in captivity. However, the inability to reproduce natural spawning conditions or to control fish reproductive cycles prevent us from mastering the dynamics of endocrine mechanisms (Lapeyre 2008). With tilapia reproduction management, there are two main hormonal manipulation objectives: (A) to induce spermiation/ovulation processes of gametes and stimulate the spawning process; and (B) to increase efficiency in spawning synchronization to improve the production of gametes and sperm.

During tilapia breeding, females generally present more problems than males, which are usually the limiting factor. For this reason, studies generally focus on females and protocols to induce permiation are lacking. However, sperm may sometimes be a limiting factor in fertilization. So in order to obtain high mature sperm concentrations, hormonal stimulation processes can be performed with tilapia males. The use of LH releasing hormone (LHRHa) seems to increase sperm count on the day after injection (Garcia-Abiado et al. 1996). Other studies reflect on human Chorionic Gonadotropin (hCG) injections inducing the expressions of both *star1* and *star2* in male tilapia testis, which would contribute to MIH production during sperm maturation and spermiation (Yu et al. 2014).

Female tilapias are very prolific animals capable of spawning all year long with a synchronous gonadal development and maternal behaviour (allocate eggs from different females). Thus, efforts of hormonal therapies focus on not only improving spawning synchrony, but also on controlling final oocyte maturation for subsequent artificial fertilization. For these purposes, a wide variety of hormones has been tested in Nile tilapia, such as GnRH and its agonists (Piamsomboon et al. 2019), pituitary extracts (Fernandes et al. 2013) and hCG (Coward and Bromage 2002, El-Gamal and El-Greisy 2005, Owusu-Frimpong 2008). A comparative study into the effectiveness of these hormones has been performed by Fernandes et al. (2013), who determined that hCG was the most potent and effective treatment for inducing synchronicity and gamete collection in Nile tilapia. However, the effects of hCG administration on gamete collection and fertilization quality varied depending on hCG concentration and number of doses (Fernandes et al. 2013). Piamsomboon et al. (2019) showed that a combination of two GnRHa injections (15 and 30 µg/kg, with an 8-hour interval) and oral dopamine antagonist administration (5 mg/kg) was also effective in successfully inducing spawning in tilapia females. In hormonal induction protocols, induction time has become increasingly more important as several studies have shown time-dependent responses of the hypothalamic-pituitary axis. The research by Rasines et al. (2013) into Senegalese sole demonstrated an effect of the time of day at which it is hormonally induced (with a GnRH analog) with the production of the obtained larvae. They carried out three inductions at three different times of the day (6 am, 12 am and 7 pm), and reported the induction of the highest values durin g larval production at 6 am compared to the groups induced at other times. These results highlight the importance of considering daytime responses when establishing hormonal protocols. Nevertheless, current research on hormonal treatments for tilapia reproduction is still scarce, hence the need to further investigate and develop new protocols that consider innovative factors like light colour and stimulation timing. In this way, the development of improved therapies will help to minimize the difficulties found in relation to artificial fertilization, and boost the genetic selection and reproductive efficiency processes of tilapia.

Acknowledgments

The present research was partially funded by projects "CHRONOLIPOFISH" (RTI2018-100678-A-I00), BLUESOLE (AGL2017-82582-C3-3-R) and CRONOFISH (RED2018-102487-T), granted by the Spanish Ministry of Science, Innovation and Universities, and co-funded by FEDER, to JFLO and FJSV. JFLO was also funded by a "*Ramón y Cajal*" research fellowship (RYC-2016-20959).

References cited

Abucay, J.S., G.C. Mair, D.O.F. Skibinski and J.A. Beardmore. 1999. Environmental sex determination: The effect of temperature and salinity on sex ratio in *Oreochromis niloticus* L. Aquaculture 173: 219–234.

Aizen, J., H. Kasuto and B. Levavi-Sivan. 2007a. Development of specific enzyme-linked immunosorbent assay for determining LH and FSH levels in tilapia, using recombinant gonadotropins. Gen. Comp. Endocrinol. 153: 323–332.

Aizen, J., H. Kasuto, M. Golan, H. Zakay and B. Levavi-Sivan. 2007b. Tilapia follicle-stimulating hormone (FSH): Immunochemistry, stimulation by gonadotropin-releasing hormone, and effect of biologically active recombinant FSH on steroid secretion. Biol. Reprod. 76: 692–700.

Baroiller, J.F., D. Desprez, Y. Carteret, P. Tacon, F. Borel, M. Hoareau, et al. 1997. Influence of environmental and social factors on the reproductive efficiency in three tilapia species, *Oreochromis niloticus, O. aureus,* and the red tilapia (red Florida strain). Proc. Int. Symp. Tilapias in Aquaculture. New York, USA 1: 238–252.

Baroiller, J. and A. Toguyeni. 2004. The Tilapiini tribe: Environmental and social aspects of reproduction and growth. Proc. Fisheries and Aquaculture. EOLSS, Oxford, UK.

Beaulieu, J.M. and R.R. Gainetdinov. 2011. The physiology, signaling, and pharmacology of dopamine receptors. Pharma. Rev. 63: 182–217.

Bhujel, R.C. 2000. A review of strategies for the management of Nile tilapia (*Oreochromis niloticus*) broodfish in seed production systems, especially hapa-based systems. Aquaculture 181: 37–59.

Biran, Jakob, O. Palevitch, S. Ben-Dor and B. Levavi-Sivan. 2012. Neurokinin Bs and Neurokinin B receptors in zebrafish-potential role in controlling fish reproduction. P. Natl. A. Sci. Biol. 109: 10269–10274.

Biran, J., M. Golan, N. Mizrahi, S. Ogawa, I.S Parhar and B. Levavi-Sivan. 2014. Direct regulation of gonadotropin release by neurokinin B in tilapia (*Oreochromis niloticus*). Endocrinology 155: 4831–4842.

Biswas, A.K., T. Morita, G. Yoshizaki, M. Maita and T. Takeuchi. 2005. Control of reproduction in Nile tilapia *Oreochromis niloticus* (L.) by photoperiod manipulation. Aquaculture 243: 229–239.

Bogomolnaya, A., Z. Yaron, V. Hilge, D. Graesslin, V. Lichtenberg and M. Abraham. 1989. Isolation and radioimmunoassay of a steroidogenic gonadotropin of tilapia. Isr. J. Aquacult. Bamid. 41: 123–136.

Campos-Mendoza, A., B.J. McAndrew, K. Coward and N. Bromage. 2004. Reproductive response of Nile tilapia (*Oreochromis niloticus*) to photoperiodic manipulation, effects on spawning periodicity, fecundity and egg size. Aquaculture 231: 299–314.

Chattoraj, A., M. Seth, A. Basu, T.G. Shrivastav, S. Porta and S.K. Maitra. 2009. Temporal relationship between the circulating profiles of melatonin and ovarian steroids under natural photo-thermal conditions in an annual reproductive cycle in carp *Catlacatla*. Biol. Rhythm Res. 40: 347–359.

Chen, C.C. and R.D. Fernald. 2008. GnRH and GnRH receptors: Distribution, function and evolution. J. Fish Biol. 73: 1099–1120.

Cornish, D.A. 1998. Seasonal steroid hormone profiles in plasma and gonads of the tilapia, *Oreochromis mossambicus*. Water SA 24: 257–263.

Cowan, M., C. Azpeleta and J.F. López-Olmeda. 2017. Rhythms of the endocrine system of fish: A review. J. Co. Physiol. B. 187: 1057–1089.

Coward, K. and N.R. Bromage. 2002. Stereological point-counting, an accurate method for assessing ovarian function in tilapia. Aquaculture 212: 383–401.

Das, K., S. Ogawa, T. Kitahashi and I.S. Parhar. 2018. Expression of neuropeptide Y and gonadotropin-releasing hormone gene types in the brain of female Nile tilapia (*Oreochromis niloticus*) during mouthbrooding and food restriction. Peptides 112: 67–77.

De Alba, G. De, N. Michele, N. Mourad, J.F. Paredes, F.J. Sánchez-Vázquez and J.F. López-Olmeda. 2019. Daily rhythms in the reproductive axis of Nile tilapia (*Oreochromis niloticus*): Plasma steroids and gene expression in brain, pituitary, gonad and egg. Aquaculture 507: 313–321.

De Silva, S.S. and K. Radampola. 1990. Effect of dietary protein level on the reproductive performance of *Oreochromis niloticus*. Asian Fish. Soc. Manila 1: 559–563.

Elizur, A., N. Zmora, I. Meiri, H. Kasuto, H. Rosenfeld, M. Kobayashi, et al. 2000. Gonadotropins – from genes to recombinant proteins. Int. Symp. Reprod. Physiol. of Fish. Norway 1: 462–465.

El-Gamal, A.E. and Z.A. El-Greisy. 2005. Effect of photoperiod, temperature and HCG on ovarian recrudescence and ability of spawning in Nile tilapia *Oreochromis niloticus* (Teleostei, Cichlidae). Egypt. J. Aquatic Res. 31: 419–431.

El-Sayed, A.F.M., C.R. Mansour and A.A. Ezzat. 2005. Effects of dietary lipid source on spawning performance of Nile tilapia (*Oreochromis niloticus*) broodstock reared at different water salinities. Aquaculture 248: 187–196.

El-Sayed, A.F.M. and M. Kawanna. 2007. Effects of photoperiod on growth and spawning efficiency of Nile tilapia (*Oreochromis niloticus* L.) broodstock in a recycling system. Aquaculture Res. 38: 1242–1247.

Falcón, J., L. Besseau, S. Sauzet and G. Boeuf. 2007. Melatonin effects on the hypothalamo–pituitary axis in fish. Trends in Endocrinol. Metabolism. 18: 81–88.

Fernandes, A.F.A., É.R. Alvarenga, D.A.A. Oliveira, C.G. Aleixo, S.A. Prado, R.K. Luz, et al. 2013. Production of oocytes of Nile tilapia (*Oreochromis niloticus*) for in vitro fertilization via hormonal treatments. Reprod. Domest. Anim. 48: 1049–1055.

Fishelson, L. 2003. Comparison of testes structure, spermatogenesis, and spermatocytogenesis in young, aging, and hybrid cichlid fish (Cichlidae, Teleostei). J. Morph. 256: 285–300.

Fujimura, K. and N. Okada. 2007. Development of the embryo, larva and early juvenile of Nile tilapia *Oreochromis niloticus* (Pisces: Cichlidae). Developmental staging system. Dev. Growth, Differ. 49: 301–324.

Garcia-Abiado, M.A.R., J.G. Merculio and G.C. Mair. 1996. The effect of luteinizing hormone-releasing hormone analogue (LHRHa) on sperm count of *Oreochromis niloticus* (L.). Aquaculture Res. 27: 95–100.

Gur, G., P. Melamed, H. Rosenfeld, A. Elizur and Z. Yaron. 2000. Mechanisms involved in the effect of GnRH, PACAP, and NPY on gonadotropin subunit mRNAs in tilapia pituitary cells. Proc. Int. Symp. Reproductive Physiol. Fish. Norway, 1: 466–468.

Hyder, M. 1972. Endocrine regulation of reproduction in Tilapia. Gen. Comp. Endocr. 3: 729–740.

Ijiri, S., H. Kaneko, T. Kobayashi, D.-S. Wang, F. Sakai, B. Paul-Prasanth, M. Nakamura and Y. Nagahama. 2008. Sexual dimorphic expression of genes in gonads during early differentiation of a teleost fish, the Nile tilapia *Oreochromis niloticus*. Biol. Reprod. 78: 333–341.

Jiang, Q., A. Lian and Q. He. 2016. Dopamine inhibits somatolactin gene expression in tilapia pituitary cells through the dopamine D2 receptors. Comp. Biochem. Physiol. A. 197: 35–42.

Jin, Y.H., J.W. Park, J. Kim and J.Y. Kwon. 2016. Neurokinin B-related peptide suppresses the expression of GnRH I, Kiss2 and tac3 in the brain of mature female Nile tilapia *Oreochromis niloticus*. Dev. Reprod. 20: 51–55.

Kasper, R.S., N. Shved, A. Takahashi, M. Reinecke and E. Eppler. 2006. A systematic immunohistochemical survey of the distribution patterns of GH, prolactin, somatolactin, β–TSH, β–FSH, β–LH, ACTH, and α–MSH in the adenohypophysis of *Oreochromis niloticus*, the Nile tilapia. Cell Tissue Res. 325: 303–313.

Kobayashi, T. and M. Nakamura. 2009. Molecular aspects of gonadal differentiation in a teleost fish, the Nile tilapia. Sex. Dev. 3: 108–117.

Kwok, H.-F., W.-K. So, Y. Wang and W. Ge. 2005. Zebrafish gonadotropins and their receptors: I. Cloning and characterization of zebrafish follicle-stimulating hormone and luteinizing hormone receptors-evidence for their distinct functions in follicle development. Biol. Reprod. 72: 1370–1381.

Lapeyre, B.A. 2008. Control of reproduction in Nile tilapia (*Oreochromis niloticus*) by manipulation of environmental factors. PhD Thesis, University of Göttingen, Germany.

Lapeyre, B.A., A. Müller-Belecke and G. Hörstgen-Schwark. 2009. Control of spawning activity in female Nile tilapia (*Oreochromis niloticus* L.) by temperature manipulation. Aquaculture Res. 40: 1031–1036.

Levavi-Sivan, B. and Z. Yaron. 1993. Intracellular mediation of GnRH action on GTH release in tilapia. Fish Physiol. Biochem. 11: 51–59.

Levavi-sivan, B., H. Safarian, H. Rosenfeld, A. Elizur, A. Avitan and I. Oceanographic. 2004. Regulation of gonadotropin-releasing hormone (GnRH)-receptor gene expression in tilapia: Effect of GnRH and dopamine. Biol. Reprod. 70: 1545–1551.

Levavi-Sivan, B., J. Biran and E. Fireman. 2006. Sex steroids are involved in the regulation of gonadotropin-releasing hormone and dopamine D2 receptors in female tilapia pituitary. Biol. Reprod. 75: 642–650.

Levavi-Sivan, B., J. Bogerd, E. Mañanos, A. Gómez and J.J. Lareyre. 2010. Perspectives on fish gonadotropins and their receptors. Gen. Comp. Endocr. 165: 412–437.

Little, D.C., D.J. Macintosh and P. Edwards. 1993. Improving spawning synchrony in the Nile tilapia, *Oreochromis niloticus* (L.). Aquaculture Res. 24: 399–405.

Liu, G., F. Luo, Q. Song, L. Wu, Y. Qiu, H. Shi, et al. 2014. Blockage of progestin physiology disrupts ovarian differentiation in XX Nile tilapia (*Oreochromis niloticus*). Biochem. Bioph. Res. Co. 473: 29–34.

Martinez-Chavez, C.C., M. Minghetti and H. Migaud. 2008. GPR54 and rGnRH I gene expression during the onset of puberty in Nile tilapia. Gen. Comp. Endocr. 156: 224–233.

Melamed, P., G. Gur, A. Elizur, H. Rosenfeld, B. Levavi-Sivan, F. Rentier-Delrue, et al. 1996. Differential effects of gonadotropin-releasing hormone, dopamine and somatostatin and their second messengers on the mRNA levels of gonadotropin IIβ subunit and growth hormone in the teleost fish, tilapia. Neuroendocrinology 64: 320–328.

Melamed, P., G. Gur, H. Rosenfeld, A. Elizur and Z. Yaron. 1997. The mRNA levels of GtH Iβ, GtH IIβ and GH in relation to testicular development and testosterone treatment in pituitary cells of male tilapia. Fish Physiol. Biochem. 17: 93–98.

Melamed, P., G. Gur, H. Rosenfeld, A. Elizur, R.W. Schulz and Z. Yaron. 2000. Reproductive development of male and female tilapia hybrids (*Oreochromis niloticus* × *O. aureus*) and changes in mRNA levels of gonadotropin (GtH) Iβ and IIβ subunits. J. Exp. Zool. 286: 64–75.

Migaud, H., A. Davie and J.F. Taylor. 2010. Current knowledge on the photoneuroendocrine regulation of reproduction in temperate fish species. J. Fish. Biol. 76: 27–68.

Miura, C., T. Higashino and T. Miura. 2007. A progestin and an estrogen regulate early stages of oogenesis in fish. Biol. Reprod. 77: 822–828.

Mizrahi, N., C. Gilon, I. Atre, S. Ogawa, I.S. Parhar and B. Levavi-Sivan. 2019. Deciphering direct and indirect effects of neurokinin B and GnRH in the brain-pituitary axis of tilapia. Front. Endocrinol. 1: 10.

Muñoz-Cueto, J.A., J.A. Paullada-Salmerón, M. Aliaga-Guerrero, M.E. Cowan, I.S. Parhar and T. Ubuka. 2017. A journey through the gonadotropin-inhibitory hormone system of fish. Front. Endocrinol. 8: 285.

Nakamura, M. and Y. Nagahama. 1989. Differentiation and development of Leydig cells, and changes of testosterone levels during testicular differentiation in tilapia *Oreochromis niloticus*. Fish Physiol. Biochem. 7: 211–219.

Oba, Y., T. Hirai, Y. Yoshiura, T. Kobayashi and Y. Nagahama. 2001. Fish gonadotropin and thyrotropin receptors: The evolution of glycoprotein hormone receptors in vertebrates. Comp. Biochem. Physiol. B. 129: 441–448.

Ogawa, S., M. Sivalingam, J. Biran, M. Golan, R.S. Anthonysamy, B. Levavi-Sivan, et al. 2016. Distribution of LPXRFa, a gonadotropin-inhibitory hormone ortholog peptide, and LPXRFa receptor in the brain and pituitary of the tilapia. J. Comp. Neurol. 524: 2753–2775.

Ortiz, J., L. Valladares, D. Muñoz, J. Caza, B. Manjunatha and R.R. Kundapur. 2017. Levels of 17β-estradiol, vitellogenin, and prostaglandins during the reproductive cycle of *Oreochromis niloticus*. Lat. Am. J. Aquat. Res. 45: 930–936.

Owusu-Frimpong, M. 2008. Controlled artificial reproduction in mouth brooding tilapia with human chorionic gonadotropin. J. Ghana Science Association 10: 70–77.

Pandit, N.P., R.K. Bhandari, Y. Kobayashi and M. Nakamura. 2015. High temperature-induced sterility in the female Nile tilapia, *Oreochromis niloticus*. Gen. Comp. Endocrinol. 213: 110–117.

Parhar, I.S. 1997. GnRH in tilapia: Three genes, three origins and three roles. GnRH neurons: Gene to Behavior, 99–122.

Parhar, I.S., T. Soga, Y. Ishikawa, Y. Nagahama and Y. Sakuma. 1998. Neurons synthesizing gonadotropin-releasing hormone mRNA subtypes have multiple developmental origins in the medaka. J. Comp. Neurol. 401: 217–226.

Parhar, I.S., T. Soga, Y. Sakuma and R.P. Millar. 2002. Spatio-temporal expression of gonadotropin-releasing hormone receptor subtypes in gonadotropes, somatotropes and lactotropes in the cichlid fish. J. Neuroendocrinol. 14: 657–665.

Parhar, I.S., S. Ogawa and Y. Sakuma. 2004. Laser-captured single digoxigenin-labeled neurons of gonadotropin-releasing hormone types reveal a novel G protein-coupled receptor (Gpr54) during maturation in cichlid fish. Endocrinol. 145(8): 3613–3618.

Park, J.W., J.H. Kim, Y.H. Jin and J.Y. Kwon. 2012. Expression profiles of Kiss2, GPR54 and GnRH receptor I mRNAs in the early life stage of Nile tilapia, *Oreochromis niloticus*. Dev. Reprod. 16: 31–38.

Park, J.W., Y.H. Jin, S.Y. Oh and J.Y. Kwon. 2016. Kisspeptin2 stimulates the HPG axis in immature Nile tilapia (*Oreochromis niloticus*). Comp. Biochem. Physiol. B. 202: 31–38.

Piamsomboon, P., N.S. Mehl, S. Sirivaidyapong and J. Wongtavatchai. 2019. Assisted reproduction in Nile tilapia *Oreochromis niloticus*: Milt preservation, spawning induction and artificial fertilization. Aquaculture 507: 139–143.

Poleo, G.A., C.G. Lutz, G. Cheuk and T.R. Tiersch. 2005. Fertilization by intracytoplasmic sperm injection in Nile tilapia (*Oreochromis niloticus*) eggs. Aquaculture 250: 82–94.

Prat, F., S. Zanuy and M. Carrillo. 2001. Effect of gonadotropin-releasing hormone analogue (GnRHa) and pimozide on plasma levels of sex steroids and ovarian development in sea bass (*Dicentrarchus labrax* L.). Aquaculture, 198: 325–338.

Rana, K. 1988. Reproductive biology and the hatchery rearing of tilapia eggs and fry. *In*: Recent Advances in Aquaculture. Springer, Dordrecht. 1: 343–406.

Rasines, I., M. Gomez, I. Martin, C. Rodríguez, E. Mañanos and O. Chereguini. 2013. Artificial fertilization of cultured Senegalese sole (*Solea senegalensis*): Effects of the time of day of hormonal treatment on inducing ovulation. Aquaculture 392: 94–97.

Ridha, M.T. and E.M. Cruz. 1999. Effect of different broodstock densities on the reproductive performance of Nile tilapia, *Oreochromis niloticus* (L.), in a recycling system. Aquaculture Res. 30: 203–210.

Ridha, M.T. and E.M. Cruz. 2000. Effect of light intensity and photoperiod on Nile tilapia *Oreochromis niloticus* L. seed production. Aquaculture Res. 31: 609–617.

Ridha, M.T. and E.M. Cruz. 2003. Effect of different schedules for broodstock exchange on the seed production of Nile tilapia *Oreochromis niloticus* (L.) in freshwater. Aquaculture Int. 11: 267–276.

Rosenfeld, H., B. Levavi-Sivan, P. Melamed, Z. Yaron and A. Elizur. 1997. The GTH β subunits of tilapia: Gene cloning and expression. Fish Physiol. Biochem. 17: 85–92.

Rosenfeld, H., B. Levavi-Sivan, G. Gur, P. Melamed, I. Meiri, Z. Yaron, et al. 2001. Characterization of tilapia FSHβ gene and analysis of its 5′ flanking region. Comp. Biochem. Physiol. B. 129: 389–398.

Rothbard, S. 1991. Hormonal profile associated with breeding behaviour in *Oreochromis nioticus*. Reprod. Physiol. Fish. 206: 15.

Saunders, D.M., M. Podaima, G. Codling, J.P. Giesy and S. Wiseman. 2015. A mixture of the novel brominated flame retardants TBPH and TBB affects fecundity and transcript profiles of the HPGL-axis in Japanese medaka. Aquat. Toxicol. 158: 14–21.

Schulz, R.W. and T. Miura. 2002. Spermatogenesis and its endocrine regulation. Fish Physiol. Biochem. 26: 43–56.

Servili, A., Y. Le Page, J. Leprince, A. Caraty, S. Escobar, I.S. Parhar, et al. 2011. Organization of two independent kisspeptin systems derived from evolutionary-ancient kiss genes in the brain of zebrafish. Endocrinology 152: 1527–1540.

Servili, A., A.V. Canario, O. Mouchel and J.A. Muñoz-Cueto. 2020. Climate change impacts on fish reproduction are mediated at multiple levels of the brain-pituitary-gonad axis. Gen. Comp. Endocrinol. 291: 113439.

Shawky, S.M., S.I. Fathalla and I.S. Abu-alya. 2018. Effect of seasonal variations (breeding and non-breeding seasons) on productive performance and reproductive hormonal profile in Nile tilapia (monosex and mixed sex). J. Life Sci. Int. 19: 1–15.

Smith, C. and S.R. Haley. 1988. Steroid profiles of the female tilapia, *Oreochromis mossambicus*, and correlation with oocyte growth and mouthbrooding behavior. Gen. Comp. Endocrinol. 69: 88–98.

So, W.K., H.-F. Kwok and W. Ge. 2005. Zebrafish gonadotropins and their receptors: II. Cloning and characterization of zebrafish follicle-stimulating hormone and luteinizing hormone subunits: Their spatial-temporal expression patterns and receptor specificity. Biol. Reprod. 72: 1382–1396.

Soga, T., S. Ogawa, R.P. Millar, Y. Sakuma and I.S. Parhar. 2005. Localization of the three GnRH types and GnRH receptors in the brain of a cichlid fish: Insights into their neuroendocrine and neuromodulator functions. J. Comp. Neurol. 487: 28–41.

Srisakultiew, P. and K.L. Wee. 1988. Synchronous spawning of Nile tilapia through hypophyzation and temperature manipulation. Proc. Int. Symp. Tilapia on Aquaculture, Philippines 1: 275–284.

Srisakultiew, P. 1993. Studies on the reproductive biology of *Oreochromis niloticus* L. Ph.D. Thesis, University of Stirling, Scotland, 267 pp.

Stickney, R.R. 2000. Tilapia culture. *In*: R.R. Stickney [ed.]. Encyclopedia of Aquaculture. John Wiley and Sons. New York, USA.

Tacon, P., J.F. Baroiller, P.Y. Le Bail, P. Prunet and B. Jalabert. 2000. Effect of egg deprivation on sex steroids, gonadotropin, prolactin, and growth hormone profiles during the reproductive cycle of the mouthbrooding cichlid fish *Oreochromis niloticus*. Gen. Comp. Endocrinol. 117: 54–65.

Tang, Z., Y. Zhou, J. Xiao, H. Zhong, W. Miao, Z. Guo, et al. 2019. Transcriptome analysis of ovary development in Nile tilapia under different photoperiod regimes. Front. Genetics 10: 894.

Tao, W., J. Yuan, L. Zhou, L. Sun, Y. Sun, S. Yang, et al. 2013. Characterization of gonadal transcriptomes from Nile tilapia (*Oreochromis niloticus*) reveals differentially expressed genes. PloS One 8: e63604.

Tokarz, J., G. Möller, M.H. de Angelis and J. Adamski. 2015. Steroids in teleost fishes: A functional point of view. Steroids 103: 123–144.

Tsutsui, K., E. Saigoh, K. Ukena, H. Teranishi, Y. Fujisawa, M. Kikuchi, et al. 2000. A novel avian hypothalamic peptide inhibiting gonadotropin release. Biochem. Piophys. Res. Co. 275: 661–667.

Van Bohemen, C.G., J.G.D. Lambert, H.T. Goos and P.G.W.J. Van Oordt. 1982. Estrone and estradiol participation during exogenous vitellogenesis in the female rainbow trout, *Salmo gairdneri*. Gen. Comp. Endocrinol. 46: 81–92.

Vilela, D.A.R., S.G.B. Silva, M.T.D. Perixoto, H.P. Godinho and L.R. França. 2003. Spermatogenesis in teleost: Insights from the Nile tilapia (*Oreochromis niloticus*) model. Fish Physiol. Biochem. 28: 187–190.

Volpato, G.L., C.R.A. Duarte and A.C. Luchiari. 2004. Environmental color affects Nile tilapia reproduction. Brazilian J. Med. Biol. Res. 37: 479–483.

Wang, D.S., B. Senthilkumaran, C.C. Sudhakumari, F. Sakai, M. Matsuda, T. Kobayashi, et al. 2005. Molecular cloning, gene expression and characterization of the third estrogen receptor of the Nile tilapia, *Oreochromis niloticus*. Fish Physiol. Biochem. 31: 255.

Wang, W., W. Liu, Q. Liu, B. Li, L. An, R. Hao, et al. 2016. Coordinated microRNA and messenger RNA expression profiles for understanding sexual dimorphism of gonads and the potential roles of microRNA in the steroidogenesis pathway in Nile tilapia (*Oreochromis niloticus*). Theriogenology 85: 970–978.

Weber, G.M., J.F.F. Powell, M. Park, W.H. Fischer, A.G. Craig, J.E. Rivier, et al. 1997. Evidence that gonadotropin-releasing hormone (GnRH) functions as a prolactin-releasing factor in a teleost fish (*Oreochromis mossambicus*) and primary structures for three native GnRH molecules. J. Endocrinol. 155: 121–132.

Weltzien, F.A., E. Andersson, Ø. Andersen, K. Shalchian-Tabrizi and B. Norberg. 2004. The brain–pituitary–gonad axis in male teleosts, with special emphasis on flatfish (Pleuronectiformes). Comp. Biochem. Physiol. A. 137: 447–477.

Yan, H., S. Ijiri, Q. Wu, T. Kobayashi, S. Li, T. Nakaseko, et al. 2012. Expression patterns of gonadotropin hormones and their receptors during early sexual differentiation in Nile tilapia *Oreochromis niloticus*. Biol. Reprod. 87: 116–119.

Yaron, Z., G. Gur, P. Melamed, H. Rosenfeld and B. Levavi-sivan. 2001. Regulation of gonadotropin subunit genes in tilapia. Comp. Biochem. Physiol. B. 129: 489–502.

Yaron, Zvi, G. Gur, P. Melamed, H. Rosenfeld, A. Elizur and B. Levavi-Sivan. 2003. Regulation of fish gonadotropins. Int. Rev. Cytology 225: 131–185.

Yaron, Z. and B. Levavi-Sivan. 2011. Endocrine regulation of fish reproduction. Encyclopedia of fish physiology: From genome to environment 2: 1500–1508.

Yu, X., L. Wu, L. Xie, S. Yang, T. Charkraborty, H. Shi, et al. 2014. Characterization of two paralogous StAR genes in a teleost, Nile tilapia (*Oreochromis niloticus*). Mol. Cell. Endocrinol. 392: 152–162.

10

Embryonic and Larval Development of Tilapias

Manuel Yúfera

Instituto de Ciencias Marinas de Andalucía (ICMAN-CSIC) Campus Universitario Río San Pedro s/n, 11519 Puerto Real, Cádiz, Spain

1. Introduction

Tilapiine cichlids, and particularly Nile tilapia (*Oreochromis niloticus*), are currently considered a group of continental fish species of primary importance for the worldwide production of protein for human consumption from aquatic origin (FAO 2018). The impressive farming success of tilapias in different continents and geographic regions is due to their versatility, notable tolerance to stress, high capacity for adaptation to a variety of environments and excellent growth rates. Furthermore, tilapias display omnivorous feeding habits that allow these fish species to accept a wide variety of feeds (MacIntosh and Little 1995, El-Sayed 1999, Omasaki et al. 2017).

As in the other teleosts, all these biological characteristics appear progressively during the ontogeny and first life stages. The developmental traits of this group of species are obviously related to the natural habitats where they inhabit as well as to the corresponding behaviour and feeding habits they display to have a successful life in those environments. With the exception of the reproductive system, final morphology and complete organ functionality are acquired at the beginning of the juvenile stage, but in tilapias the principal features appear early in the development as corresponding to species with fast and precocious development.

On the other hand, due to the above-mentioned adaptive characteristics and their short reproductive cycle, tilapias have been utilized as model species, gaining increasing importance as a laboratory animal for physiological, osmoregulation, genetic and evolutionary studies (Wood et al. 1994, Johanning and Specker 1995, Wright and Pohaydak 2001, Santini and Bernardi 2005), and therefore, there is a justified interest for a deep knowledge of their ontogeny.

In spite of the importance of this group of species, the studies on its embryonic and larval development are not as extensive as would be expected. Although there are many publications on these early stages, the current knowledge is very

---

*Corresponding author: manuel.yufera@icman.csic.es

fragmentary and partially distributed among species, strains and lineages obtained in studies that have been performed with different scientific objectives and under different experimental conditions. The available information on the ontogeny of embryos and larvae process has been mainly obtained from studies with Nile tilapia and Mozambique tilapia (*Oreochromis mossambicus*), with punctual studies performed in other species and hybrids.This chapter shows different aspects of the developmental process of tilapias that may be relevant for the larvae and fry rearing, taking as a basis the published information on these two tilapia species.

2. Basic biological characteristics in relation to environmental factors

There are some fundamental biological characteristics that are inexorably defining the developmental pattern of tilapias and that is necessary to mention. Firstly, as in most fish species, two main environmental factors may influence the developmental history in this group, water temperature and salinity. In consequence, their anatomy and physiology are adapted to respond properly to the expected range of variation of these variables in their original habitats.

Both Nile and Mozambique tilapia are continental and warm water species with African origin that adapt very well to subtropical and temperate waters. Tilapias grow and reproduce at temperatures over 21-23 °C (Rana 1990, Anken et al. 1993, Popma and Lovshin 1996), having optimum growth performance in the range of 25-32 °C (Popma and Lovshin 1996, Campinho et al. 2004), though both species can tolerate a wider range of temperatures between 8 and 42 °C (Philippart and Ruwet 1982). As a consequence, the temperature for incubation and larval rearing in the farming practices oscillates primarily between 23 and 30 °C because they are more dependent on the natural environmental conditions, while the experimental studies have usually been performed between 27 and 30 °C.

In relation to water salinity, although recognised as freshwater fishes living in continental waters, it is a fact that tilapias have certain salinity tolerance and can also be considered as eurihaline fishes, at least some of them (Chervinski 1982, Popma and Lovshin 1996). Nile tilapia is a moderate salinity-tolerant species that can grow in waters with low salinity, though its reproductive performance declined progressively at increasing salinities (Watanabe 1985, Suresh and Lin 1992). The eggs are able to tolerate rearing salinities up to 20 ppt (Fridman et al. 2012a). But Mozambique tilapia is the species showing the highest salinity tolerance, being able to live in seawater conditions (Chervinski 1982, Suresh and Lin 1992, Uchida et al. 2000). Larvae of the hybrid red tilapia also showed a high tolerance to changes in salinity (Rahmah et al. 2020).

On the other hand, these species are continuous breeders with multiple spawns during the reproductive stage, as long as the key environmental conditions as temperature and salinity keep within the favourable range. This feature has a particular relevance because the changes in maternal age and body size during the reproductive stage may affect in some degree the fecundity and characteristics of eggs and larvae (MacIntosh and Little 1995, Tsadik 2008). In teleost, egg volume depends on the amount of yolk accumulated during the vitellogenesis that in last instance will determine the embryonic developmental time and size of the hatched larvae (Rana 1985, Polo et al. 1991).

Regarding the characteristics of feeding function, the digestive tract of tilapia species at the adult stage shows a clearly recognizable Y-shaped stomach with three sections (Caceci et al. 1997, Morrison and Wright 1999) that has a high acidification capacity (Fish 1960, Moriarty 1973, Caceci et al. 1997, Hlophe et al. 2014), and a long intestine separated from the rectum by the ileocecal valve (Morrison and Wright 1999). This structure allows to digest different types of food further than macrophytes, particularly from lower trophic levels, which correspond to teleosts with omnivorous feeding habits.

Other important detail to take into account when talking about ontogeny in tilapias is that these species, as other freshwater fish, are considered to have a fast and precocious development; their larvae have a relatively well-developed gut and accept prepared feeds from first feeding (Drossou et al. 2006) and few days after they attained the main juvenile morphology and physiological features.

3. Developmental staging

The terminology of the different successive developmental steps has been diverse in the studies on tilapias. Furthermore, some authors just prefer to refer the period of time in hour or days from fertilization (hpf, dph) or days from hatching (dph) to explain the temporal sequence of each particular event without giving any specific stage name. These different criteria may induce certain confusion in identifying the different stages. Nevertheless, classic early nomenclature as embryonic stage (from fertilization to hatching), early larva stage or yolksac larva (from hatching to the end of yolk reabsorption) and exogenous feeding larva or just larva (from the onset of the exogenous feeding to the full acquirement of juvenile features), will always serve as clear references for the purposes of this chapter, although in this last stage it is difficult to establish the moment in which the transition to juvenile has been definitively completed. In the present review, the first 24 h corresponding to the day of fertilization or to the day of hatching is considered as 0 dpf and 0 dph, respectively, and therefore, the numerical order has been accordingly corrected in those studies in which the first 24 h was considered as day 1.

In Nile tilapia, Galman and Avtalion (1989) established 15 developmental stages from egg fertilization to hatching (occurring at 72 hpf) and other 9 larval stages from hatching to the complete yolk reabsorption and start of feeding (attained at 6 dph) by observing changes in external morphological characters with scanning microscopy in eggs and larvae incubated at 24-26 °C. Morrison et al. (2001) also studied the embryonic development in Nile tilapia from fertilization to the onset of the exogenous feeding and the inflation of swim bladder in eggs and larvae maintained at 28 °C. This study, that describes the succession of events day by day, is based primarily on the examination of internal organs and some glands at histological level. In this study, the staging is organized according to classic embryonic steps (zygote, cleavage, blastula, gastrula, segmentation and pharyngula periods) that ended about 100 hpf, followed by a hatching period which includes the events occurring during the emerging of free larvae from the chorion that occurs along the fourth day after fertilization, and finally by an early larval period during the reabsorption of the vitelline reserves up to start of exogenous feeding. Fujimura and Okada (2007) established 32 stages along the embryo and larval development, based primarily on external features and the number of caudal fin ray elements. These stages were grouped in three main developmental periods. The first two, like in Morrison et al. (2001) study, include the period from the embryo development up to hatching (stages 1 to 18), and the free swimming larval period from hatching up to the end of yolk resorption and onset of exogenous feeding (stages 19 to 25). Furthermore, this staging system includes stages beyond the yolksac larval stage including the one named as early juvenile period (stages 26 to 32). However, from stages 28 to 32, the only noticeable change is the increasing number of caudal fin ray elements. Obviously, length and weight of larvae continue increasing progressively during these last stages. All the 32 stages were defined over the development observed during a month from fertilization at 28 °C. Figure 1, taken from Piamsomboon et al. (2019) supplementary data, shows a detailed and explicative description of the different developmental stages in Nile tilapia according to the staging defined by Fujimura and Okawa (2007).

On the other hand, 22 developmental stages were established from fertilization to end of yolk resorption in Mozambique tilapia (Cattin 1989), also based on externally observable features in experiments performed at 28 °C. The first 16 stages were defined in the period from fertilization to hatching at 95 hpf, and the following ones from hatching to the complete yolk depletion and active swimming larva that occurred at 7 dph.

4. Embryos and hatching

Tilapia exhibits maternal mouth-brooder incubation. Therefore, the first observable step after the egg fertilization is the presence of large ovoid orange eggs in the oropharyngeal cavity of female. However, most of the studies on embryonic development have been carried out under artificial incubation conditions in aquaria or jars after removing the eggs from the female's mouth. Morrison et al. (2001) reported that, in Nile tilapia, the hatching glands, the glands responsible for secreting the proteases able to hydrolyse the fibrillar proteins constituting the chorion, are profusely distributed over the different parts of embryo and the nearest yolk surface. These hatching glands are more evident the day before hatching. Concordantly, the hatching enzymes start to digest the inner layers of the chorion, that appears already partially digested the day before hatching (Morrison et al. 2003). In agreement with these observations, the expression of RNAm codifying these hatching enzymes (low choriolytic hatching enzyme, *TLCE* and high choriolytic enzyme, *THCE*) has been detected at the gastrula stage and progressively increasing up to hatching (He et al. 2017). The gene expression of these enzymes declines in the hatched larvae but is still noticeable. This would indicate an additional function in the developing larvae other than just breaking the chorion.

Eggs and larvae characteristics, size and quality may depend on many different biotic and abiotic factors. Thus, egg diameter and weight have been related to tilapia species and lineages, size and/or age of broodstock, water temperature and maternal

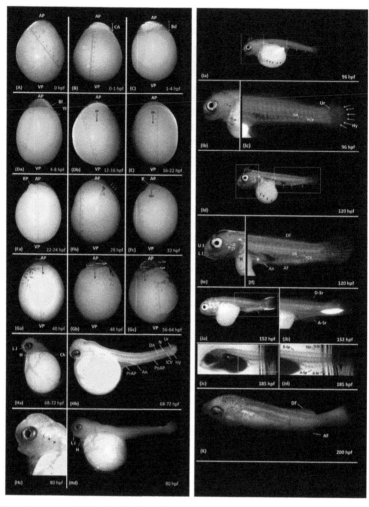

Fig. 1. Development of Nile tilapia *Oreochromis niloticus* embryo incubated at 30±1 °C. (Left panel) The embryonic development phase. (A) non-fertilized egg, AP; animal pole, VP; vegetal pole. (B) zygote stage, CA; cytoplasmic accumulation. (C) cleavage stage, Bd; blastodisc. (Da) early blastula stage, Bl; blastoderm layer, Yl; Yolk syncytial layer. (Db) late blastula stage, EP; epibody. (E) gastrula stage. (Fa-Fc) segmentation stage, BP; Brain primordium, red arrow; embryonic axis, B; brain, OpC; optic cup. (Ga-Gc) pharyngula stage, H; heart, L; Len, OpP; optic primordium, Ce; cerebellum. (Ha-Hd) hatching stage, LJ; lower jaw, Ch; chorion, PrAP; preanal plexus, PoAP; postanal plexus, An; anus, DA; dorsal aorta, N; notochord, ICV; inferior caudal vein, Ur; urostylar artery and vein, Hy; loops in hypural region, I, II, III; 1st, 2nd and 3rd gill arch. (Right panel) Larval development and the early juvenile phase. (Ia-If) Early larva period, Arrow; melanophores, black arrow head; yolk capillaries, At; atrium, Ve; ventricle, AF; anal fin, DF; dorsal fin, PCV; profundal caudal vein, white asterisk; intestinal content. (Ja-Jd) Late larva period, D-Sr; dorsal soft ray of dorsal fin, A-Sr; anal soft ray of anal fin, D-Sp; dorsal spines of dorsal fin, A-Sp; anal spines of anal fin, TM; Tilapia mark. (K) Early juvenile phase, hpf; hour(s) post fertilization. (Reprinted with permission from Piamsomboon et al. (2019), Aquaculture 507 (2019) 139–143)

nutritional conditions among other factors (Rana 1985, Tsadik 2008, Sarmento et al. 2018). Therefore, a wide range of mean egg diameters, between 1.4 and 2.9 mm, has been reported in the different studies. Considering the ovoid shape of tilapia eggs, egg diameter has been usually provided as the mean of the long and short axis (Coward and Bromage 1999).

Most of the information has been obtained in Nile tilapia in experiments performed between 28 and 30 °C of water temperature. Thus, Gunasekera et al. (1996) reported mean egg diameters in the range 2.28-2.43 mm, without statistical variation among the successive spawning, or with the maternal body weight or the dietary protein level of maternal feed, although this last factor induced a subtle change in the protein content of eggs (between 273 and 280 g/kg). Lu and Takeuchi (2004) obtained eggs from 2.10 to 2.30 mm (2.0 to 2.50 mg dry weight), again without variation among successive spawning or feed type. Likewise, Morrison et al. (2003) reported mean values in the range 2.3-2.9 mm, Ng and Wang (2011) in the range 2.38-2.44 mm, and Fernandes et al. (2013) in the range 2.6-2.8 mm. Campos-Mendoza et al. (2004) also reported similar mean diameter values in the range 2.36-2.47 mm. In their experiment, the largest eggs were obtained from fish maintained under a 12L:12D photoperiod coinciding with the lowest number of spawns and total fecundity. Sarmento et al. (2018) described an increase in Nile tilapia egg diameter from 2.3 to 2.6 mm (long axis) and weight from 3.5 to 4.5 mg by increasing the vitamin C content in the maternal diet from 0 to 942 mg/kg diet. Bombardelli et al. (2017) found an increase in the weight of eggs from 5.9 to 6.5 mg as well as in the corresponding freshly hatched larvae from 7.2 to 7.9 mg when the broodstock was fed on diets containing lower digestible protein and higher digestible energy. Lower egg size values were observed by Biswas et al. (2005), with mean diameters between 1.75 and 2.36 mm, and Tsadik (2008) with mean diameters increasing from 1.8 to 2.1 mm (0.29 to 0.31 mg dry weight) with the increase of the maternal age and body weight. Still lower egg size has been reported in Nile tilapia by Piamsomboon et al. (2019), with only 1.42-1.68 mm for the major axis. Similarly, Smitherman et al. (1988) found egg diameters from 1.55 to 1.75 mm in three different regional strains of Nile tilapia without noticeable differences among them.

In a study about the effect of dietary lipid source in the broodstock feed, Hajizadeh et al. (2008) reported mean egg diameters between 1.9 and 2.5 mm, and egg dry weights between 1.6 and 3.6 mg. These authors found that the dietary lipid sources had no significant effect on egg diameter, egg volume and egg dry weight. However, both weight and length of hatched larvae were slightly lower when only palm oil was used in the feed formulation in comparison with a more complete lipid composition. A wide mean egg volume variation has also been described in Mozambique tilapia, ranging from 2.42 to 5.39 mm^3 (0.80 to 2.00 mg dry weight) in different egg clutches (Rana 1985).

As observed in the above mentioned studies, variability in egg size and volume, and therefore in the amount of vitelline reserves, is a matter of fact in tilapia that will affect to some extend the posterior development. In teleost, the vitelline reserves of the yolk sac constitute the source of nutrients and energy for the developing embryo and larva up to the onset of exogenous feeding, when the yolk has been completely or almost-completely exhausted (Yúfera and Darias 2007). As occurs during the vitellogenesis, in which the biotic and abiotic circumstances are defining the egg volume at spawning, the efficiency in which the yolk is utilized depends on genetic characteristics and environmental factors (Polo et al. 1991, Wang et al. 2015). The size of larvae at the start of exogenous feeding is strongly dependent on yolk volume and composition, but also on the efficiency of yolk utilization.

Nile tilapia larvae typically hatch at 4 dpf at 28 °C (Rana 1990, Morrison et al. 2001, Lu and Takeuchi 2004, Fujimura and Okada 2007). This developmental time changes with the incubation temperature from 6 dpf at 20 °C to 2 dpf at 35 °C (Rana 1990). The hatched larvae require between 6 and 8 additional days to completely reabsorb the yolk reserves (Macintosh and Little 1995, Lu and Takeuchi 2004, Fujimura and Okada 2007, Hajizadeh et al. 2008). Nevertheless, Galman and Avtalion (1989) reported only 3 dpf for hatching and other seven days to the end of yolk reabsorption at 24-26 °C. In Mozambique tilapia, time for hatching at 28 °C was also 4 dpf according to Rana (1985), but it ranges from 2 dpf at 32 °C to 4 dpf at 22 °C in the study by Campinho et al. (2004).

These time ranges and values mentioned above correspond to experiments with specific strains and different maternal feeding conditions that may change among studies performed under similar temperatures. Illumination conditions also affect the embryonic development (Wang et al. 2019). Working with Nile tilapia eggs incubated at 28 °C, these authors found that both photoperiod and light intensity alters the incubation time up to hatching, the hatching rate, the yolk utilization and the size of larvae at the end of yolk resorption. Overall, most favourable conditions were 13-14 hours of day-length and 1000-1100 lux of intensity.

A similar variability has also been observed in the size of larvae emerging from the eggs. Thus, total lengths between 4.8 and 6.2 mm in freshly hatched larvae and between 7.3 and 9.0 mm at first feeding have been reported in Nile tilapia (Morrison et al. 2001, Fujimura and Okada 2007), though a larval total length of 8 mm at hatching in eggs naturally incubated within the female mouth at 25 °C has also been reported (Khalil et al. 2011). In Mozambique tilapia, the larvae at first feeding were reported to have a total length of 6-7 mm with independence from the egg size (Rana 1985). However, the larvae derived from the smaller eggs attained the irreversible starvation at 15-16 dph, while in the larvae derived for larger eggs this point-of-no-return occurred at 21 dph.

5. Ontogenesis of systems and key structures

5.1. Skeletal ontogenesis

The development of cartilaginous and bony structures of the fish skeleton allows the normal functionality of crucial structures necessary for adequate breathing, feeding and swimming.

Some aspects of the skeleton ontogeny in Nile tilapia have been studied by Fujimura and Okada (2007, 2008), paying particular attention to the cranial development due to the importance of the feeding apparatus at the start of feeding in a group of fish showing variated trophic specialization and adaptation of feeding behaviours to diverse environments.

According to these authors, the primordia of the pharyngeal arches were clearly detected at the pharyngula period, but the morphogenesis of the pharyngeal skeleton occurred mainly during the hatching period. More specifically, the cartilage differentiation in the jaw structures of Nile tilapia started in stage 17 (about 4 dpf at 28 °C), just the day before hatching, being extensive in stages 20 and 21 (about 2-3 dph). At this stage 17 pectoral fin rudiments and the differentiation of the gill arches are also detected. In stage 18 (at hatching), the gill arches develop the cartilaginous rod and gill filaments and the caudal fin rays start to form. On the other hand, the chondrocytes corresponding to the incipient gill arches were also detected at the end of yolk reabsorption at 25 °C in the posterior region of the buccopharynx by Khalil et al. (2011). The osteoid for the dermal bones were observed at stage 20, time at which the calcium uptake changes from the skin to the gills. Finally, the mineralization of the skeleton begins at stage 25 (4-6 dph).

Skeletal ontogenesis has also been studied in detail in Mozambique tilapia (Campinho et al. 2004). In this species, the chondrogenesis starts just after hatching (between at 0.5 and 3 dph at 27 °C) and before the notochord flexion according to the following order, the cranium, the elements pectoral fin, neural and haemal arches, elements of the caudal, dorsal and anal fins, and finally the pleural ribs and the pelvic fin. The chondrogenesis is completed firstly in those structures related to feeding and respiration, by the end of the yolksac larval stage. In the elements related to swimming and manoeuvrability, the chondrogenesis occured much later, at the transitiion from larva to juvenile. The ossification of these structures starts coinciding with the end of the chondrogenesis, starting by the cranial region and followed by the neural and haemal arches and the hypurals. In this study, the ossification was not yet completed after two months of experiment. A more recent study (Weigele et al. 2015) described similar sequences for chondrogenesis and ossification but with some variations in the developmental time of some events and structures. This study shows a detailed description of the sequential expression pattern of mRNA codifying the bone matrix proteins in fish, osteopontin-like protein (OP-L) and osteonectin (SPARC), in the different emerging skeletal elements. Both OP-L and SPARC appear simultaneously with the calcification of the bony skeleton elements of the cranium, including tooth, just after hatching, while they were detected before calcification in the fin rays.

5.2. Gut development

Ontogeny of the digestive tract has been studied mostly in Nile tilapia (Morrison et al. 2001, Khalil et al. 2011). Overall, the characteristics of the anatomical development follow the rules described for many fish species but particularly those related to freshwater species with large eggs and a precocious development at hatching. During the larval stage of teleosts, the gastrointestinal tract undergoes a transformation from a simple closed tube to a more complex system with a segmented tract with differentiated luminal mucosa in each section, and annexed glands as in the juveniles.

The initial development was described more in detail by Morrison et al. (2001) during 8 dpf up to the opening of the mouth observing eggs and larvae maintained at 28 °C of temperature. According to these authors, the commencement of the gut development occurs at the embryo stage before hatching, at 48 hpf approximately.

The incipient gut can be observed in the embryo as a closed tube with detectable lumen below the notochord at the completion of the epiboly. The formation of the liver is also detected at this stage between the yolk and gut. Later in the pharyngula stage, the developing pharynx and oesophagus show the lumen with the epithelium lined with cuboidal cells, the intestine lined with columnar epithelium, and an anlage at the end of the oesophagus corresponding to the emerging stomach.

At hatching, the larvae exhibits a digestive tract already forming a small loop surrounding the pancreas, and maintaining the same segments, the buccopharynx and oesophagus from one side, and the intestine from the other side lined with different epithelium types. A small cecum in one side of the gut indicates the primordial stomach. Goblet cells are not still present. The pancreas, a well-vascularized liver and the gallbladder lined with cuboidal cells are also present at this stage.

The following days during yolk resorption, this still rudimentary digestive tract becomes more clearly segmented in four sections, buccopharynx, oesophagus, stomach and intestine, with differentiated tissues like in the precedent days. The formation of the pyloric sphincter separates the small stomach and the intestine. Goblet cells appear in the epithelium of the buccopharyngeal cavity and the lumen of the oesophagus. Small goblet cells also appear in the intestine. At the end of this stage, the mouth is still covered by an oropharyngeal membrane.

At eight days after fertilization, taste buds appear in the epithelium of the pharynx and a noticeable constriction between the still small stomach and the intestine marks the pyloric sphincter. The wall of the oesophagus is surrounded by a muscular layer lined with a stratified squamous epithelium. In the developing stomach, the epithelium is formed by columnar cells as in the intestine. The rectum is lacking goblet cells. At this is time, the mouth becomes open and some prey can be detected in the lumen of the gut.

Similar developmental schedule has been described by Fabillo et al. (2004), although in this study the experimental temperature was not provided. Interestingly, these authors indicated that the evaginations in the anterior intestine corresponding to the emerging pyloric caeca appeared at 5 dph. This is an important developmental event that marks the final steps in the morphological transition to juvenile-adult digestive tract.

The gut development after the commencement of feeding has been studied by Khalil et al. (2011) in larvae reared at 25 °C, although this description focused only on the features developed at days 4 and 35 after hatching. Thus, according to these authors, at day 4 after hatching the yolk is completely consumed. At this time, the first few taste buds appear scattered in the epithelium of the buccal cavity. Moreover, mucus-secreting cells are already evident. The mucosa of the oesophagus presents several layers of saccular mucous cells containing acid and neutral mucopolysaccharides, while the submucosa is constituted by a loose connective tissue, a circular tunica muscularis and a thin serosa. The stomach is starting to develop with some few gastric glands observable in the columnar epithelium, although mucopolyssacharides secretion is not yet detectable. The submucosa presents an areolar connective tissue with lymph spaces and blood vessels, while the tunica muscular presents two layers with longitudinal and circular muscle fibres, respectively. The intestine has a relatively thick wall, with still short mucosa folds and a narrow luminal cavity, and a well-developed muscular layer. Mucous cells containing neutral mucopolysaccharides are abundant in this segment of the gut.

At 35 dph, all these structures are completely formed. The buccal cavity has a stratified epithelium with several cell layers and numerous mucous cells that secrete neutral mucopolysaccharides. The taste buds also proliferate mainly in the anterior part of the cavity. The submucosa also has several layers including a fibrous connective tissue layer, stratum compactum and an areolar capillarized connective tissue. The oesophagus, presenting a wide lumen, has a folded mucosa covered of abundant mucous cells, with acid and neutral mucopolyssacharides, as well as carboxylated and sulphated glycoproteins, a circular striated muscular layer and an areolar connective tissue. A thin epithelium forms the serosa. The stomach appears well formed, with a folded mucosa constituted by a simple layer of columnar cells, in which proliferates numerous gastric glands with neutral mucopolyssacharides. The intestine maintains the same structure of columnar cells combined with mucous cells but showing a wider lumen and with numerous short and curved mucosal folds. The submucosa is formed by a loose areolar connective tissue rich in blood vessels. The intestine is also surrounded by a circular and a longitudinal layer of muscle fibres.

This study of Khalil et al. (2011) also examines the potential benefit of treating the broodstock with thyroid hormones. These authors found that the larvae whose mothers were previously injected with thyroxine grew better and the gut structures developed faster than those derived from untreated control individuals. This positive effect of a maternal treatment with thyroxine was also observed in larvae of Mozambique tilapia (Lam 1980).

The annexed glands also undergo a rapid differentiation during these early days of development. After 4 dph, the liver consists in basophilic hepatocytes organized around hepatic sinusoids nearby the remaining yolk sac. These hepatocytes present centred nuclei, and reduced cytoplasm with lipid vacuoles. At 35 dph, the liver turns in a mass of polygonal hepatocytes with irregular boundaries arranged in cords around the veins. The hepatic cells present a large nucleus with one nucleolus and nuclear membrane, while the cytoplasm shows a granular appearance and contains vacuoles of glycogen. On the other hand, the pancreas is extrahepatic with separate exocrine and endocrine regions. The exocrine pancreas is arranged in acini and ducts with cells containing acidophilic zymogen granules. In the endocrine pancreas, the cells are arranged around capillaries and grouped in islets of Langerhans.

Although not always clearly specified, it seems that Nile tilapia has a period of mixed feeding between the strict endogenous and exogenous feeding (Fujimura and Okada 2007, Khalil et al. 2011). The digestive tract of teleost fish is already functional with the first food ingestion. Clear symptoms of digestion are easily observable even in species with altricial development in which the gut is still more undeveloped than in Tilapia (Yúfera and Darias 2007, Rønnestad et al. 2013). In fact, some digestive enzymes are already available in the developing gut during the yolksac stage before the opening of the mouth. Thus, Pereira et al. (2018) analysed the activity of gastric, pancreatic and intestinal digestive enzymes in embryo and fasted larvae from hatching to 10 dph in Nile tilapia while trying to find the potential effect of changing the dietary crude protein content in the maternal feed. They found that overall activity of pancreatic and gastric enzymes was very low or negligible during the egg cleavage phase, increasing quickly after hatching. The fact that acid protease also appeared during the first days after hatching confirms the precocious development pattern of this species. Intestinal brush border enzymes also increased after hatching was clearly noticeable at 7 dph. Tengjaroenkul et al. (2002), using histochemical techniques, also detected the activity of maltase, leucine aminopeptidase, dipeptidyl aminopeptidase IV, non-specific esterases, and alkaline phosphatase in the brush border of the intestine at hatching, while the lipase appeared at 3 dph. Likewise, first signs of nutrient absorption in the form of the lipid storage in the liver was observed two days after first feeding (Khalil et al. 2011). The pancreatic trypsin activity was also measured a couple of days after the onset of exogenous feeding (Drossou et al. 2006), probably as a consequence of the sampling schedule of this study. Interestingly in this later study, different levels of activity were observed in relation to the type of feed, being higher in larvae fed on feeds with poorer nutritional quality (Drossou et al. 2006). This response has also been observed in other freshwater fish, while in marine altricial larvae this feed-effect appears usually later in the development (Yúfera et al. 2018).

In Mozambique tilapia, the start of the activity of digestive proteases was examined in larvae maintained at 25-28 °C (Lo and Weng 2006). These authors found pepsin activity for the first time on 3 dph, the day of the opening of the mouth, and the activity continues increasing the following days. However, trypsin and chymotrypsin were detected only by 5 dph, the day of the commencement of feeding.

There is practically no information about the start of functionality of digestion at molecular level in Nile tilapia. Obviously, the mRNA of proteases precursors was already clearly expressed by 10 days after the commencement of feeding (Silva et al. 2019). In Mozambique tilapia, pepsinogen, trypsinogen and chymotrypsinogen transcripts have been cloned and their expression measured by semi-quantitative methods by RT-PCR and southern blotting analysis (Lo and Weng 2006). The expression of trypsinogen and chymotrypsinogen was detected on 1 dph, while the pepsinogen expression was detected by 2 dph, that is, before the opening of the mouth and the start of the hydrolytic activity in all cases.

Furthermore, the expression of peptides transported has also been examined in Mozambique tilapia during the embryonic development and yolk reabsorption (Con et al. 2019). Interestingly, the three isoforms PepT1a (*slc15a1a*), PepT1b (*slc15a1b*), and PepT2 (*slc15a2*) were expressed before the start of the exogenous feeding but not simultaneously. PepT1a and PepT1b were mainly overexpressed first from 3 to 10 dpf, while PepT2 were mainly expressed from 11 dpf onwards. The results would indicate the readiness for peptides absorption of enterocytes before the onset of feeding.

5.3. Swimming bladder

Swimming bladder is an indispensable structure for maintaining the larvae in the water column, and for developing an adequate swimming and manoeuvrability that allow to catch living prey and floating edible particles with increasing efficiency.

There is a critical window of time for the inflation of the bladder after which it cannot occur. In both Nile and Mozambique tilapia, swimbladder inflation occurs between 7 and 10 dph at temperatures ranging 26-28 °C (Fishelson 1966, Doroshev and Cornacchia 1979, Lingling and Qianru 1981, Morrison et al. 2001, Fujimira and Okada 2007) before the complete yolk reabsorption.

The ontogeny of this structure has been described for Mozambique tilapia (Doroshev and Cornacchia 1979). The primordial bladder before the inflation is a compressed sac with an outer layer with a squamous epithelium and an internal layer with columnar cells. This inner layer changes into a flattened epithelium after the inflation of the bladder by removing unnecessary structures. According to these authors, there is no connection between swim bladder and the digestive tract and therefore this species does not swallow atmospheric gas for the initial inflation of the bladder. Contrarily in Nile tilapia, the pneumatic duct connecting the swimbladder to the digestive tract was described by Morrison et al. (2001). According to these authors, the inflation of the swimbladder occurs together with the commencement of feeding at 3 dph.

5.4. Ontogeny of osmoregulatory capacity

Although living in freshwater environments, tilapias have a notable tolerance for higher salinities (see Chapter 6 of the present book). The ion exchanges occurred in specialized rich-mitochondria cells, named chloride cells or ionocytes. These cells, responsible for this osmoregulatory capacity, are mainly situated in epithelium, gills, kidney and intestine. The ontogeny of the chloride cells and osmoregulation capacity has been addressed in studies on several fish species. In teleosts, these cells, and therefore the osmoregulation, appear early during the embryonic development (Alderdice 1988, Varsamos et al. 2005), mainly on the epithelium of the yolksac and other skin surfaces. With the development of gills, digestive tract and urinary organs at the juvenile stage, the fish attains a complete osmoregulatory capacity. As already mentioned above, both Nile and Mozambique tilapias have significant salinity tolerance, mainly the second one.

In Mozambique tilapia incubated at 26-28 °C, the first chloride cells appear in the skin of embryos as early as 48 hpf and just after hatching (96 hpf), the larvae shows chloride cells with apical crypts facing the water (Hwang et al. 1994). These authors indicate that these cells constitute the main site of active ion transport up to 10 dph. The dense presence of chloride cells in the yolksac integument of Mozambique tilapia embryos and larvae was verified by Ayson et al. (1994). These authors also demonstrated that these cells are activated when the larvae are transferred to seawater, indicating their role in ion trans-epithelial transport and chloride-secreting activity (Ayson et al. 1994, Kaneko and Shirashi 2001). Likewise, Na^+, K^+-ATPase activity in the yolksac membrane declines as the yolk-sac is being absorbed and the gills are developing (Kosztowny et al. 2008). Yanagie et al. (2009) found that embryos and larvae of this species are able to regulate the osmolality of body fluid in a range of 300 to 370 mOsm/kg, although in adults these levels are lower and with less fluctuations. The highest level was measured at two days before hatching, declining quickly up to hatching and more progressively during the yolksac stage up to 8 dph. Figure 2, taken from this study, illustrates very well the ontogenetical pattern of these structures related to the osmoregulation during these early days. The gills develop quickly after hatching in Mozambique tilapia showing numerous chlorides cells at 3 dph, increasing further at 10 dph with the development of secondary lamellae and becoming a functional ionoregulatory organ even before they start the gas exchanging (Li et al. 1995).

In Nile tilapia, the osmoregulatory capacity has also been reported at hatching time, due to the presence of chloride cells on the yolksac epithelium and other skin areas (Fridman et al. 2011). During the yolk absorption, the chloride cells on the outer opercula and tail teguments increase their size while the chloride cells of the yolksac epithelium decrease in size. Concordantly, the osmoregulatory ability also increased during development, being constant from 2 dph until yolksac absorption (Fridman et al. 2012b). In Nile tilapia, the primordial opercula start covering the developing gills at hatching, and four gill arches with short filaments and vascularized lamellae are clearly observable at 1 dph. When the yolksac is practically exhausted, the operculum becomes completely developed covering the gill filaments (Fujimura and Okada 2007, Fridman et al. 2011).

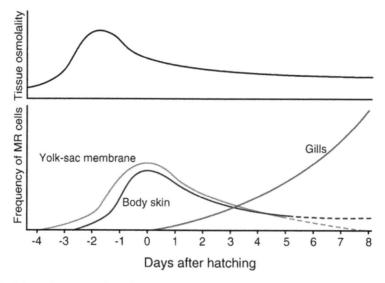

Fig. 2. Schematic presentation of ontogenic changes in tissue osmolality and developmental sequence of mitochondria-rich (MR) cells in Mozambique tilapia developing in freshwater (Reprinted with permission from Yanagie et al. Comp. Biochem. Physiol. (2009) 154A: 263–269).

6. Concluding remarks

As corresponds to most freshwater species and to cichlids in particular, tilapias exhibit a quick development attaining most of the characteristics of the juvenile stages few days after the opening of the mouth. The digestive tract is almost completely developed only few days after hatching at first feeding and becomes fully developed in an adult-like gut including the stomach few days after. This early high gut functionality allows the tilapias to nourish on a wide variety of feeds during the larval rearing that otherwise turns quickly in juvenile rearing. Likewise, the ionocites system spreads throughout the body surface at hatching and the subsequent quick gill development allows the larvae to cope with changes in water salinity from very early ages, practically from hatching. Obviously, some of these abilities are reinforced during post-larva/juvenile growth.

As perceived from this short review, there is still some missing information on the development of this group of species in comparison with other farmed fish families or groups. Most of the information on ontogeny was obtained years ago and there are some aspects that probably need an updating. There are of course some new very detailed studies of some structures such as the development of the inner ear (Weigele et al. 2017). However, studies at molecular and genomic level during embryonic development and first larval stages are still scarce or not detailed enough. The importance of some hybrids and polyploid strains is evident in this group and there are no particular comparative studies on the ontogeny. Variability in results from these different strains is difficult to evaluate because the strain characteristic is usually not typified. Probably the high adaptive plasticity of tilapias and the success of these species around the world is preventing to have a better global view of their ontogeny.

References cited

Alderdice, D.F. 1988. Osmotic and ionic regulation in teleost eggs and larvae. pp. 163–242. *In*: Hoar, W.S. and D.J. Randall (eds.). The Physiology of Developing Fish: Eggs and Larvae. Fish Physiology, Vol. 11A. Academic Press, London.

Anken, R.H., T. Kappel, K. Slenzka and H. Rahmann. 1993. The early morphogenetic development of the cichlid fish, *Oreochromis mossambicus* (Perciformes, Teleostei). Neth. J. Zool. 231: 1–10.

Ayson, F.G., T. Kaneko, S. Hasegawa and T. Hirano. 1994. Development of mitochondrion rich cells in the yolk-sac membrane of embryos and larvae of tilapia, *Oreochromis mossambicus*, in fresh water and seawater. J. Exp. Zool. 270: 129–135.

Bombardelli, R.A., E.S.R. Goes, S.M.N. Sousa, M.A. Syperreck, M.D. Goes, A.C.O. Pedreira, et al. 2017. Growth and reproduction of female Nile tilapia fed diets containing different levels of protein and energy. Aquaculture 479: 817–823.

Caceci, T., H.A. El-Habback, S.A. Smith and B.J. Smith. 1997. The stomach of *Oreochromis niloticus* has three regions. J. Fish Biol. 50: 939–952.

Campinho, M.A., K.A. Moutou and D.M. Power. 2004. Temperature sensitivity of skeletal ontogeny in *Oreochromis mossambicus*. J. Fish Biol. 65: 1003–1025.

Campos-Mendoza, A., B.J. McAndrew, K. Coward and N. Bromage. 2004. Reproductive response of Nile tilapia (*Oreochromis niloticus*) to photoperiodic manipulation: Effects on spawning periodicity, fecundity and egg size. Aquaculture 231: 299–314.

Cattin, P.M. 1989. The ontogeny and morphology of the upper pharyngeal pad of *Oreochromis mossambicus* (Peters) and its possible role in the rearing of young. PhD Thesis. University of Johannesburg, 188 p.

Con, P., T. Nitzan, T. Slosman, S. Harpaz and A. Cnaani. 2019. Peptide transporters in the primary gastrointestinal tract of pre-feeding Mozambique tilapia larva. Front. Physiol. 10: 808.

Coward, K. and N.R. Bromage. 1999. Spawning periodicity, fecundity and egg size in laboratory-held stocks of a substrate-spawning tilapiine, *Tilapia zillii* (Gervais). Aquaculture 171: 251–267.

Doroshev, S.I. and J.W. Cornacchia. 1979. Initial swim bladder inflation in the larvae of *Tilapia mossambica* (Peters) and *Moronesaxatilis* (Walbaum). Aquaculture 16: 57–66.

Drossou, A., B. Ueberschär, H. Rosenthal and K-H. Herzig. 2006. Ontogenetic development of the proteolytic digestion activities in larvae of *Oreochromis niloticus* fed with different diets. Aquaculture 256: 479–488.

El-Sayed, A.-F.M. 1999. Alternative dietary protein sources for farmed tilapia, *Oreochromis* spp. Aquaculture 179: 149–168.

Fabillo, M.D., A.A. Herrera and J.S. Abucay. 2004. Effects of delayed first feeding on the development of the digestive tract and skeletal muscles of Nile tilapia, *Oreochromis niloticus* L. Proceedings 6th International Symposium on Tilapia in Aquaculture Philippine International Convention Center. Roxas Boulevard, Manila, Philippines, pp. 301–315.

FAO. 2018. State of Fisheries and Aquaculture in the world. Meeting the Sustainable Development Goals. Rome, 210 pp.

Fernandes, A.F., É.R. Alvarenga, D.A. Oliveira, C.G. Aleixo, S.A. Prado, R.K. Luz, et al. 2013. Production of oocytes of Nile tilapia (*Oreochromis niloticus*) for *in vitro* fertilization via hormonal treatments. Reprod. Domest. Anim. 48: 1049–1055.

Fish, G.R. 1960. The comparative activity of some digestive enzymes in the alimentary canal of Tilapia and perch. Hydrobiologia 15: 161–178.

Fridman, S., J.E. Bron and K.J. Rana. 2011. Ontogenetic changes in location and morphology of chloride cells during early life stages of the Nile tilapia (*Oreochromis niloticus* (L.)) adapted to freshwater and brackish water. J. Fish Biol. 79: 597–614.

Fridman, S., J.E. Bron and K.J. Rana, 2012a. Influence of salinity on embryogenesis, survival, growth and oxygen consumption in embryos and yolk-sac larvae of the Nile tilapia. Aquaculture 334: 182–190.

Fridman S., J.E. Bron and K.J. Rana. 2012b. Ontogenic changes in the osmoregulatory capacity of the Nile tilapia *Oreochromis niloticus* and implications for aquaculture. Aquaculture 356–357: 243–249.

Fujimura, K. and N. Okada. 2007. Development of the embryo, larva and early juvenile of Nile tilapia *Oreochromis niloticus* (Pisces: Cichlidae). Developmental staging system Develop. Growth Differ. 49: 301–324.

Fujimura, K. and N. Okada. 2008. Bone development in the jaw of Nile tilapia *Oreochromis niloticus* (Pisces: Cichlidae) develop. Growth Differ. 50: 339–355.

Galman, O.R. and R.R. Avtalion. 1989. Further study of the embryonic development of *Oreochromis niloticus* (Cichlidae, Teleostei) using scanning electron microscopy. J. Fish Biol. 34: 653–664.

Gunasekera, R.M., K.F. Shim and T.J. Lam. 1996. Effect of dietary protein level on spawning performance and amino acid composition of eggs of Nile tilapia, *Oreochromis niloticus*. Aquaculture 146: 121–134.

Hajizadeh, A., K. Jauncey and K. Rana. 2008. Effects of dietary lipids source on egg and larvae quality of Nile tilapia, *Oreochromis niloticus* (L.). 8th International Symposium on Tilapia in Aquaculture 2008: 965–977.

He, J., Y. Liang, C. Xie, H. Qiao and P. Xua. 2017. Expression patterns of the hatching enzyme genes during embryonic development of tilapia. Aquaculture 479: 845–849.

Hlophe, S.N., N.A.G. Moyo and I. Ncube. 2014. Postprandial changes in pH and enzyme activity from the stomach and intestines of *Tilapia rendalli* (Boulenger, 1897), *Oreochromis mossambicus* (Peters, 1852) and *Clarias gariepinus* (Burchell, 1822). J. Appl. Ichthyol. 30: 35–41.

Hwang, P.P., Y.N. Tsai and Y.C. Tung. 1994. Calcium balance in embryos and larvae of the freshwater-adapted teleost, *Oreochromis mossambicus*. Fish Physiol. Biochem. 13: 325–333.

Johanning, K.M. and J.L. Specker. 1995. Characterization of yolk proteins during oocyte development of tilapia, *Oreochromis mossambicus*. Comp. Biochem. Physiol. 112B(2): 177–189.

Lam, T.J. 1980. Thyroxine enhances larval development and survival in *Sarotheroden* (*Tilapia*) *mossambicus* Ruppell. Aquaculture 21: 287–291.

Li, J., J. Eygensteyn, R.A.C. Lock, P.M. Verbost, A.J.H. van der Heijden, S.E. Wendelaar Bonga, et al. 1995. Branchial chloride cells in larvae and juveniles of freshwater tilapia *Oreochromis mossambicus*. J. Exp. Biol. 198: 2177–2184.

Lo, M.J. and C.F. Weng. 2006. Developmental regulation of gastric pepsin and pancreatic serine protease in larvae of the euryhaline teleost, *Oreochromis mossambicus*. Aquaculture 261: 1403–1412.

Lu, J. and T. Takeuchi. 2004. Spawning and egg quality of the tilapia *Oreochromis niloticus* fed solely on raw Spirulina throughout three generations. Aquaculture 234: 625–640.

Kaneko, T. and K. Shirashi. 2001. Evidence for chloride secretion from chloride cells in the yolk-sac membrane of Mozambique tilapia larvae adapted to seawater. Fish. Sci. 67: 541–543.

Khalifa, N.S.A., I.E.H. Belal, K.A. El-Tarabily, S. Tariq and A.A. Kassab. 2018. Evaluation of replacing fish meal with corn protein concentrate in Nile tilapia *Oreochromis niloticus* fingerlings commercial diet. Aquac. Nutr. 24: 143–152.

Khalil, N.A., H.M.M.K. Allah and M.A. Mousa. 2011. The effect of maternal thyroxine injection on growth, survival and development of the digestive system of Nile tilapia, *Oreochromis niloticus*, larvae. Adv. Bioscience Biotechnol. 2: 320–329.

Kosztowny, A., T. Hirano and G. Grau. 2008. Developmental changes in Na$^+$, K$^+$ -atpase activity in Mozambique tilapia (*Oreochromis mossambicus*) embryos and larvae in various salinities. 8th International Symposium on Tilapia in Aquaculture 2008: 447–457.

MacIntosh, D.J. and D.C. Little. 1995. Nile tilapia (*Oreochromis niloticus*). pp. 277–320. *In*: Bromage, N.R. and R.J. Roberts (eds.). Broodstock Management and Egg and Larval Quality. Oxford-Blackwell Science.

Moriarty, D.J.W. 1973. The physiology of digestion of blue–green algae in the cichlid fish, *Tilapia nilotica*. J. Zool. 171: 25–39.

Morrison, C.M. and J.R. Wright Jr. 1999. A study of the histology of the digestive tract of the Nile tilapia. J. Fish Biol. 54: 597–606.

Morrison, C.M., T. Miyake and J.R. Wright Jr. 2001. Histological study of the development of the embryo and early larva of *Oreochromis niloticus* (Pisces: Cichlidae). J. Morph. 247: 172–195.

Morrison, C.M., B. Pohajdak, M. Henry and J.R. Wright Jr. 2003. Structure and enzymatic removal of the chorion of embryos of the Nile tilapia. J. Fish Biol. 63: 1439–1453.

Ng, W.-K. and Y. Wang. 2011. Inclusion of crude palm oil in the broodstock diets of female Nile tilapia, *Oreochromis niloticus*, resulted in enhanced reproductive performance compared to broodfish fed diets with added fish oil or linseed oil. Aquaculture 314: 122–131.

Omasaki, S., K. Janssen, M. Besson and H. Komen. 2017. Economic values of growth rate, feed intake, feed conversion ratio, mortality and uniformity for Nile tilapia. Aquaculture 481: 124–132.

Pereira, M.M., M.M. Evangelista, E. Gisbert and E. Romagosa. 2019. Nile tilapia broodfish fed high-protein diets: Digestive enzymes in eggs and larvae. Aquacult. Res. (in press).

Piamsomboon, P., N.S. Mehl, S. Sirivaidyapong and J. Wongtavatchai. 2019. Assisted reproduction in Nile tilapia *Oreochromis niloticus*: Milt preservation, spawning induction and artificial fertilization. Aquaculture 507: 139–143.

Polo, A., M. Yúfera and E. Pascual. 1991. Effect of temperature on egg and larval development of *Sparusaurata* L. Aquaculture 92: 367-375.

Rahmah, S., H.J. Liew, N. Napi and S.A. Rahmata. 2020. Metabolic cost of acute and chronic salinity response of hybrid red tilapia *Oreochromis* sp. Larvae. Aquac. Rep. 16: 100233.

Rana, K.J. 1985. Influence of egg size on growth, onset of feeding, point-of-no-return, and survival of unfed *Oreochromis mossambicus* fry. Aquaculture 46: 119–131.

Rana, K.J. 1990. Influence of incubation temperature on *Oreochromis niloticus* (L.) eggs and fry 1: Gross embryology, temperature tolerance and rates of embryonic development. Aquaculture 87: 165–181.

Santini, S. and G. Bernardi. 2005. Organization and base composition of tilapia Hox genes: Implications for the evolution of Hox clusters in fish. Gene 346: 51–61.

Sarmento, N.L.A.F., E.F.F. Martins, D.C. Costa, C.C. Mattioli, G.S.C. Julio, L.G. Figueiredo, et al. 2018. Reproductive efficiency and egg and larvae quality of Nile tilapia fed different levels of vitamin C. Aquaculture 482: 96–102.

Silva, W.S., L.S. Costa, J.F. López-Olmeda, N.C.S. Costa, W.M. Santos, P.A.P. Ribeiro, et al. 2019. Gene expression, enzyme activity and performance of Nile tilapia larvae fed with diets of different CP levels. Animal 13(7): 1376–1384.

Smitherman, R.O., A.A. Khateri, N.I. Cassell' and R.A. Dunham. 1988. Reproductive performance of three strains of *Oreochromis niloticus*. Aquaculture 70: 29–37.

Suresh, A.V. and C.K. Lin, 1992. Tilapia culture in saline water: A review. Aquaculture 106: 201–226.

Tengjaroenkul, B., B.J. Smith, S.A. Smith and U. Chatreewongsin. 2002. Ontogenic development of the intestinal enzymes of cultured Nile tilapia, *Oreochromis niloticus* L. Aquaculture 211: 241–251.

Tsadik, G.G. 2008. Effects of maternal age on fecundity, spawning interval, and egg quality of Nile tilapia, *Oreochromis niloticus* (L.). J. World Aquacult. Soc. 39(5): 671–677.

Uchida, K., T. Kaneko, H. Miyazaki, S. Hasegawa and T. Hirano. 2000. Excellent salinity tolerance of Mozambique tilapia (*Oreochromis mossambicus*): Elevated chloride cell activity in the branchial and opercular epithelia of the fish adapted to concentrated seawater. Zool. Sci. 17(2): 149–160.

Varsamos, S., C. Nebel and G. Charmantier. 2005. Ontogeny of osmoregulation in postembryonic fish: A review. Comp. Bioche. Physiol. 141A: 401–429.

Wang, H., G. Liang, J. Liu, H. Yang, J. Qiang and P. Xu. 2015. Combined effects of temperature and salinity on yolk utilization in Nile tilapia (*Oreochromis niloticus*). Aquacult. Res. 46: 2418–2425.

Wang, H., W. Shi, L. Wang, C. Zhu, Z. Pan and N. Wu. 2019. Light conditions for commercial hatching success in Nile tilapia (*Oreochromis niloticus*). Aquaculture 509: 112–119.

Watanabe, W.O. and C.-M. Kuo. 1985. Observations on the reproductive performance of Nile tilapia (*Oreochromis niloticus*) in laboratory aquaria at various salinities. Aquaculture 49: 315–323.

Weigele, J., T.A. Franz-Odendaal and R. Hilbig. 2015. Expression of SPARC and the osteopontin-like protein during skeletal development in the cichlid fish *Oreochromis mossambicus*. Dev. Dyn. 244(8): 955–972.

Weigele, J., T.A. Franz-Odendaal and R. Hilbig. 2017. Formation of the inner ear during embryonic and larval development of the cichlid fish (*Oreochromis mossambicus*), Connect Tissue Res. 58(2): 172–195.

Wood, C.M., H.L. Bergman, P. Laurent, J.N. Maina, A. Narahara and P.J. Walsh. 1994. Urea production, acid-base regulation and their interactions in the Lake Magadi tilapia, a unique teleost adapted to a highly alkaline environment. J. Exp. Biol. 189: 13–36.

Wright, J.R. Jr. and B. Pohajdak. 2001. Cell therapy for diabetes using piscine islet tissue. Cell Transplant. 10: 125–143.

Yanagie, R., K.M. Lee, S. Watanabe and T. Kaneko. 2009. Ontogenic change in tissue osmolality and developmental sequence of mitochondria-rich cells in Mozambique tilapia developing in freshwater. Comp. Biochem. Physiol. 154A: 263–269.

Yúfera, M. and M.J. Darias. 2007. The onset of exogenous feeding in marine fish larvae. Aquaculture 268: 53–63.

Yúfera, M., G. Martínez-Rodríguez and F.J. Moyano. 2018. The digestive function in developing fish larvae and fry: From molecular gene expression to enzymatic activity. pp. 51–86. *In*: M. Yúfera (ed.). Emerging Issues in Fish Larvae Research. Springer International Publishing, Cham.

Tilapia Larviculture

Ronald Kennedy Luz* and Gisele Cristina Favero

Laboratório de Aquacultura, Universidade Federal de Minas Gerais, Escola de Veterinária,
Departamento de Zootecnia. Avenida Antônio Carlos, n° 6627, CEP - 30161-970,
Minas Gerais, Brazil

1. Introduction

Larviculture of tilapia, *Oreochomis niloticus*, is performed in several regions of the world with various production systems being used with success. Such production systems include fertilized earthen pond systems, hatchery systems, hapas installed inside earthen ponds, hapas installed inside greenhouses, concrete tanks installed inside greenhouses with water heating, green water systems, recirculating aquaculture systems, aquariums in hatcheries or small tanks and biofloc systems (Gall and Bakar 1999, Dan and Little 2000, Sanches and Hayashi 2001, El-Sayed 2002, Little et al. 2003, El-Sayed and Kawanna 2004, Little and Edwards 2004, El-Sayed 2006, El-Sayed and Kawanna 2008, Luz et al. 2012b, Nasr-Allah et al. 2014, Ferdous et al. 2014, Hui et al. 2019b, Mirzakhani et al. 2019). However, nutritional needs and management for the different types of systems can vary depending on growing conditions and must be constantly improved to obtain more sustainable larviculture. This, in turn, promotes animal welfare and more efficient production in terms of the quantity and quality of animals for success in later stages of the tilapia production chain.

2. Sex inversion of tilapia larvae

The use of monosex populations in fish production can provide one or more of the following advantages: high growth rates, control of unwanted reproduction, reduced sexual/territorial behaviors, reduced variation in lot size at the end of production and reduced risk of environmental impacts due to the escape of exotic fish (Beardmore et al. 2001). Individuals of male monosex populations have higher somatic growth rates because most of the energy that would be needed for gonadal development and reproduction can be stored (Gale et al. 1999).

*Corresponding author: luzrk@yahoo.com

Masculinization of tilapia larvae is advantageous because males can grow 20% faster than females (Golan and Levavi-Sivan 2014) and have more uniform lots in terms of size (Ugonna et al. 2018). Several studies have confirmed improved responses in growth, such as weight gain and specific growth rate, with male monosex lots of tilapia (Chakraborty and Banerjee 2010, Siddik et al. 2014, Singh et al. 2017) compared to mixed-sex lots.

There are two techniques for the direct production of monosex tilapia - sexual inversion by hormonal induction, which is the most widely used technique in commercial production, and masculinization by water temperature manipulation. There are yet other masculinization techniques in addition to these, such as genetic manipulation, which is performed prior to or soon after fertilization.

2.1. Masculinization by hormonal induction

Masculinization of larvae through hormonal induction has been performed with a large number of fish species and is considered even more efficient for cichlids due to their early sexual differentiation (Phelps and Popma 2000). The hormones that induce sexual inversion allow genetic females to phenotypically express male characteristics and genetic males to remain phenotypically male (Pandian and Varadaraj 1988).

There are numerous studies on the use of steroidal and non-steroidal hormones during gonadal differentiation of tilapia larvae (Abdelghany 1996, Abucay and Mair 1997, Beardmore et al. 2001, Bombardelli and Hayashi 2005, Dan and Little 2000, Desprez et al. 1995, 2003, El-Sayed et al. 2012, Gale et al. 1999, Johnstone et al. 1983, Kamaruzzaman et al. 2009, Leonhardt et al. 1999, Melo et al. 2019, Sayed et al. 2016, Wassermann and Afonso 2003), some of which were published as early as the 1960s and 1970s (Clemens and Inslee 1968, Eckstein and Spira 1965, Nakamura 1975). However, administration protocols vary widely, such as the type, nature, dosage and route of administration of the hormone used, period of initiation and duration of treatment, stocking densities of the larvae and the management practices adopted for these procedures, which can result in different responses and influence the success of masculinization.

17α-methyltestosterone is classified as a synthetic steroid. It is a methylated derivative of the natural hormone testosterone (El-Greisy and El-Gamal 2012, Piferrer and Donaldson 1991) and is the most commonly used hormone for sexual inversion of several species of fish (Devlin and Nagahama 2002, Sayed et al. 2016). The most common administration procedure, and with the best results (El-Sayed 2006), is oral administration after dissolving in alcohol and mixing with larval diet (Beardmore et al. 2001). A less costly technique than diet supplementation with hormone (Pandian and Sheela 1995) is administration by immersion baths, which involves periodic or continuous exposure of larvae in solution containing the masculinizing hormone.

Studies involving the administration of 17α-methyltestosterone in diet and/or by immersion bath, including dosages, duration of treatment and the percentage of masculinization obtained, are provided in Tables 1 and 2. Studies using other hormones, such as 17α-methydihydrotestosterone and 17α-ethynyltestosterone (Wassermann and Afonso 2003) and 11α-hydroxyandrostenedione (Desprez et al.

2003) for the sexual inversion of *O. niloticus* and Florida red tilapia, respectively, are also provided in Tables 1 and 2.

The hormones 17α-methydihydrotestosterone and 17α-methyltestosterone are of restricted use in the USA (Wassermann and Afonso 2003) and the use of synthetic androgens is prohibited in some European countries due to the impact they can have on the environment by leaving residues (Desprez et al. 2003). Some studies with tilapia have highlighted that, in addition to environmental impacts, the use of synthetic hormones such as 17α-methyltestosterone can also be harmful to the health of the animals themselves, with possible genotoxic effects on red blood cells (Sayed et al. 2016), histological changes in the liver (Gayão et al. 2013) and the presence of hormonal residues in the muscle of masculinized tilapia (Chu et al. 2006). Thus, the use of natural androgens, such as 11α-hydroxyandrostenedione (Desprez et al. 2003), and the use of plant extracts (Felix and Oscar 2019, Gabriel et al. 2017, Ghosal et al. 2016, Ghosal and Chakraborty 2014, Mukherjee et al. 2018, Sani et al. 2019, Ugonna et al. 2018) have been presented as alternatives to alleviate these problems while being effective at masculinization.

2.2. Masculinization by manipulating water temperature

Sex determination in tilapia is genetically controlled by sex and autosomal chromosomes; however, environmental factors such as temperature can strongly influence this determination (D'Cotta et al. 2001a, Kwon et al. 2002). High water temperatures (between 32 and 36.5 °C) during early larval development in tilapia have been shown to significantly increase the proportion of males in a population (Baroiller et al. 1995, 2009), as found for *O. niloticus* (Abucay et al. 1999, Baras et al. 2001, Borges et al. 2005), *O. aureus* (Desprez and Mélard 1998) and *O. mossambicus* (Wang and Tsai 2000). However, water temperature must be manipulated before and during gonadal sex differentiation (Borges et al. 2005), that is, beginning about 10 days post-fertilization (dpf) and lasting for at least 10 days (Baroiller et al. 2009). The ideal period for Nile tilapia is 10 to 30 dpf (Nivelle et al. 2019).

The use of high temperatures (close to 37 °C) during the first weeks of exogenous feeding of *O. niloticus* larvae resulted in greater than 90% males (Baras et al. 2001). Maintenance of Chitralada strain tilapia in water at a temperature of around 35 °C from the 10th day after hatching resulted in about 72% males, while the control group (temperature around 27 °C) had about 62% males (Borges et al. 2005).

Temperature acts on the tilapia masculinization process via a cascade of sexual differentiation that prevents ovarian differentiation and redirects testicular development (D'Cotta et al. 2001a). The structure and functions of many proteins and other macromolecules are strongly influenced by changes in temperature, allowing sexual determination as males or females (Devlin and Nagahama 2002). This phenomenon may be related to an unknown gene, called MM20C, whose expression is increased when elevated temperature stimulates the masculinization of gonads of *O. niloticus* (D'Cotta et al. 2001a). Another study with *O. niloticus* larvae evaluated the role of a protein known as cyp19a1a, whose decrease or inhibition in expression prevents the conversion of androgens to estrogens and, consequently, blocks ovarian

Table 1. Masculinization in tilapia larvae by inclusion of the inductor in the diet. Inductors, concentrations in the diet, duration of treatment and male percentages (Adapted from El-Sayed 2006)

Species	Inductor	Concentration in diet	Duration (days)	Male (%)	References
Oreochromis sp.	17α-methyltestosterone	60 mg/kg	65	92 (RAS[1])	David-Ruales et al. 2019
Oreochromis sp.	17α-methyltestosterone	60 mg/kg	65	66 (BFT[2])	David-Ruales et al. 2019
O. niloticus	*Buccholzia coriacea* seed	4 g/kg	30	83.33	Felix and Oscar 2019
O. niloticus	Fadrozole nanoparticles	350 and 500 ppm	30	100	Joshi et al. 2019
O. niloticus	*Mucuna pruriens* seed	0.2 g/kg	30	93.79	Mukherjee et al. 2018
O. niloticus	*Asparagus racemosus* roots	0.2 g/kg	30	92.24	Mukherjee et al. 2018
Red tilapia	17α-methyltestosterone	50 mg/kg	30	100	Singh et al. 2018
O. niloticus	*Carica papaya* seed meal	4.27 g/kg	28	82.19	Ugonna et al. 2018
Red tilapia	17α-methyltestosterone	50 mg/kg	30	100	Basavaraja and Raghavendra 2017
O. niloticus	*Aloe vera* powder	4% in diet	30	67.62	Gabriel et al. 2017
O. niloticus	Dried carp testes	100% in diet	30	88.29	Shrivastav et al. 2016 and 2017
O. niloticus	*Tribulus terrestris* seed extract	15 g/kg	30	76.6	Ghosal et al. 2016
O. niloticus	17α-methyltestosterone	60 mg/kg	21	93.3	Jensi et al. 2016
O. mossambicus	Letrozole + 17α-methyltest.	100 mg letroz. + 25 mg 17α-methyltest			Das et al. 2012
O. niloticus	17α-methyltestosterone	60 mg/kg	28	97	El-Sayed 2012
O. niloticus	17α-methyltestosterone	1200 mg/kg	28	>95	Phelps and Okoko 2011
O. niloticus	17α-methyltestosterone	60 mg/kg	28	98	Moreira et al. 2010
O. niloticus	17α-methyltestosterone	5 mg/kg	21	75	Kamaruzzaman et al. 2009
O. niloticus	17α-methyltestosterone	50 and 60 mg/kg	30	96.66	Rouf et al. 2008

(Contd.)

Table 1. (*Contd.*)

Species	Inductor	Concentration in diet	Duration (days)	Male (%)	References
O. niloticus	17α-methyltestosterone	60 mg/kg	30	98.73	Neumann et al. 2009
Red tilapia	17α-methyltestosterone	60 mg/kg	30	89.46	Neumann et al. 2009
O. niloticus	17α-methyltestosterone	60 mg/kg	29	100	Meurer et al. 2008
O. niloticus	17α-methyltestosterone	60 mg/kg	30	98.33	Tachibana et al. 2008
O. niloticus	17α-methyltestosterone	70 mg/kg	25	95.4	Mateen and Ahmed 2007
O. niloticus	17α-methyltestosterone	50 mg/kg	23	100	Bhandari et al. 2006
O. niloticus	17α-methyltestosterone	60 mg/kg	33	100	Boscolo et al. 2005a
O. niloticus	17α-methyltestosterone	60 mg/kg	28	100	Boscolo et al. 2005c
Red tilapia	11α -hydroxyandrostenedione	50 mg/kg	28	100	Desprez et al. 2003

[1]Recirculation Aquaculture System. [2]Biofloc Technology.

Table 2. Masculinization in tilapia larvae by immersion bath. Inductors, concentrations in water, duration of treatment and male percentages (Adapted from El-Sayed 2006)

Species	Inductor	Concentration in water	Duration	Male (%)	References
Tilapia zilli	*Nigella sativa* seed extract	0.15 g/L of extract	2 h once a week	86.67	Sani et al. (2019)
Oreochromis spp.	17α-methyltestosterone	300 μg/L	12 h	90	Singh et al. (2018)
O. niloticus	*Tribulus terrestris* seed extract	0.15 g/L of extract	4 times (once weekly) during 30 days	81.4	Ghosal et al. (2016)
O. niloticus	*Basela alba* leaf extract	0.1 g/L	4 times (once weekly) during 30 days	70.3	Ghosal and Chakraborty (2014)
O. niloticus	17α-methyltestosterone	6 mg/L	3 h (1st bath, 6 DAH[1]) 3 h (2nd bath, 10 DAH[1])	64.46	Dias-Koberstein et al. (2007)
O. niloticus	17α-methyltestosterone	2 mg/L	36 h	85.19 (15 DAH[1])	Bombardelli and Hayashi (2005)
O. niloticus	17α-methyltestosterone	1800 μg/L	4 h	86	Wassermann and Afonso (2003)
O. niloticus	17α-methyldihydrotestosterone	1800 μg/L	4 h	90	Wassermann and Afonso (2003)
O. niloticus	17α-ethynyltestosterone	1800 μg/L	4 h	86.7	Wassermann and Afonso (2003)

[1]Days After Hatching.

differentiation, allowing testicular development (Wang et al. 2017). As a result, these authors found a decrease in the expression of this protein in females and a less decrease in expression in males, both submitted to treatments with increased water temperature.

Although several studies have confirmed the effects of water temperature on sexual determination in tilapia (Abucay et al. 1999, Baras et al. 2001, Baroiller et al. 1995, Borges et al. 2005, D'Cotta et al. 2001a, b, Desprez and Mélard 1998, Nivelle et al. 2019, Shen and Wang 2014, Vernetti et al. 2013, Wang et al. 2017, Wang and Tsai 2000), the technique is influenced by many variables, and there are simpler masculinization techniques that present better results (Fuentes-Silva et al. 2013).

2.3. Masculinization by genetic manipulation – Triploidy

The triploidy technique consists of the production of sterile fish (Piferrer et al. 2009), with the aim of avoiding reproduction in the production environment and improving growth and carcass quality (Peruzzi et al. 2007, Piferrer et al. 2009). This technique has been generating interest, especially in tilapia production, since females spend a large portion of their energy on egg production and not on somatic growth (Mair 1993, Tebaldi and Junior 2009, Tiwary et al. 2004).

The process of triploidy induction begins soon after ovulation and sperm entry into the oocyte (El-Sayed 2006). The second meiotic division of the fertilized egg is suppressed (Maxime 2008) through physical shock treatments, such as by temperature (hot or cold) [in *O. aureus* (Byamungu et al. 2011), in *O. niloticus* (Hussain et al. 1991, Razak et al. 1999) and in hybrid *O. mossambicus* × *O. niloticus* (Pradeep et al. 2012)] or by hydrostatic pressure [in *O. niloticus* (El-Gamal et al. 1999, Hussain et al. 1991), in *O. mossambicus* (Varadaraj and Pandian 1988)]. Chemical induction treatments [in *O. niloticus* (El-Gamal et al. 1999), in *O. mossambicus* (Varadaraj and Pandian, 1988)] and, more recently, electric induction [in hybrid *O. mossambicus* × *O. niloticus* (Hassan et al. 2018)] are other treatments for generating triploidy.

Physical shock treatments (temperature and hydrostatic pressure) are more widely used for triploidy than chemical treatments (Teskeredzic et al. 1993), with temperature shocks being the most common, least expensive and most applicable to large scale fish production with satisfactory results (Tiwary et al. 2004). Despite being very effective, shock treatments by hydrostatic pressure are more expensive because they require adequate equipment and are applicable only to small and medium scale commercial production (Teskeredzic et al. 1993). Triploidy induction by electric shock has recently been presented as a viable alternative, but it remains a virtually unexplored technique (Hassan et al. 2018), especially in the production of tilapia.

Triploidy is considered a more economical, practical and effective method for large-scale production of sterile fish (Maxime 2008). This method may be very important for controlling reproduction in tilapia, but conclusions on whether it can be used to improve growth performance of these fish are difficult to make (El-Sayed 2006).

3. Androgenesis

Androgenesis is a genetic manipulation technique that results in exclusively paternal DNA in a diploid organism (Horváth and Orban 1995, Myers et al. 1995). This production of homogametic super-males (YY genotype) occurs with species for which sex determination in males is XY, as is the case with the *O. niloticus* and *O. mossambicus* (Marengoni and Onoue 1998). A cross between a super-male (YY) and a normal female (XX) will produce 100% normal male offspring (XY) (Marengoni and Onoue 1998, Shelton 2002).

The process of androgenesis occurs by inactivating the genetic material of females by ultraviolet (UV), gamma or X ray irradiation (Karayücel et al. 2002, Komen and Thorgaard 2007). Subsequent fertilization of the oocyte of a female whose DNA was denatured by sperm of a normal male forms an egg. Physical shock treatments of temperature or pressure are then applied to the egg during the first mitotic division (Karayücel et al. 2002) and cause the sperm's genetic material to duplicate and form a diploid nucleus, thus producing animals with a totally paternal DNA. Temperature shocks are simpler to perform than hyperbaric shocks, which are more difficult to standardize and apply to large quantities of eggs; temperature shocks of up to 42 °C can be used for tilapia (Komen and Thorgaard 2007).

4. Nutrition in tilapia larviculture

Nutrition in tilapia larviculture has been studied for decades because it is fundamental for successful production. In this sense, the ability to use formulated diets beginning with the first feeding has facilitated the larviculture of this species. This has also enabled studies to develop diets with better quality in relation to protein and energy levels, as well as to search for alternative foods that can favor good development and lower cost. During this phase, the optimum temperature for best growth, feed conversion and survival is 28 °C (El-Sayed and Kawanna 2008).

The use of diets for the first exogenous feeding is made possible by the enzymatic system present in the larvae even before the first feeding, which is important for the digestion and absorption of nutrients in food. The enzymes maltase, leucine aminopeptidase, dipeptidyl aminopeptidase IV, lipase, non-specific esterases, and intestinal alkaline phosphatase are present in the intestine of tilapia larvae before the start of exogenous feeding. However, only lipase was detected at three days post-hatching. Besides, due to the localization of the different enzyme in the intestine, undifferentiated enterocytes may be involved in enzyme production, especially peptidase and alkaline phosphatase, before the first exogenous feeding (Tengjaroenkul et al. 2002).

Table 3 summarizes the requirements for Nile tilapia larvae. Tilapia larvae fed diets containing 3,000, 4,000 or 5,000 kcal of crude energy (EC)/kg of diet and 30, 35, 40, 45 or 50% crude protein (CP) for four weeks showed lower growth rates and high mortality in low energy diets, regardless of the level of protein. There was no improvement in growth and feed conversion rates for 50% protein at any energy level. A positive correlation was found between protein levels and body lipid content, with diets with 45% CP and 4000 kcal of energy being recommended (El-Sayed and Teshima 1992). Next to that recommendation, greater growth and a better feed conversion ratio were found for tilapia larvae fed diets containing 40% of digestible protein (DP) for 98 days, while a worse feed conversion ratio was obtained for 20% DP. With respect to protein efficiency rate, there was a decrease as CP levels increased from 20 to 50% (Siddiqui et al. 1988).

Likewise, for isoenergetic (4000 kcal/kg of diet), isocalcic and isophosphoric diets with 32.7, 37.0, 41.3, 45.7 and 50.0% CP, the best weight obtained after 28 days of feeding was with 38.56% DP. Survival rate decreased with increasing levels of DP, which can be explained by increased excretion of nitrogenous residues due to the use of diets with higher levels of protein, and thus the ratio of 105 mg of DP/kcal of digestible energy (DE) is recommended (Hayashi et al. 2002).

The use of diets containing 25, 35, or 45% CP for larvae fed to satiety twice a day for six days a week did not affect survival. However, greater growth and a better feed conversion ratio were achieved with 45% CP, while less growth, a higher hepatosomatic index, lower protein content and higher lipid content in the carcass were observed with the use of 25% CP. In addition, all physiological variables and activities of the enzymes aspartate aminotransferase and alanine aminotransferase in serum, liver and muscle increased significantly with increasing levels of protein in the diet, showing the importance of protein to the physiology of larvae.

In addition to performance and survival, another way of assessing animal quality as a function of diet is the use of stress resistance tests. In this sense, Nile tilapia larvae were fed diets containing 32, 40 and 55% CP. Diets containing 32 and 40% CP were not specific for larvae, so they were ground to present an appropriate pellet size similar to the 55% CP diet (commercial diet). After 30 days of feeding, the animals were submitted to the air exposure test for 7 and 10 minutes. The lowest resistance to stress due to air exposure was for animals who were fed the diet containing 55% CP. For the levels of 32 and 55%, there was a reduction in the resistance to stress with increased time of exposure to air, confirming that 40% CP in the diet may be the best for this phase. According to the results of resistance to stress, the variables of weight and length were greater at the end of larviculture for larvae fed with 40% CP (Luz et al. 2012a).

As well as assessing growth performance and resistance to stress, understanding the functioning of the enzyme system during larviculture, depending on the levels of crude protein in the diet, can also help in the search for more efficient management at this stage. Using diets with four levels of CP (30, 36, 42 and 48%) and 3400 kj/g of diet of DE for the GIFT strain did not affect survival rates. After 10 days of feeding, length and weight were reduced and the relative expression of pepsinogen was higher in fish fed 30% CP. However, at the end of the experiment (30 days of feeding), weight was higher for diets with 30% and 42% CP. Relative expression of chymotrypsinogen, carboxypeptidase, NPY and CCK were higher in fish fed 42% CP. No differences were registered in the relative expression of trypsinogen and lipase in the first 30 days of exogenous feeding. The activities of digestive enzymes, such as trypsin, pepsin and chymotrypsin, were similar among diets on both 10 and 20 days of feeding, whereas at 30 days of feeding, trypsin activity was higher for 30 and 42% CP while pepsin activity was less for 30% CP. The 42% CP level showed the highest chymotrypsin activity, which demonstrated variation in enzyme activity during the first 30 days of feeding due to CP level. At the end of the 30 days, the test of resistance to stress due to exposure to air for 7 minutes revealed that fish fed 36% CP had a higher survival rate (Silva et al. 2019). In this sense, the authors related that commercial larviculture uses the same protein level throughout the first 30 days and recommended a new strategy using 36 to 48% CP for the first 10 days of exogenous feeding, 30 and 48% CP for days 11 to 20, and levels between 30 and 42% CP for days 21 to 30.

Although these studies demonstrate that diets with more than 50% CP are not efficient for Nile tilapia larvae, feed industries continue to offer such diets to producers. These diets can lead to poor performance and worsening water quality due to excess water protein and greater excretion of nitrogenous products, which harm the environment, in addition to producing animals less resistant to stress as previously demonstrated.

In addition to protein, energy requirement is also important for good development of larvae. An energy/protein ratio of 3.300 kcal of DE/kg of diet and 85.49 mg of PD/kcal of DE is recommended since higher levels of energy can lead to worse performance due to a low rate of food intake due to the greater inclusion of soy oil (Boscolo et al. 2005c).

Table 3. Protein requirements of Nile tilapia larvae

Nile tilapia	Average weight (g) or age	Requirement (%)	References
Larvae	0.012	45% CP[1]	El-Sayed and Teshima (1992)
	0.838	40% CP	Siddiqui et al. (1988)
	Three days post-hatch	38,6% DP[2]	Hayashi et al. (2002)
	0.4 – 0.5	45% CP	Abdel-Tawwab et al. (2010)
	0.010	40% CP	Luz et al. (2012)
	0.010	36 to 48% CP - First 10 days of initial exogenous feeding	Silva et al. (2019)
	0.26	30 to 42% CP - 11 to 30 days of initial exogenous feeding	Silva et al. (2019)

[1] Crude protein
[2] Digestible protein

Knowledge of vitamin and mineral requirements is also essential in this phase of tilapia larviculture. Vitamin C is important as an antioxidant and improves stress resistance, which is a fundamental factor especially during the early stages of fish life. During the sexual inversion phase, estimated optimum values for vitamin supplementation for diets with 40% CP and 3,600 kcal of DE/kg of diet were: 859.5 mg for weight, 765.0 mg for length and 685.7 mg of vitamin C for survival rate. Regarding sexual inversion, the different levels of vitamin C supplementation had no effect on the masculinization rate, which was estimated to be between 89 and 91% (Toyama et al. 2000). The estimated recommendation for mineral and vitamin supplementation for better performance of GIFT strain larvae is 1.88% (Pessini et al. 2015).

The use of the enzyme phytase (BASF-Natuphos 5000®) for Nile tilapia larvae during sexual inversion, at levels of 500, 1000, 2000 or 4000 FU/kg of diet with diets containing 3143 kcal of DE and 30% CP, did not lead to differences in growth performance or survival. However, the percentage of calcium in the carcass increased with increasing phytase levels, and there was a better percentage of phosphorus in the carcass for an estimated level of 1990 UF/kg of diet (Furuya et al. 2004).

Another important point to be considered for tilapia larviculture is the use of probiotics and prebiotics. The addition of 0.1% probiotic (commercial product guaranteeing 1010 live cells of *Saccharomyces cerevisiae* per gram) during sexual inversion did not improve performance, survival or the hepatosomatic index of the animals. However, bacteria of the genus *Bacillus* sp. were present in the intestine of larvae with the use of probiotics, while bacteria of the genera *Bacillus* sp. and *Enterococcus* sp. were present in the animals fed a diet without probiotics (Meurer et al. 2008). The use of yeast (*S. cerevisiae*, BFP, Dock Road, Felixstone, UK) at 1.0–5.0 g/kg of diet resulted in better growth, no difference in survival rate after 12 weeks of feeding, a greater number of red blood cells and higher hematocrit values (Abdel-Tawwab et al. 2008).

The use of mananoligosaccharides (MOS; commercial product SAF-Mannan® containing 23% β-glucan and 21% α-mananes) as a prebiotic, in the proportions of 0, 0.15, 0.30, 0.45, 0.60 and 0.75% in the diet and replacing cornmeal, had no effect on performance, survival, moisture, protein, ether extract, ash and calcium and phosphorus content in tilapia carcasses during masculinization. However, the best feed conversion ratio was estimated at 0.34% MOS. In addition, the use of MOS had a positive influence on intestinal villus density and length, resulting in a mucosa with greater integrity, which may have been due to reduced colonization by bacteria that act by inhibiting their adherence to the enterocyte through the connection with the glycocalix (Schwarz et al. 2011).

Alternative foods have also been evaluated for Nile tilapia larviculture. Diets with 20, 40 or 50% of cowpea protein concentrate had higher survival rates (96%); however, 20 and 30% of the protein concentrate had greater growth, indicating greater nutritional value for Nile tilapia larvae. Increased cowpea protein concentrate in the diet also reduced the amount of water in the carcass, while protein content was improved (Olvera-Novoa et al. 1997). The use of different vegetable protein (soybean meal and corn gluten) and supplementation with lysine and methionine or calcium and phosphorus in order to obtain isoaminoacidic, isocalcic and isophosphoric diets produced no improvement in performance during sexual inversion compared to diets with different animal protein (fish meal and poultry viscera meal) (Meurer et al. 2005). However, the inclusion of up to 40% of poultry viscera meal, replacing fish meal, was effective in sexual inversion, and led to better performance, which may have been related to the increase in the amount of lipids and the decrease in the amount of starch in the diets (Boscolo et al. 2005b).

The results presented here show the possibilities of using different dietary ingredients in tilapia larviculture. However, there remains a wide variety of products and by-products that must be evaluated for the development of diets that cost less, are more efficient for animal development and cause less pollution to the environment. Despite the ease of using formulated diets for Nile tilapia larviculture, supplementation with natural foods may provide additional alternatives. The use of the microalgae *Spirulina platensis*, either in natura or bioencapsulated in copepods, during sexual inversion resulted in better growth than did diet formulated for Nile tilapia larvae, and did not affect survival and sexual inversion rates (greater than 90% masculinization) (Moreira et al. 2010). In fact, *S. platensis* was more accepted by tilapia larvae than were *Euglena gracilis* and *Chlorella vulgaris* (Lu et al. 2004). However, tilapia larvae grown with *Chlorella* sp. in formulated diets experienced better performance than when in a green water system of a fish farm or when fed a clear water diet (Costa et al. 2011). In addition to microalgae, *Artemia* nauplii and the combination of *Chlorella* + *Artemia* nauplii + formulated diet with 50% CP are also good alternatives for better growth and survival in tilapia larviculture (Al-Shamsi et al. 2006).

Therefore, it is evident that diets formulated for Nile tilapia larviculture can be supplemented with live food resulting in improved final productivity.

In addition to nutrition, there are other important management practices to be considered for successful Nile tilapia larviculture.

5. Management during tilapia larviculture

Various management practices adopted during larviculture can affect growth, survival and water quality, and thus must be taken into account for successful production of fingerlings. Among management practices, feeding level, stocking density, water salinity, feeding frequency, photoperiod and luminosity, as well as biofloc technology, deserve special attention.

5.1. Feeding management

Feeding during the early stages of development of Nile tilapia is facilitated by the ability to use commercial diet beginning with the first feeding, which also facilitates management. However, differences among several studies in feeding management practices have been recorded: 30-45% of body weight (Santiago et al. 1987), 30% of body weight (El-Sayed 2002), or 30-20% of body weight (El-Sayed and Kawanna 2004). Some studies reduce the amount during the larviculture phase, such as initial feeding at 25% of body weight and gradually reducing to 10% (Ferdous et al. 2014), or a feeding at 20% of body weight for the first 10 days and 10% for the subsequent 20 days (Silva et al. 2019).

5.2. Stocking density

Adequate stocking density is fundamental to successful larviculture because it maximizes growth, water use and production structure. A comparison of densities of 2 and 10 larvae/liter during the sexual inversion phase in hapas installed in a green water system revealed similar survival rates, but reduced weight, length and feed conversion ratio of the larvae of the higher stocking density (Sanches and Hayashi 1999). A similar result was found, also for the sexual inversion phase, in net cage systems with densities of 1, 3 and 5 larvae/L (Vera Cruz and Mair 1994). Variables such as survival, masculinization efficiency and condition factor were found to be similar among stocking densities of 1, 3, 5 and 7 larvae/L for 30 days in a recirculation aquaculture system, but the higher stocking densities led to less larvae growth (Tachibana et al. 2008). Despite reduced growth, higher densities can result in greater production per area (Sanches and Hayashi 1999, Tachibana et al. 2008). Survival was also found to be similar for densities between 3 and 20 larvae/L in a recirculation aquaculture system, but there was also reduced growth with increasing density, except for densities of 3 and 5 larvae/L, which did not differ (El-Sayed 2002).

The use of small (30 × 15 × 30 cm) or large (30 × 30 × 30 cm) compartments and stocking densities of 20, 60, 120 and 200 larvae/L for 56 days with continuous flow of water did not affect larval body size, despite reduced dissolved oxygen levels with increasing density (Gall and Bakar 1999). Densities of 1, 10, 20 and 30 larvae/L also did not affect performance of Nile tilapia larvae after 28 days of exogenous feeding while being reared in water salinity of 2 g/L, despite increases in turbidity and total water ammonia, which are directly related to increased stocking density (Luz et al. 2012b).

Stocking density is an important consideration of management, mainly with regards to maximizing production area and feasibility of intensive cultivations. However, the different results of the studies presented here may be due to the different cultivation systems used and the adopted management practices, which makes the search for greater productive efficiency necessary. Nonetheless, it is clear that production can be intensified without loss of performance.

5.3. Water salinity

Salinity has been an important aspect of freshwater fish management because it is related to osmoregulation and improves the conditions of the farming environment. It has also been widely used in the transport of live animals and for the prevention, control and treatment of some diseases.

The mechanisms of tolerance of freshwater larvae to different gradients of salinity are still poorly understood. However, the drinking rate for *O. mossambicus* in the early stages of life increased from 2 to 10 days after hatching, while larvae in seawater (33.5 g of salt/L) imbibed more water than those in freshwater. In addition, water permeability was lower in embryos and larvae adapted to seawater, while chloride ion turnover rates in seawater were 50 to 100 times higher than in freshwater (Miyazaki et al. 1998).

The optimal temperature/salinity/pH combination for Nile tilapia fertilization was found to be 27.6 °C/9.3 ppt/7.5, and for hatching 27.1 °C/9.2 ppt/7.4 (Hui et al. 2014), while the temperature/salinity combination of 28–30 °C and 4–6 ppt was recommended for the period of sac fry rearing (Hui et al. 2015). Reduced hatching rates were registered at 15 and 20 ppt, and survival at yolk-sac absorption was inversely related to increasing salinity with higher mortality with salinities of 15, 20 and 25 ppt. Besides, oxygen consumption of larvae between three- and six-days post-hatching in freshwater was always higher than those in 7.5–25 ppt. Salinity also had an inverse effect on larval standard length, producing shorter larvae from hatching until six days post-hatching (Fridman et al. 2012). However, exposure of newly hatched Nile tilapia larvae (up to 24 h post-hatching) to 4 and 6 g of salt (NaCl)/L for 72 h caused greater osmoregulation expenditure, morphological alterations to yolk-sac surface and histological damage to skeletal muscle compared to freshwater or 2 g of salt (NaCl)/L (Melo et al. 2019).

Nile tilapia larviculture performed with a salinity of 25 g of salt (NaCl)/L and using acclimatization with 5 g of salt (NaCl)/L per day for 28 days resulted in sexual inversion failure, probably because of the stress caused by acclimatization and the high salinity, which may have inhibited masculinization (Moreira et al. 2011). The direct transfer of larvae at the time of the first feeding to different salinities (up to 6 g of salt (NaCl)/L) led to total mortality in the maximum salinity of 6 g of salt (NaCl)/L before 10 days of feeding (Luz et al. 2013). Furthermore, according to these authors, performance and survival were lower for 4 g of salt (NaCl)/L than for fresh water and to 2 g of salt (NaCl)/L, which were similar at the end of 30 days of feeding. However, transferring freshwater larvae of hybrid red tilapia, *Oreochromis* sp., to different salinities found acute exposure to 20 g of salt (NaCl)/L to be lethal due to high metabolic cost, while 55.6% of larvae survived 24 h of exposure to 16 g/L (Rahmah et al. 2020). All larvae survived 30 days of exposure to salinities of 0 to 12 g/L with similar metabolic oxygen consumption and growth performance.

Studies have shown wide variation in the tolerance of Nile tilapia larvae to different salinity gradients, which may be due to different Nile tilapia strains (Basiao et al. 2005, Oca et al. 2015). In addition, different forms of acclimatization need to be better evaluated for the initial stages of tilapia cultivation with the aim of reducing osmotic shock, since early exposure to low salinity could confer pre-adaptation for exposure to higher salinities for further growth (Fridman et al. 2012).

5.4. Feeding frequency

The correct feeding frequency is important for reducing diet loss and improving the efficiency of the use of feed. Growth during the sexual inversion phase in hapas installed in green water was reduced with a feeding frequency of twice a day, with feed frequencies of four to five times a day being the most adequate (Sanches and Hayashi 2001). Another study with tilapia larvae grown in hapas obtained best growth and survival with a feeding frequency of five times a day (Ferdous et al. 2014). A feeding frequency of three times a day for 60 days in an indoor recirculation system was successful (El-Sayed and Kawanna 2004). However, using automatic feeders for larvae stored in hapas and comparing three feeding frequencies – 12 times/day, once an hour; 24 times/day, once every 30 minutes; and 48 times/day, once every 15 minutes – revealed the best treatment to be with the feeding frequency of 48 times/day (Schäfer 2015).

5.5. Photoperiod and luminosity

Photoperiod and luminosity are directly related to the biological rhythms of animals. Tilapia eggs and fry are responsive to light and are influenced by light and photic phase (Hui et al. 2019a, b). The optimal photoperiod/intensity combination for hatching was 13.45 h/1020 lx (12.85 μmolm^{-2} s^{-1}) (Hui et al. 2019a), while a photoperiod of >18 h resulted in the best growth for larvae of Nile tilapia (GIFT strain, 7 dph); however, the optimum combination of 14–16 h/1650–1900 lx (20.79–23.94 μmolm^{-2} s^{-1}) was better for larviculture (Hui et al. 2019b). These authors also state that continuous lighting is not recommended. The best growth for Nile tilapia larvae after 60 days was with 24L:0D and 18L:6D with a constant light intensity of 2500 lux, with18L:6D being recommended; however, the performance of fingerlings reared for 90 days was not affected by photoperiod (El-Sayed and Kawanna 2004). In a similar study, opposite results were observed during the sexual inversion phase (28 days) with a photoperiod of 12.5 light having better results, whereas in the post inversion phase (more than 46 days), longer photoperiods promoted greater growth of chitralized tilapia fry (Bezerra et al. 2008). These results indicate that the best photoperiod varies with the development of fish (El-Sayed and Kawanna 2004, Bezerra et al. 2008).

6. Biofloc technology

Biofloc technology (BFT) has been gaining more and more prominence in the production of tilapia, as well as in larviculture. The daily net uptake of microbial protein by tilapia from biofloc suspension is about 25% of the normal protein ration given to this species (Avnimelech and Kochba 2009). Biofloc also contains bioactive components such as essential fatty acids, vitamins, minerals, carotenoids and phytosteroids (Toledo et al. 2016).

The application of BFT to Nile tilapia broodstock tanks and during Nile tilapia larviculture can improve the quality and performance of produced larvae compared to larvae originating from the control broodstock tanks (without the addition of molasses) (C). Post-challenge (*Streptococcus agalactiae* suspension) survival of tilapia larvae originating from BFT broodstock was higher than that of larvae originating from C broodstock. Besides, survival of larvae from C broodstock and larviculture in BFT was 36% higher than that of larvae from the C broodstock and larviculture in water (without the post-challenge addition of molasses). Larvae originating from BFT broodstock had higher survival of a salinity stress test (35 g/L NaCl) than did larvae from C broodstock (Ekasari et al. 2015), showing the benefits of this management practice for reproduction and larviculture.

Although larvae of BFT had greater final length, greater gain in length and a higher condition factor during sexual inversion than larvae in a recirculating aquaculture system, the percentage of males was higher in RAS (92%), since settleable solid concentrations of above 35 cm in the BFT decreased the masculinization percentage (66%) (David-Ruales et al. 2019).

Nile tilapia larvae (initial average body weight of 2.7 g) in biofloc showed an increase in weight gain from 71.8 to 319.9% greater than the control group. In addition, biofloc larvae experienced increased length and diameter of intestinal villi, which may have resulted in higher nutrient absorption and improvements in humoral immune responses (lysozyme activity, total protein, albumin, globulin and hemolytic activity) (Mirzakhani et al. 2019).

New technologies to improve the efficiency of sexual inversion with biofloc need to be evaluated since BFT provides better performance and higher animal quality, and is friendlier to the environment due to less water exchange (see Chapter 14 of the present book for further details on BFT technology).

7. Health of Nile tilapia larvae

As production systems intensify, the chance for health problems increases. In this sense, it is always important to employ prevention through good production practices, water quality maintenance and management (nutrition, stocking densities, food management and environmental conditions, etc.) that provide better performance and animal welfare. Nonetheless, it is important to know the products that are available to prevent and control diseases.

As preventive management, supplementation with different levels of baker´s yeast (*S. cerevisiae*) (0.25 to 5.0 g/kg of diet) for 12 weeks showed decreased mortality 10 days after intraperitoneal injection with the bacterium *Aeromonas hydrophila* with increased yeast supplementation. This finding indicates that, in addition to improving fish growth, supplementation with yeast also improves resistance to infection, with the optimal level of yeast in the diet being 1.0 g/kg (Abdel-Tawwab et al. 2008). Nile tilapia larvae fed with Epi-1 transgenic *Artemia* cysts acquired enhanced immunity toward bacterial challenge with Gram-positive *Streptococcus iniae* or Gram-negative *Vibrio vulnificus* (204), and so this feed should be considered a potential functional feed for tilapia larvae (Ting et al. 2018). The use of the truncated growth hormone produced in the supernatant of *Pichia pastoris* culture stimulates innate immune parameters (lysozyme and lectins) in tilapia larvae (Acosta et al. 2008). Salinity (>5 ppt) could be effective in improving the survival of embryos and larvae and help prevent parasites, bacteria and viruses in commercial hatcheries (Hui et al. 2014). However, there are some tilapia strains that are more sensitive to salinity than others.

Treatments of formalin 1:40,000 and sodium chloride 2.5 g/L plus formalin 1:40,000 eliminated 100% of the parasites (cutaneous monogenoid *Gyrodactylus* sp. and cutaneous protozoa *Trichodina* sp. and *Ichtyophthirius* sp.) of larvae of common tilapia submitted to long-term baths (10 days) (Silva et al. 2009).

The tilapia lake virus (TiLV) has become a major worldwide concern for tilapia production. Collections carried out in Thailand from 2012 to 2017 from four different hatcheries located in four provinces found the majority of tested samples of fertilized eggs, yolk-sac larvae and larvae to be TiLV positive (Dong et al. 2017). According to these authors, over 40 countries worldwide have imported tilapia larvae and fingerlings, requiring care to control and prevent further spread of TiLV.

8. Considerations and perspectives

Tilapia larviculture has been experiencing rapid growth globally to meet the demand of the growing production of the species. The constant search for new technologies and biotechnologies for more sustainable production that provides better well-being is fundamental for continuity in the availability and quantity and quality of fingerlings. In this sense, studies of nutrition, macronutrients and the use of probiotics and prebiotics that are efficient at producing fingerlings that are more resistant to post-larviculture stress (harvesting, fasting, transport and release in a new environment) are important. Another important consideration is the continued search for better health control measures in hatcheries where fingerlings are sold to different fish farms to prevent the spread of diseases, such as tilapia lake virus (TiLV), among others.

Acknowledgments

The authors are grateful to Conselho Nacional de Desenvolvimento Científico e Tecnológico (CNPq-Brazil), and Coordenação de Aperfeiçoamento de Pessoal de Nível Superior. R.K. Luz received a research grant from the Conselho Nacional de Desenvolvimento Científico e Tecnológico (CNPq No. 308547/2018-7).

References cited

Abdelghany, A.E. 1996. Effects of feeding 17α-methyltestosterone and withdrawal on feed utilization and growth of Nile tilapia, *Oreochromis niloticus* L., fingerlings. J. Appl. Aquacult. 5: 67–76.

Abdel-Tawwab, M., A.M. Abdel-Rahman and N.E.M. Ismael. 2008. Evaluation of commercial live baker's yeast, *Saccharomyces cerevisiae* as a growth and immunity promoter for Fry Nile tilapia, *Oreochromis niloticus* (L.) challenged in situ with *Aeromonas hydrophila*. Aquaculture 280: 185–189.

Abucay, J.S. and G.C. Mair. 1997. Hormonal sex reversal of tilapias: Implications of hormone treatment application in closed water systems. Aquacult. Res. 28: 841–845.

Abucay, J.S., G.C. Mair, D.O.F. Skibinski and J.A. Beardmore. 1999. Enviromental sex determination: The effect of temperature and salinity on sex ratio in *Oreochromis niloticus* L. Aquaculture 173: 219–234.

Acosta, J., Y. Carpio, V. Besada, R. Morales, A. Sánchez, Y. Curbelo, et al. 2008. Recombinant truncated tilapia growth hormone enhances growth and innate immunity in tilapia fry (*Oreochromis* sp.). Gen. Comp. Endocrinol. 157: 49–57.

Al-Shamsi, L., W. Hamza and A.-F. El-Sayed. 2006. Effects of food sources on growth rates and survival of Nile tilapia (*Oreochromis niloticus*) fry. Aquat. Ecosyst. Health Manage. 9: 447–455.

Avnimelech, Y. and M. Kochba. 2009. Evaluation of nitrogen uptake and excretion by tilapia in bio floc tanks, using 15N tracing. Aquaculture 287: 163–168.

Baras, E., B. Jacobs and C. Mélard. 2001. Effect of water temperature on survival, growth and phenotypic sex of mixed (XX-XY) progenies of Nile tilapia *Oreochromis niloticus*. Aquaculture 192: 187–199.

Baroiller, J-F., C. Fredéric and G. Eddy. 1995. Temperature sex determination in two tilapia, *Oreochromis niloticus* and the red tilapia (red Florida strain): Effect of high or low temperature. pp. 158–160. *In*: F. Goetz and P. Thomas [eds.]. The Reproductive Physiology of Fish. University of Texas, Austin, USA.

Baroiller, J.F., H. D'Cotta, E. Bezault, S. Wessels and G. Hoerstgen-Schwark. 2009. Tilapia sex determination: Where temperature and genetics meet. Comp. Biochem. Physiol. A: Mol. Integr. Physiol. 153: 30–38.

Basavaraja, N. and C.H. Raghavendra. 2017. Hormonal sex reversal in red tilapia (*Oreochromis niloticus* and *Oreochromis mossambicus*) and inheritance of body colour in *O. mossambicus* and red tilapia. Aquacult. Int. 25: 1317–1331.

Basiao, Z.U., R.V. Eguia and R.W. Doyle. 2005. Growth response of Nile tilapia fry to salinity stress in the presence of an 'internal reference' fish. Aquac. Res. 36: 712–720.

Beardmore, J.A., G.C. Mair and R.I. Lewis. 2001. Monosex male production in finfish as exemplified by tilapia: Applications, problems and prospects. Aquaculture 197: 283–301.

Bezerra, K.S., A.J.G. Santos, M.R. Leite, A.M. Silva and M.R. Lima. 2008. Crescimento e sobrevivência da tilápia chitralada submetida a diferentes fotoperíodos. Pesqui. Agropecu. Bras. 43: 737–743.

Bhandari, R.K., M. Nakamura, T. Kobayashi and Y. Nagahama. 2006. Suppression of steroidogenic enzyme expression during androgen-induced sex reversal in Nile tilapia (*Oreochromis niloticus*). Gen. Comp. Endocrinol. 145: 20–24.

Bombardelli, R.A. and C. Hayashi. 2005. Masculinização de larvas de tilápia do Nilo (*Oreochromis niloticus* L.) a partir de banhos de imersão com 17α-metiltestosterona. Rev. Bras. Zootec. 34: 365–372.

Borges, A.M., J.O.C. Moretti, C. McManus and A.S. Mariante. 2005. Produção de populações monosexo macho de tilápia-do-nilo da linhagem Chitralada. Pesqui. Agropecu. Bras. 40: 153–159.

Boscolo, W.R., C. Hayashi, F. Meurer, A. Feiden, R.A. Bombardelli and A. Reidel. 2005a. Farinha de resíduos da filetagem de tilápias na alimentação de tilápia-do-Nilo (*Oreochromis niloticus*) na fase de reversão sexual. Rev. Bras. Zootec. 34: 1807–1812.

Boscolo, W.R., F. Meurer, A. Feiden, C. Hayashi, A. Reidel and A.L. Genteline. 2005b. Farinha de vísceras de aves em rações para a tilápia do Nilo (*Oreochromis niloticus* L.) durante a fase de reversão sexual. Rev. Bras. Zootec. 34: 373–377.

Boscolo, W.R., A. Signor, A. Feiden, R.A. Bombardelli, A.A. Signor and A. Reidel. 2005c. Energia digestível para larvas de tilápia-do-Nilo *(Oreochromis niloticus)* na fase de reversão sexual. Rev. Bras. Zootec. 34: 1813–1818.

Boscolo, W.R., A. Signor, A.A. Signor, A. Feiden and A. Reidel. 2008. Inclusão de amido em dietas para larvas de tilápia-do-nilo. Rev. Bras. Zootec. 37: 177–180.

Byamungu, N., V.M. Darras and E.R. Kuhn. 2001. Growth of heat-shock induced triploids of blue tilapia, *Oreochromis aureus*, reared in tanks and in ponds in Eastern Congo: Feeding regimes and compensatory growth response of triploid females. Aquaculture 198: 109–122.

Chakraborty, S.B. and S. Banerjee. 2010. Comparative growth performance of mixed-sex and monosex Nile tilapia population in freshwater cage culture system under Indian perspective. Int. J. Biol. 2: 44–50.

Chu, P-S., M. Lopez, S. Serfling, C. Gieseker and R. Reimschussel. 2006. Determination of 17α-methyltestosterone in muscle tissues of tilapia, rainbow trout, and salmon using liquid chromatography-tandem mass spectrometry. J. Agric. Food Chem. 54: 3193–3198.

Clemens, H.P. and T. Inslee. 1968. The production of unisexual broods by Tilapia mossambica sex-reversed with methyl testosterone. Trans. Am. Fish. Soc. 97: 18–21.

Costa, F.T.M., F.R.C. Reis, J.M.S. Santos, S.M. Maciel, T.S. Biserra, R.L. Moreira, et al. 2011. *Chlorella* sp. como suplemento alimentar durante a larvicultura de tilápia do Nilo. Rev. Bras. Saúde Prod. Anim. 12: 1103–1115.

D'Cotta, H., A. Fostier, Y. Guiguen, M. Govoroun and J-F. Baroiller. 2001a. Search for genes involved in the temperature-induced gonadal sex differentiation in the tilapia, *Oreochromis niloticus*. J. Exp. Zool. 290: 574–585.

D'Cotta, H., A. Fostier, Y. Guiguen, M. Govoroun and J-F. Baroiller. 2001b. Aromatase plays a key role during normal and temperature-induced sex differentiation of tilapia *Oreochromis niloticus*. Mol. Reprod. Dev. 59: 265–276.

Dan, N.C. and D.C. Little. 2000. The culture performance of monosex and mixed-sex new-season and overwintered fry in three strains of Nile tilapia (*Oreochromis niloticus*) in northern Vietnam. Aquaculture 184: 221–231.

Das, R., M.A. Rather, N. Basavaraja, R. Sharma and U.K. Udit. 2012. Effect of nonsteroidal aromatase inhibitor on sex reversal of *Oreochromis mossambicus* (Peters, 1852). Isr. J. Aquacult. 64: 1–6.

David-Ruales, C.A., E.M. Betancur-Gonzalez and R.D. Valbuena-Villareal. 2019. Sexual reversal with 17α-methyltestosterone in *Oreochromis* sp.: Comparison between Recirculation Aquaculture System (RAS) and Biofloc Technology (BFT). J. Agric. Sci. Technol. A 9: 131–139.

Desprez, D., C. Mélard and J.C. Philippart. 1995. Production of a high percentage of male offspring with 17α-ethynylestradiol sex-reversed *Oreochromis aureus*. II. Comparative reproductive biology of females and F2 pseudofemales and large-scale production of male progeny. Aquaculture 130: 35–41.

Desprez, D. and C. Mélard. 1998. Effect of ambient water temperature on sex determinism in the blue tilapia *Oreochromis aureus*. Aquaculture 162: 79–84.

Desprez, D., E. Géraz, M.C. Hoareau, C. Mélard, P. Bosc and J.F. Baroiller. 2003. Production of a high percentage of male offspring with a natural androgen, 11ß-hydroxyandrostenedione (11ßOHA4), in Florida red tilapia. Aquaculture 216: 55–65.

Devlin, R.H. and Y. Nagahama. 2002. Sex determination and sex differentiation in fish: An overview of genetic, physiological, and environmental influences. Aquaculture 208: 191–364.

Dias-Koberstein, T.C.R., A.G. Neto, M.V. Stéfani, E.B. Malheiros, M.F. Zanardi and M.A. Santos. 2007. Reversão sexual de larvas de tilápia do Nilo (*Oreochromis niloticus*) por meio de banhos de imersão em diferentes dosagens hormonais. Rev. Acad. 5: 391–395.

Dong, H.T., G.A. Ataguba, P. Khunrae, T. Rattanarojpong and S. Senapin. 2017. Evidence of TiLV infection in tilapia hatcheries from 2012 to 2017 reveals probable global spread of the disease. Aquaculture 479: 579–583.

Eckstein, B. and M. Spira. 1965. Effect of sex hormones on gonadal differentiation in cichlid, *Tilapia aurea*. Biol. Bull. 129: 482–489.

El-Gamal, A-R.A., K.B. Davis, J.A. Jenkins and E.L. Torrans. 1999. Induction of triploidy and tetraploidy in Nile tilapia, *Oreochromis niloticus* (L.). J. World Aquac. Soc. 30: 269–275.

El-Greisy, Z.A. and A.E. El-Gamal. 2012. Monosex production of tilapia, *Oreochromis niloticus* using different doses of 17α-methyltestosterone with respect to the degree of sex stability after one year of treatment. Egypt. J. Aquat. Res. 38: 59–66.

Ekasari, J., D.R. Rivandi, A.P. Firdausi, E.H. Surawidjaja, M.J. Zairin, P. Bossier, et al. 2015. Biofloc technology positively affects Nile tilapia (*Oreochromis niloticus*) larvae performance. Aquaculture 441: 72–77.

El-Sayed, A.-F.M. and S. Teshima. 1992. Protein and energy requirements of Nile tilapia, *Oreochromis niloticus*, fry. Aquaculture 103: 55–63.

El-Sayed, A.-F.M. 2002. Effects of stocking density and feeding levels on growth and feed efficiency of Nile tilapia (*Oreochromis niloticus*) fry. Aquacult. Res. 33: 621–626.

El-Sayed, A.-F.M. and M. Kawanna. 2004. Effects of photoperiod on the performance of farmed Nile tilapia *Oreochromis niloticus*: Growth, feed utilization efficiency and survival of fry and fingerlings. Aquaculture 231: 393–402.

El-Sayed, A.-F.M. 2006. Reproduction and seed production. pp. 125–131. *In*: El-Sayed, A.-F.M. [ed.]. Tilapia Culture. CABI Publishing, Cambridge, USA.

El-Sayed, A.-F.M. and M. Kawanna. 2008. Optimum water temperature boosts the growth performance of Nile tilapia (*Oreochromis niloticus*) fry reared in a recycling system. Aquacult. Res. 39: 670–672.

El-Sayed, A-F.M., E.-S.H. Abdel-Aziz and H. Abdel-Ghani. 2012. Effects of phytoestrogens on sex reversal of Nile tilapia (*Oreochromis niloticus*) larvae fed diets treated with 17α-methyltestosterone. Aquaculture 360–361: 58–63.

Felix, E. and E.V. Oscar. 2019. Effect of wonderful Kola seed meal (*Buchholzia coriacea*) on growth, masculinizing potency and gonadal gross morphology of the Nile tilapia (*Oreochromis niloticus*, Linnaeus 1758). J. Aquacult. Fish Health. 8: 191–198.

Ferdous, Z., N. Nahar, S. Hossen, K.R. Sumi and M.M. Ali. 2014. Performance of different feeding frequency on growth indices and survival of monosex tilapia, *Oreochromis niloticus* (Teleostei: Cichlidae) fry. Int. J. Fish. Aquat. Stud. 1: 80–83.

Fridman, S., J. Bron and K. Rana. 2012. Influence of salinity on embryogenesis, survival, growth and oxygen consumption in embryos and yolk-sac larvae of the Nile tilapia. Aquaculture 334–337: 182–190.

Fuentes-Silva, C., G.M. Soto-Zarazúa, I. Torres-Pacheco and A. Flores-Rangel. 2013. Male tilapia production techniques: A mini-review. Afr. J. Biotechnol. 12: 5496–5502.

Furuya, W.M., P.R. Neves, L.C.R. Silva, D. Botaro, C. Hayashi and E.S. Sakaguti. 2004. Fitase na alimentação da tilápia do Nilo (*Oreochromis niloticus*), durante o período de reversão de sexo. Acta Sci. Anim. Sci. 26: 299–303.

Gabriel, N.N., J. Qiang, X.Y. Ma, J. He, P. Xu and E. Omoregie. 2017. Sex-reversal effect of dietary *Aloe vera* (Liliaceae) on genetically improved farmed Nile tilapia fry. North Am. J. Aquacult. 79: 100–105.

Gale, W.L., M.S. Fitzpatrick, M. Lucero, W.M. Contreras-Sánchez and C.B. Schreck. 1999. Masculinization of Nile tilapia (*Oeochromis niloticus*) by immersion in androgens. Aquaculture 178: 349–357.

Gall, G.A.E. and Y. Bakar. 1999. Stocking density and tank size in the design of breed improvement programs for body size of tilapia. Aquaculture 173: 197–205.

Gayão, A.L.B.A., H. Buzollo, G.C. Favero, A.A.S. Junior, M.C. Portella, C. Cruz, et al. 2013. Histologia hepática e produção em tanques-rede de tilápia-do-nilo masculinizada hormonalmente ou não masculinizada. Pesqui. Agropecu. Bras. 48: 991–997.

Ghosal, I. and S.B. Chakraborty. 2014. Effects of the aqueous leaf extract of *Basella alba* on sex reversal of Nile tilapia, *Oreochromis niloticus*. J. Pharm. Biol. Sci. 9: 162–164.

Ghosal, I., D. Mukherjee, C. Hancz and S.B. Chakraborty. 2016. Efficacy of *Basella alba* and *Tribulus terrestris* extracts for production of monosex Nile tilapia, *Oreochromis niloticus*. J. App. Pharmacol. Sci. 5: 152–158.

Golan, M. and B. Levavi-Sivan. 2014. Artificial masculinization in tilapia involves androgen receptor activation. Gen. Comp. Endocrinol. 207: 50–55.

Hassan, A., V.T. Okomoda and P.J. Pradeep. 2018. Triploidy induction by electric shock in red hybrid tilapia. Aquaculture 495: 823–830.

Hayashi, C., W.R. Boscolo, C.M. Soares and F. Meurer. 2002. Exigência de proteína digestível para larvas de tilápia do Nilo (*Oreochromis niloticus*), durante a reversão sexual. Rev. Bras. Zootec. 31: 823–828.

Horváth, L. and L. Orbán. 1995. Genome and gene manipulation in the common carp. Aquaculture 129: 157–181.

Hui, W., Z. Xiaowen, W. Haizhen, Q. Jun, X. Pao and L. Ruiwei, 2014. Joint effect of temperature, salinity and pH on the percentage fertilization and hatching of Nile tilapia (*Oreochromis niloticus*). Aquacult. Res. 45: 259–269.

Hui, W., L. Guodong, L. Jiahui, Y. Hongshuai, Q. Jun and X. Pao. 2015. Combined effects of temperature and salinity on yolk utilization in Nile tilapia (*Oreochromis niloticus*). Aquacult. Res. 46: 2418–2425.

Hui, W., S. Wenjing, W. Long, Z. Chuankun, P. Zhengjun and W. Nan. 2019a. Light conditions for commercial hatching success in Nile tilapia (*Oreochromis niloticus*). Aquaculture 509: 112–119.

Hui, W., S. Wenjing, W. Long, Z. Chuankun, P. Zhengjun, C. Guoliang, et al. 2019b. Can larval growth be manipulated by artificial light regimes in Nile tilapia (*Oreochromis niloticus*)? Aquaculture 506: 161–167.

Hussain, M.G., A. Chatterji, B.J. McAndrew and R. Johnstone. 1991. Triploidy induction in Nile tilapia, *Oreochromis niloticus* L. using pressure, heat and cold shocks. Theor. App. Genet. 81: 6–12.

Jensi, A., K.K. Marx, M. Rajkumar, R.J. Shakila and P. Chidambaram. 2016. Effect of 17α-methyl testosterone on sex reversal and growth of Nile tilapia (*Oreochromis niloticus* L., 1758). Eco. Env. & Cons. 22: 1493–1498.

Johnstone, R., D.J. Macintosh and R.S. Wright. 1983. Elimination of orally administered 17α-methyltestosterone by *Oreochromis mossambicus* (tilapia) and *Salmo gairdneri* (rainbow trout) juveniles. Aquaculture 35: 249–257.

Joshi, H.D., V.K. Tiwari, S. Gupta, R. Sharma, W.S. Lakra and U. Sahoo. 2019. Application of nanotechnology for the production of masculinized tilapia, *Oreochromis niloticus* (Linnaeus, 1758). Aquaculture 511: 734206.

Kamaruzzaman, N., N.H. Nguyen, A. Hamzah and R.W. Ponzoni. 2009. Growth performance of mixed sex, hormonally sex reversed and progeny of YY male tilapia of the GIFT strain, *Oreochromis niloticus*. Aquacult. Res. 40: 720–728.

Karayücel, S., I. Karayücel, D. Penman and B. McAndrew. 2002. Production of androgenetic Nile tilapia, *Oreochromis niloticus* L.: optimization of heat shock duration and application time to induce diploidy. Isr. J. Aquacult. 54: 145–156.

Komen, H. and G.H. Thorgaard. 2007. Androgenesis, gynogenesis and the production of clones in fishes: A review. Aquaculture 269: 150–173.

Kwon, J.Y., B.J. McAndrew and D.J. Penman. 2002. Treatment with an aromatase inhibitor suppresses high-temperature feminization of genetic male (YY) Nile tilapia. J. Fish Biol. 60: 625–636.

Leonhardt, J.H. and Urbinati, E.C. 1999. Estudo comparativo do crescimento entre machos de tilápia do Nilo, *Oreochromis niloticus*, sexados e revertidos. Bol. Inst. Pesca. 25: 19–26.

Little, D.C., R.C. Bhujel and T.A. Pham. 2003. Advanced nursing of mixed-sex and mono-sex tilapia (*Oreochromis niloticus*) fry, and its impact on subsequent growth in fertilized ponds. Aquaculture 221: 265–276.

Little, D.C. and P. Edwards. 2004. Impact of nutrition and season on pond culture performance of mono-sex and mixed-sex Nile tilapia (*Oreochromis niloticus*). Aquaculture 232: 279–292.

Lu, J., T. Takeuchi and H. Satoh. 2004. Ingestion and assimilation of three species of freshwater algae by larval tilapia *Oreochromis niloticus*. Aquaculture 238: 437–449.

Luz, R.K., P.A.P. Ribeiro, A.L. Ikeda, A.E.H. Santos, R. Melillo Filho, E.M. Turra, et al. 2012a. Performance and stress resistance of Nile tilapias fed different crude protein levels. Rev. Bras. Zootec. 41: 457–461.

Luz, R.K., W.S. Silva, R. Melillo Filho, A.E.H. Santos, L.A. Rodrigues, R. Takata, et al. 2012b. Stocking density in the larviculture of Nile tilapia in saline water. Rev. Bras. Zootec. 41: 2385–2389.

Luz, R.K., A.E.H. Santos, R. Melillo Filho, E.M. Turra and E.A. Teixeira. 2013. Larvicultura de tilápia em água doce e água salinizada. Pesqui. Agropecu. Bras. 48: 1150–1153.

Mair, G.C. 1993. Chromosome-set manipulation in tilapia – Techniques, problems and prospects. Genet. Aquacult. 111: 227–244.

Marengoni, N.G. and Y. Onoue. 1998. Ultraviolet-induced androgenesis in Nile tilapia, *Oreochromis niloticus* (L.), and hybrid Nile H blue tilapia, *O. aureus* (Steindachner). Aquacult. Res. 29: 359–366.

Mateen, A. and I. Ahmed. 2007. Effect of androgen on sex reversal and growth of Nile tilapia (*Oreochromis niloticus*). Pak. J. Agri. Sci. 44: 272–276.

Maxime, V. 2008. The physiology of triploid fish: Current knowledge and comparisons with diploid fish. Fish Fish. 9: 67–78.

Melo, L.H., Y.S. Martins, R.M.C. Melo, P.S. Prado, R.K. Luz, N. Bazzoli, et al. 2019. Low salinity negatively affects early larval development of Nile tilapia, *Oreochromis niloticus*: Insights from skeletal muscle and molecular biomarkers. Zygote 27: 375–381.

Meurer, F., C. Hayashi, W.R. Boscolo, C.R. Schamber and R.A. Bombardelli. 2005. Fontes protéicas suplementadas com aminoácidos e minerais para a tilápia do Nilo durante a reversão sexual. Rev. Bras. Zootec. 34: 1–6.

Meurer, F., C. Hayashi, M.M. Costa, A.S. Mascioli, L.M.S. Colpiniand and A. Freccia. 2008. Levedura como probiótico na reversão sexual da tilápia-do-Nilo. Rev. Bras. Saude Prod. Anim. 9: 804–812.

Miyazaki, H., T. Kaneko, S. Hasegawa and T. Hirano. 1998. Developmental changes in drinking rate and ion and water permeability during early life stages of euryhaline tilapia, *Oreochromis mossambicus*, reared in fresh water and seawater. Fish Physiol. Biochem. 18: 277–284.

Mirzakhani, N., E. Ebrahimi, S.A.H. Jalali and J. Ekasari. 2019. Growth performance, intestinal morphology and nonspecific immunity response of Nile tilapia (*Oreochromis niloticus*) fry cultured in biofloc systems with different carbon sources and input C:N ratios. Aquaculture 512: 734235.

Moreira, R.L., J.M. Costa, R.V. Queiroz, P.S. Moura and W.R.L. Farias. 2010. Utilização de *Spirulina platensis* como suplemento alimentar durante a reversão sexual de tilápias do Nilo. Rev. Caatinga 23: 134–141.

Moreira, R.L., R.R.O. Martins and W.R.L. Farias. 2011. Utilização de *Spirulina platensis* como suplemento alimentar durante a reversão sexual da tilápia-do-Nilo (Var. chitralada) em água salina. Cienc. Anim. Bras. 12: 76–82.

Mukherjee, D., I. Ghosal, C. Hancz and S.B. Chakraborty. 2018. Dietary administration of plants extracts for production of monosex tilapia: Searching a suitable alternative to synthetic steroids in tilapia culture. Turk. J. Fish. Aquatic Sci. 18: 267–275.

Myers, J.M., D.J. Penman, K.J. Rana, N. Bromage, S.F. Powell and B.J. McAndrew. 1995. Applications of induced androgenesis with tilapia. Aquaculture 137: 149–160.

Nakamura, M. 1975. Dosage-dependent changes in the effect of oral administration of methyltestosterone on gonadal sex differentiation in Tilapia mossambica. Bull. Fish. Sci. Hokkaido Univ. 26: 99–108.

Nasr-Allah, A.M., M.W. Dickson, D.A.R. Al-Kenawy, M.F.M. Ahmed and G.O. El-Naggar. 2014. Technical characteristics and economic performance of commercial tilapia hatcheries applying different management systems in Egypt. Aquaculture 426–427: 222–230.

Neumann, E., T.C.R. Dias-Koberstein and F.M.S. Braga. 2009. Desempenho de três linhagens de tilápia submetidas ao tratamento com 17-α-metiltestosterona em condições ambientais não controladas. Rev. Bras. Zootec. 38: 973–979.

Nivelle, R., V. Gennotte, E.J.K. Kalala, B. Ngoc, M. Muller, C. Mélard, et al. 2019. Temperature preference of Nile tilapia (*Oreochromis niloticus*) juveniles induces spontaneous sex reversal. PLoS One 14: 1–19.

Oca, G.A.R-M., J.C. Román-Reyes, A. Alaniz-Gonzalez, C.O. Serna-Delval, G. Muñoz-Cordova and H. Rodríguez-González. 2015. Effect of salinity on three tilapia (*Oreochromis* sp.) strains: Hatching rate, length and yolk sac size. Int. J. Aquat. Sci. 6: 96–106.

Olvera-Novoa, M.A., F. Pereira-Pacheco, L. Olivera-Castillo, V. Pkrez-Flores, L. Navarro and J.C. Sámano. 1997. Cowpea (*Vigna unguiculata*) protein concentrate as replacement for fish meal in diets for tilapia (*Oreochromis niloticus*) fry. Aquaculture 158: 107–116.

Pandian, T.J. and K. Varadaraj. 1988. Techniques for producing all-male and all-triploid *Oreochromis mossambicus*. The Second International Symposium on Tilapia in Aquaculture, ICLARM Conf. Proc. Thailand. 15: 243–249.

Pandian, T.J. and S.G. Sheela. 1995. Hormonal induction of sex reversal in fish. Aquaculture 138: 1–22.

Pessini, J.E., A. Signor, E.B. Moro, D.R.A. Fernandes, A. Feiden, W.R. Boscolo, et al. 2015. Suplementação de micronutrientes em dietas para larvas de tilápia do Nilo (*Oreochromis niloticus*). Rev. Biociênc. 21: 86–92.

Peruzzi, S., A. Kettunen, R. Primicerio and G. Kauric. 2007. Thermal shock induction of triploidy in Atlantic cod (*Gadus morhua* L.). Aquacult. Res. 38: 926–932.

Phelps, R.P. and T.J. Popma. 2000. Sex reversal of tilapia. pp. 34–59. *In*: Costa-Pierce, B.A. and J.E. Rakocy [eds.]. Tilapia Aquaculture in the Americas. The World Aquaculture Society, Baton Rouge, Louisiana, USA.

Phelps, R.P. and M. Okoko. 2011. A non-paradoxical dose response to 17α-methyltestosterone by Nile tilapia *Oreochromis niloticus* (L.): Effects on the sex ratio, growth and gonadal development. Aquacult. Res. 42: 549–558.

Piferrer, F. and E.M. Donaldson. 1991. Dosage-dependent differences in the effect of aromatizable and nonaromatizable androgens on the resulting phenotype of Coho salmon (*Oncorhynchus kisutch*). Fish Physiol. Biochem. 9: 145–150.

Piferrer, F., A. Beaumont, J-C. Falguière, M. Flajshans, P. Haffray and L. Colombo. 2009. Polyploid fish and shellfish: Production, biology and applications to aquaculture for performance improvement and genetic containment. Aquaculture 293: 125–156.

Pradeep, P.J., T.C. Srijaya, A. Papini and A.K. Chatterji. 2012. Effects of triploidy induction on growth and masculinization of red tilapia [*Oreochromis mossambicus* (Peters, 1852) × *Oreochromis niloticus* (Linnaeus, 1758)]. Aquaculture 344–349: 181–187.

Rahmah, S., H.J. Liew, N. Napi and S.A. Rahmat. 2020. Metabolic cost of acute and chronic salinity response of hybrid red tilapia *Oreochromis* sp. larvae. Aquaculture Reports (in press).

Razak, S.A., G.-L. Hwang, M.A. Rahman and N. Maclean. 1999. Growth performance and gonadal development of growth enhanced transgenic tilapia *Oreochromis niloticus* (L.) following heat-shock-induced triploidy. Mar. Biotechnol. 1: 533–544.

Rouf, M.A., R.I. Sarder, F. Rahman and M.F. Islam. 2008. Optimization of dose of methyl testosterone (MT) hormone for sex reversal in tilapia (*Oreochromis niloticus* L.). Bangladesh J. Fish. Res. 12: 135–142.

Sanches, L.E.F. and C. Hayashi. 1999. Densidade de estocagem no desempenho de larvas de tilápia-do-Nilo (*Oreochromis niloticus* L.), durante a reversão sexual. Acta Sci. 21: 619–625.

Sanches, L.E.F. and C. Hayashi. 2001. Effect of feeding frequency on Nile tilapia, *Oreochromis niloticus* (L.) fries performance during sex reversal in hapas. Acta Sci. 23: 871–876.

Sani, K.A., I.O. Obaroh, T. Yahaya, S. Aliyu and S.K. Faruku. 2019. Masculinine effects of *Nigella sativa* on tilapia. Int. J. Fish. Aquat. Stud. 7: 478–481.

Santiago, C.B., M.B. Aldaba and O.S. Reyes. 1987. Influence of feeding rate and diet form on growth and survival of Nile tilapia (*Oreochromis niloticus*) fry. Aquaculture 64: 277–282.

Sayed, A.E.-D.H., U.M. Mahmoud and I.A. Mekkawy. 2016. Erythrocytes alterations of monosex tilapia (*Oreochromis niloticus*, Linnaeus, 1758) produced using methyltestosterone. Egyptian J. Aquat. Res. 42: 83–90.

Schäfer, M.R. 2015. Otimização do arraçoamento no cultivo de tilápias GIFT em sistema automatizado de alimentação. Ph.D. Thesis, Universidade Federal do Paraná, Palotina, Brazil.

Schwarz, K.K., W.M. Furuya, M.R.M. Natali, M.C. Gaudezi and P.A.G. Lima. 2011. Mananoligossacarídeo em dietas para larvas de tilápia. Rev. Bras. Zootec. 40: 2634–2640.

Shelton, W.L. 2002. Monosex tilapia production through androgenesis. pp. 1–9. *In*: McElwee, K., Lewis, K., Nidiffer, M. and Buitrago, P. [eds.]. Nineteenth Annual Technical Report. Pond Dynamics/Aquaculture. Oregon.

Shen, Z-G. and H-P. Wang. 2014. Molecular players involved in temperature-dependent sex determination and sex differentiation in Teleost fish. Genet. Sel. Evol. 46: 1–21.

Shrivastav, R., M.K. Shrestha, N.B. Khanal and N.P. Pandit. 2016 and 2017. Assessment of dried carp testes for success on hormonal sex reversal in Nile tilapia (*Oreochromis niloticus*). Nepal. J. Aquacult. Fish. 3 and 4: 21–25.

Siddik, M.A.B., A. Nahar, E. Ahsan, F. Ahamed and Y. Hossain. 2014. Over-wintering growth performance of mixed-sex and mono-sex Nile tilapia *Oreochromis niloticus* in the northeastern Bangladesh. Croatian J. Fish. 72: 70–76.

Siddiqui, A.Q., M.S. Howlader and A.A. Adam. 1988. Effects of dietary protein levels on growth, feed conversion and protein utilization in fry and young Nile tilapia, *Oreochromis niloticus*. Aquaculture 70: 63–73.

Silva, A.L., F.C.M. Alves, E.F. Andrade Talmelli, C.M. Ishikawa, M.K. Nagata and N.E.T. Rojas. 2009. Utilização de cloreto de sódio, formalina e a associação destes produtos no controle de ectoparasitas em larvas de tilápia (*Oreochromis niloticus*). Bol. Inst. Pesca 35: 597–608.

Silva, W.S., L.S. Costa, J.F. López-Olmeda, N.C.S. Costa, W.M. Santos, P.A.P. Ribeiro, et al. 2019. Gene expression, enzyme activity and performance of Nile tilapia larvae fed with diets of different CP levels. Animal 13: 1376–1384.

Singh, E., V.P. Saini, M.L. Ojha and H.K. Jain. 2017. Comparative growth performance of monosex and mixed sex red tilapia (*O. niloticus* L.). J. Entomol. Zool. Stud. 5: 1073–1075.

Singh, E., V.P. Saini and O.P. Sharma. 2018. Orally administered 17α-methyltestosterone at different doses on the sex reversal of the red tilapia (*Oreochromis niloticus*). Int. J. Fish. Aquat. Sci. 6: 301–305.

Tachibana, L., A.F.G. Leonardo, C.F. Correa and L.A. Saes. 2008. Densidade de estocagem de pós-larvas de tilápia-do-Nilo (*Oreochromis niloticus*) durante a fase de reversão sexual. Bol. Inst. Pesca. 34: 483–488.

Tebaldi, P.C. and H.A. Junior. 2009. Produção de tetraplóides de tilápia do Nilo (*Oreochromis niloticus*) através da aplicação de choque térmico. Rev. Eletronica Vet. 10: 1–13.

Tengjaroenkul, B., B.J. Smith, S.A. Smith and U. Chatreewongsin. 2002. Ontogenic development of the intestinal enzymes of cultured Nile tilapia, *Oreochromis niloticus* L. Aquaculture 211: 241–251.

Teskeredzic, E., E.M. Donaldson, Z. Teskeredzic, I.I. Solar and E. McLean. 1993. Comparison of hydrostatic pressure and thermal shocks to induce triploidy in Coho salmon (*Oncorhynchus kisutch*). Aquaculture 117: 47–55.

Ting, C.-H., Y.-C. Chen and J.-Y. Chen. 2018. Nile tilapia fry fed on antimicrobial peptide Epinecidin-1-expressing Artemia cyst exhibit enhanced immunity against acute bacterial infection. Fish Shellfish Immunol. 81: 37–48.

Tiwary, B.K., R. Kirubagaran and A.K. Ray. 2004. The biology of triploid fish. Rev. Fish Biol. Fish. 14: 391–402.

Toledo, T.M., B.C. Silva, F.D.N. Vieira, J.L.P. Mouriño and W.Q. Seiffert. 2016. Effects of different dietary lipid levels and fatty acids profile in the culture of white shrimp *Litopenaeus vannamei* (Boone) in biofloc technology: Water quality, biofloc composition, growth and health. Aquacult. Res. 47: 1841–1851.

Toyama, G.N., J.E. Corrente and J.E.P. Cyrino. 2000. Suplementação de vitamina c em rações para reversão sexual da tilápia do Nilo. Sci. Agric. 57: 221–228.

Ugonna, B.O., S.G. Solomon, S.O. Olufeagba and V.T. Okomoda. 2018. Effect of pawpaw carica papaya seed meal on growth and a natural sex-reversal agent for Nile tilapia. North Am. J. Aquaculture. 80: 278–285.

Varadaraj, K. and T.J. Pandian. 1988. Induction of triploids in *Oreochromis mossambicus* by thermal, hydrostatic pressure and chemical shocks. Proc. Aquacult. Int. Congress. Canada 98: 531–535.

Vera Cruz, E.M. and G.C. Mair. 1994. Conditions for effective androgen sex reversal in *Oreochromis niloticus* (L). Aquaculture 122: 237–248.

Vernetti, C.H.M.M., M.D.N. Rodrigues, H.J.P. Gutierrez, C.P. Calabuig, G.A. Moreira, A.A. Nlewadim, et al. 2013. Genes involved in sex determination and the influence of temperature during the sexual differentiation process in fish: A review. Afr. J. Biotechnol. 12: 2129–2146.

Wang, L-H. and C-L. Tsai. 2000. Effects of temperature on the deformity and sex differentiation of tilapia, *Oreochromis mossambicus*. J. Exp. Zool. 286: 534–537.

Wang, Y.Y., L.X. Sun, J.J. Zhu, Y. Zhao, H. Wang, H.J. Liu, et al. 2017. Epigenetic control of cyp19a1a expression is critical for high temperature induced Nile tilapia masculinization. J. Therm. Biol. 69: 76–84.

Wassermann, G.J. and L.O.B. Afonso. 2003. Sex reversal in Nile tilapia (*Oreochromis niloticus* Linnaeus) by androgen immersion. Aquacult. Res. 34: 65–71.

CHAPTER
12

Biological Rhythms in Tilapia

José Fernando López-Olmeda*, Francisco Javier Sánchez-Vázquez
and Luisa María Vera
Department of Physiology, Faculty of Biology, University of Murcia, 30100 Murcia, Spain

1. A brief introduction to biological rhythms

Life on Earth is constantly facing cyclic environmental changes due to rotational and translational movements of the planet, which generate the day/night cycle, tides and seasons. Most of these cycles are predictable and have, thus, fostered organisms to develop biological clocks that keep track of them. Through these clocks, organisms can anticipate environmental changes and time physiological processes, such as locomotion, feeding and reproduction, to occur at specific times of the day or year, increasing their efficiency and survival, and minimizing energy expenditure (DeCoursey 2004).

Therefore, living organisms present many processes that occur in a periodic and repeatable manner that are called biological rhythms. The first observations of the presence of biological rhythms date back to ancient Egypt and Greece, where naturalists noted the rhythms of birdsongs and movements of the leaves of some plants. In the 4th century BC, Androsthenes, a naturalist who travelled with Alexander the Great, described the daily rhythms of leaves opening and closing, although he assumed that they were responses to environmental stimuli and changes in astronomic cycles (DeCoursey 2004, Madrid 2006). The first scientific report of a biological rhythm was provided by French astronomer Jean Jacques d'Ortous de Mairan who, in 1729, reported the follicular movements of the plant *Mimosa pudica*. de Mairan observed that opening and closing movements in leaves remained under constant darkness (DD) conditions for several days, with a period close to the 24-hour periodicity observed in a normal light/dark (LD) cycle (de Mairan 1729).

This persistence under constant environmental conditions was first described by de Mairan, and is one of the main characteristics of biological rhythms. In addition, biological rhythms can be classified according to their periodicity: circadian, with a period of approximately 24 h; ultradian, with a period shorter than 24 h (i.e. tidal rhythms or the firing rate of some neurons and pacemaker cells); infradian, with a

*Corresponding author: jflopez@um.es

period lasting more than 24 h, which can range from some days to one month (lunar rhythms), and even to a whole year (López-Olmeda 2017). The latter are also called seasonal rhythms, which are quite common in vertebrates and involve processes such as migration and hibernation, and especially reproduction (Goldman et al. 2004, Chemineau et al. 2007). Of all biological rhythms, the most prominent, and thus the most widely studied, are daily or circadian rhythms.

2. The circadian system

Circadian rhythms are controlled by the circadian system, which is usually divided into three different components: input signals, a pacemaker and output pathways (Pando and Sassone-Corsi 2002, López-Olmeda 2017). In the summarized scheme of the generation of biological rhythms (Figure 1), the animal receives an input from a synchronizer or *zeitgeber*, which can have an external (light and temperature cycles, tides) or internal origin (nutrients, hormones, nervous signals). The synchronizer acts on a pacemaker or oscillator by entraining the phase of biological rhythms. An oscillator is a system of components that can generate a rhythm even in the absence of external cues. This oscillator constitutes a functional anatomical region capable of sustaining its own oscillations and entraining other oscillations, and is called a pacemaker. The outputs of pacemakers are the biological rhythms (also called overt rhythms), which can be defined as rhythms in an observable characteristic that is directly or indirectly linked with, and controlled by, a pacemaker (Pando and Sassone-Corsi 2002, Johnson et al. 2004, López-Olmeda 2017).

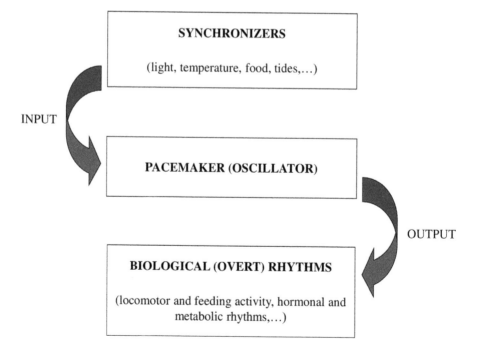

Fig. 1. Summarized scheme of the circadian system.

In mammals, the master pacemaker of the circadian system is the suprachiasmatic nucleus (SCN), a brain area located in the hypothalamus (Dibner et al. 2010, Welsh et al. 2010). Light (through LD cycles) is the most important input signal that synchronizes the SCN. Hence, this pacemaker is also called light entrainable oscillator, or LEO (Reppert and Weaver 2002). In addition to mammals, the LD cycle is considered the main synchronizer for the rhythms of all animals. Nevertheless, in most species from invertebrates to vertebrates, biological rhythms can be synchronized by other cyclic factors, such as temperature, food and tides (Rensing and Ruoff 2002, Mistlberger 2009, Tessmar-Raible et al. 2011).

In the circadian system of fish, some decades ago it was thought that the master pacemaker was located in the pineal gland (Edwards 1988, Dunlap 1999). However, in the last years, the idea of a few centralized structures that generate all circadian rhythms in fish has been challenged. Firstly, many rhythms are maintained in pineal-ablated fish (Sánchez-Vázquez et al. 2000, Yadu and Shedpure 2002). Secondly, data from invertebrates and vertebrates, excluding mammals, indicate that the circadian system may be scattered throughout the organism, and each cell possesses an autonomous clock (Plautz et al. 1997, Whitmore et al. 1998, 2000, Giebultowicz et al. 2000).

2.1. The molecular clock

The discovery of molecular mechanisms that control circadian rhythms merited the 2017 Nobel Prize in Physiology or Medicine, awarded to Drs. Jeffrey C. Hall, Michael Rosbash and Michael W. Young (Ibáñez 2017). The generation of all circadian rhythms relies on the molecular clock, present in each cell, which consists of a self-sustainable mechanism formed by positive and negative loops (Figure 2) (Pando and Sassone-Corsi 2002, Vatine et al. 2011). In vertebrates, including fish, the positive elements are the genes *clock* and *bmal* (also called *arntl*), while the negative elements are the genes *period* (*per*) and *cryptochrome* (*cry*) (Pando and Sassone-Corsi 2002, Vatine et al. 2011). The positive elements Clock and Bmal present a DNA binding domain of the helix-loop-helix (bHLH) type and a protein-protein interaction domain named PAS. Clock and Bmal heterodimerize by means of their PAS domains in the cytosol and constitute the Clock: Bmal dimer, which is an active transcription factor. Then, Clock:Bmal is translocated to the nucleus where it induces the expression of a diverse array of genes by binding to E-boxes, a sequence present in the promoter region of these genes. The E-box sequence is CANNTG with a canonical sequence of CACGTG (Gekakis et al. 1998, Hogenesch et al. 1998, Reppert and Weaver 2002). The genes activated by the Clock:Bmal transcription factor are also known as Clock-Controlled Genes (CCGs) which, in mammals, seem to account for up to 10% of the transcriptome (Panda et al. 2002). Besides these genes, Clock:Bmal stimulates the transcription of other clock genes (Figure 2): *per*, *cry*, *reverbα* and *rora* (Jin et al. 1999, Ripperger et al. 2000, Vatine et al. 2011).

In the negative loop, the Clock:Bmal complex activates the transcription of *per* and *cry*. Then, the Per and Cry proteins enter the nucleus and inhibit Clock:Bmal by closing the negative loop (Kume et al. 1999, Shearman et al. 2000). In addition, the Clock:Bmal complex activates the transcription of *reverbα* and *rora*, which form

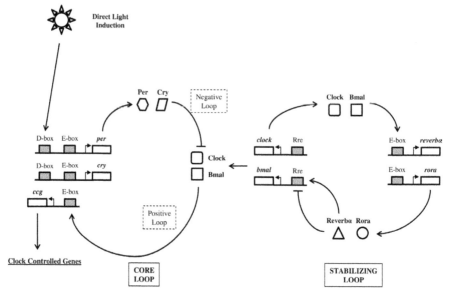

Fig. 2. Scheme of the molecular clock of teleost fish. Clock and Bmal are the positive elements of the clock, and activate the expression of clock-controlled genes (ccks) by binding to the E-box in the promoter. It also activates the expression of *per* and *cry*, which are the negative elements and inhibit Clock:Bmal. In addition, Rev-erbα and Rora constitute the stabilizing loop, which provides robustness to the main loops. The expression of *per* and *cry* may also be stimulated by direct light exposure. Adapted from Vatine et al. (2011).

the so-called stabilizing loop (Preitner et al. 2002, Reppert and Weaver 2002, Vatine et al. 2011). The elements Rev-erb α and Rora act mainly on the control of the *bmal* transcript through its binding to the Rev-erb/Ror responsive elements (Rre) in the promoter (Ueda et al. 2002, Reppert and Weaver 2002). The stabilizing loop confers robustness and stability to both the positive and negative loops (Vatine et al. 2011). Moreover, in some fish species, such as the zebrafish, *per* and *cry* expression can be induced directly by light exposure. Zebrafish cells from all tissues present non-visual opsins that can be stimulated by light (Whitmore et al. 1998, 2000, Vatine et al. 2011). The activation of these opsins after receiving light stimulates, in turn, factors like Tef, which bind to a specific region, known as D-box, of the promoter of some genes such as *per* and *cry* (Ben-Moshe et al. 2010, Mracek et al. 2012).

In Nile tilapia (*Oreochromis niloticus*), many clock genes have been identified and the daily rhythmic oscillations in these genes have been reported (Costa et al. 2016a, Wu et al. 2018). The genes from the positive (*clock1* and *bmal1*), negative (*per1b*, *per3*, *cry1a*, *cry2* and *cry5*) and stabilizing loops (*reverba*) display daily variations with peaks at similar LD cycle times as other teleost fish, such as zebrafish (*Danio rerio*), goldfish (*Carassius auratus*), medaka (*Oryzias latipes*), gilthead sea bream (*Sparus aurata*) or Senegalese sole (*Solea senegalensis*) (Cahill 2002, Velarde et al. 2009, Martín-Robles et al. 2012, Vera et al. 2013, Cuesta et al. 2014, Gómez-Boronat et al. 2019). In general, the genes from the positive loop peak around the end of the light phase, whereas the genes from the negative and stabilizing loops present a peak in the antiphase, with the highest expression appearing at the end of the dark phase or the beginning of the light phase (Figure 3) (Costa et al. 2016a).

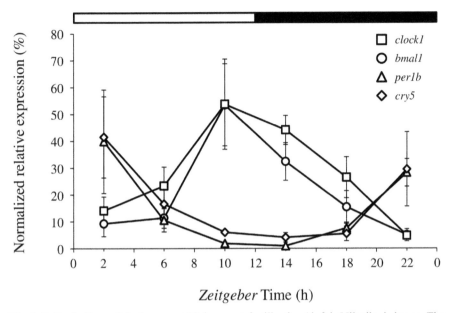

Fig. 3. Daily rhythms of clock genes at 18 days post-fertilization (dpf) in Nile tilapia larvae. The analyzed genes are *clock1* (squares), *bmal1* (circles), *per1b* (triangles) and *cry5* (rhombuses). The white and black bars above the graph indicate the light and dark phases, respectively. Modified from A.H. Espirito Santo et al. (unpublished data).

In tilapia, clock gene rhythms have been reported in several tissues, such as the brain, liver and muscle (Costa et al. 2016a, Wu et al. 2018). Rhythms in clock genes can also be detected in whole larvae homogenates early in development (Figure 3) (Espirito Santo et al. unpublished data). Of the factors that synchronize clock gene rhythms in tilapia, the LD cycle seems the most important one, especially for central tissues like the brain (Costa et al. 2016a). In addition, food can influence the rhythm of clock genes in peripheral tissues like the liver. In the experiment by Costa et al. (2016a), two tilapia groups were fed once a day at different fixed times: in the middle of the light phase (ML) and in the middle of the dark phase (MD). The phase of clock genes in the liver of the fish fed at ML was the same as in the brain, and the same as that observed in animals fed at random times, which was the expected phase in fish. However, shifting the feeding phase to MD induced a partial 6- to 8-hour shift of several clock genes in the liver (Costa et al. 2016a). This partial shift due to feeding time has been reported in the peripheral tissues of other fish species like zebrafish and gilthead sea bream (López-Olmeda et al. 2010, Vera et al. 2013), and indicates that food plays an important role as a synchronizer in peripheral tissues, unlike central tissues like the brain, where no effect due to feeding time was observed. In addition to feeding time, nutritional state can affect the rhythms of clock genes in peripheral tissues. For instance in skeletal muscle, starvation induces changes in some clock genes by shifting their phase, or by even suppressing their oscillations

(Wu et al. 2018). These changes induced by starvation are also reflected in changes in the rhythms of myogenic factors (Wu et al. 2018). Finally, clock genes have been shown to be involved in other processes in Nile tilapia, such as response to parasite infection (Ellison et al. 2018), response to cold stress (Li et al. 2020) and autophagy (Wu et al. 2020).

3. Rhythms of behavior of tilapia

In most vertebrates, the existence of a rest-activity rhythm regulated by the circadian clock and synchronized to environmental cycles is an evolutionary conserved phenomenon. Rhythms in behavioral patterns have been described in a wide variety of freshwater and marine fish species under an LD cycle. Fish can be classified into several types according to when they show most of their daily swimming activity: diurnal, nocturnal, crepuscular, and combinations of these.

In Nile tilapia, Vera et al. (2009) found wide variability in behavioral patterns in both male and female individuals. Under an LD cycle, 75% of males showed daily rhythms of locomotor activity, with 45% of them being diurnal. However, only 55% of female tilapia displayed rhythmicity, and 85% of those were diurnal. These results contrasted those with previously accepted knowledge showing that Nile tilapia was a diurnal species by not only revealing the existence of inter-individual and sex-dependent differences in their rhythmicity, but also supporting the hypothesis that freshwater fish species possess a plastic and flexible circadian system that may allow them to adapt to a relatively unstable environment (Reebs 2002). The reproductive activity of female Nile tilapia seems to clearly affect their daily locomotor rhythms. This species is a mouthbreeder, and spawning events interfere with activity patterns. Some female tilapia showed less activity levels from the start of egg incubation to several days later, whereas this decreased activity in others started a few days before spawning and was accompanied by loss of rhythmicity (Vera et al. 2009).

In Nile tilapia aquaculture, feed costs account for a high percentage of total operational costs. Therefore, feeding rhythms should be considered when designing feeding strategies to optimize these costs and to improve production efficiency. Fortes-Silva et al. (2010) used self-feeders to record feeding activity patterns in male Nile tilapia under an LD cycle, and found that feeding occurred mostly at night, with up to 93% of food demands in the dark phase and most feeding events concentrated immediately after lights went off. However, locomotor activity remained diurnal, with 67% of activity displayed in the daytime with a peak at around ML. These results showed the existence of phase independence between both behavioral rhythms in Nile tilapia, and further support the hypothesis that this fish species may possess a plastic circadian system which would allow it to display dual phasing behavior. As in other species, this flexibility might be an adaptation to potential food availability variations in their natural environment (Boujard and Leatherland 1992). Actually, Reebs (2002) proposed that foraging success could be the key factor to explain such plasticity of behavioral patterns in fish, which would be controlled by food availability, inter- and intraspecific competition, or even by the ability of fish to detect food in their habitat depending on other environmental conditions (e.g. lighting).

Different light regimes have been used to investigate the effect of photoperiod on locomotor and feeding activities in Nile tilapia. Veras et al. (2013) tested five different photoperiods (0L:24D, 6L:18D, 12L:12D, 18L:6D, 24L:0D), and observed a strong effect of light treatments on both locomotor and feeding activity levels in fingerlings. Locomotor activity was maximum when fish were kept in a 12L:12D photoperiod (around 4,500 registers/day), whereas exposure to continuous darkness (0L:24D) lowered fish activity levels to around 1,000 registers/day. Feed intake was also affected by photoperiod and showed minimal levels under the 0L:24D photoperiod, which coincided with the lowest locomotor activity levels. However, no differences appeared in the number of feeding events between the other photoperiods investigated in that study. The authors linked the reduced feed intake in the tilapia kept in continuous darkness with their lower energy demands under such conditions, when locomotor activity levels were also minimal. These findings suggested that Nile tilapia is a visual forager and, therefore, feed intake would increase when fish are exposed to longer light phases as they are able to better locate food (Dabrowski 1975, Veras et al. 2013). However, the data obtained by Fortes et al. (2010), who used self-feeders, do not support this theory.

Although the LD cycle has been considered the most powerful synchronizer of behavioral rhythms in animals, imposing fixed feeding schedules can also act as a strong *zeitgeber*, and it is well-known that animals have developed mechanisms that allow them to synchronize their behavioral rhythms to food availability, which improves food acquisition and utilization. In fact, numerous fish species show food anticipatory activity (FAA), which is characterized by increased locomotor activity before mealtime (Davidson et al. 2003). Previous research conducted in juvenile Nile tilapia has shown that feeding cycles are able to synchronize activity rhythms in this species. Consequently, the fish fed at ML showed a diurnal pattern of activity, whereas those fed at MD presented a disrupted pattern of locomotor activity in which the differences in activity levels between day and night were considerably smaller. All this reveals that feeding time had a strong synchronizing effect on this species (Guerra-Santos et al. 2017). However, in that study, tilapia did not become nocturnal when fed at night, as seen in other teleosts (Sánchez and Sánchez-Vázquez 2009, Sánchez et al. 2009, Vera et al. 2013), but the relative power of light and food signals to synchronize biological rhythms vary between fish species depending on diurnal/ nocturnal activity patterns and feeding habits (Zhdanova and Reebs 2006).

In order to determine whether behavioral rhythms in Nile tilapia are under endogenous control, constant environmental conditions (DD) and ultradian LD pulses have been employed (Vera et al. 2009). In DD, 50% of fish displayed self-sustained circadian rhythmicity and locomotor activity free-ran with τ ranging from 23.5 to 25.2 h. However, when 45-minute LD pulses were used, 58% of individuals showed circadian activity rhythms with τ ranging from 21.0 to 24.7 h in this case (Figure 4). With ultradian LD pulses, fish displayed their activity mostly in dark phases in both the tilapias that had displayed diurnal and nocturnal patterns under a 12L:12D photoperiod. This finding suggests that the circadian rhythm can be split into two components or could result from a negative light-masking phenomenon.

The endogenous control of activity patterns in tilapia has also been investigated after reversing the photoperiod from 12L:12D to 12D:12L. Under these conditions,

33% of tilapia showed gradual resynchronization to the new phase, characterized by the existence of transient cycles of activity. The rest of the fish either changed their activity rhythm phase immediately after reversing the photoperiod or did not resynchronize at all (Figure 4) (Vera et al. 2009). Altogether, these results indicate the existence of a circadian clock in Nile tilapia, but also suggest that this clock might not be as robust as in other vertebrate species, which agrees with the flexibility and wide inter-individual variability observed when behavioral rhythms are displayed under LD conditions. In fish, multiple tissues have photoreception ability and can synchronize to the environmental photocycle (Whitmore et al. 2000). Therefore, the coupling between the various oscillators could differ between individuals and would depend on physiological and environmental conditions (Iigo et al. 1994).

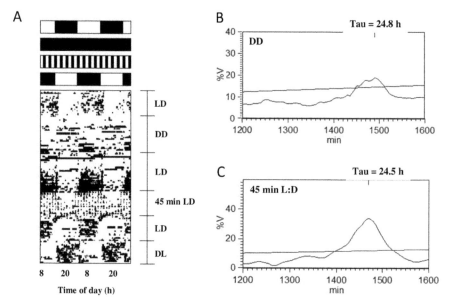

Fig. 4. Locomotor actogram from tilapia exposed to different subsequent photoperiods (A) and the periodogram analysis of tilapia subjected to DD (B) and ultradian light pulses (45:45 min LD) (C). Actogram is double-plotted for better visualization purposes. The white and black bars at the top of each graph indicate the light and dark periods, respectively. The free-running rhythm (*tau*) period is indicated above the periodograms (Adapted from Vera et al. 2009).

4. Rhythms of the digestive system of tilapia

The rhythmicity of digestive physiology has been widely investigated in mammals, and more recently, numerous studies have also focused on fish. The rhythmic secretion of digestive enzymes has been described in several species (Vera et al. 2007, Montoya et al. 2010a, López-Olmeda et al. 2012, Lazado et al. 2017), as has the existence of feed anticipatory activity in other digestive factors, such as gastric acidification (Yúfera et al. 2012, Lazado et al. 2017) and gastric evacuation (da Silva Reis et al. 2019). These rhythms are entrained by food availability cycles and driven by changes in the expression of clock genes which, in turn, regulate the expression of the genes involved in lipid metabolism and other digestive functions (Vera et al. 2013, Betancor et al. 2014, Paredes et al. 2014, 2015, Betancor et al. 2020) to improve the gut's adaptability in the daytime and to ensure optimal food breakdown and nutrient absorption and utilization. In fact, Nile tilapia producers from Brazil have not traditionally changed fish daily mealtimes in the production cycle given the general idea that doing so might result in diet waste and, consequently, economic loss. Caldini et al. (2013) investigated the effects of minor changes on tilapia's on-going feeding strategy to show that delaying the fish feeding schedule for only 1 h halfway through the production cycle resulted in major negative effects on tilapia survival and growth performance. The most dramatic result that these authors found was a 60% reduction in fish survival when the three meals they received per day were delayed for 1 h. In addition, the tank yield in this group was 43% lower than in the tilapia subjected to the same feeding schedule throughout the whole study, which was probably due to more feed waste which led to fish malnutrition and poorer growth performance. Altogether, these experimental results supported the feeding strategy followed by Brazilian farmers in tilapia production and proved that imposing fixed mealtimes is a key factor to ensure successful feeding management in aquaculture as it is linked with not only the appearance of food anticipatory activity in fish, but also with the subsequent optimization of feed intake and digestion. The optimization of nutrient utilization relies on the synchronization of numerous digestive processes.

Early works into Nile tilapia from the wild described a diet feeding behavior pattern that resulted in stomach fullness variations in the daytime, with a higher proportion of empty stomachs at night between 00:00-06:00 h. In addition, stomach pH values were lower in the afternoon when stomach fullness was greater. A decrease in pH was found when ingestion started and continued to fall as stomach contents increased, which suggested acid secretion (Getachew 1989). The acid phase of digestion induces the conversion of pepsinogen into pepsin and allows very efficient extracellular protein hydrolysis (Yúfera et al. 2012). As feeding activity depends on daily and seasonal cycles of food availability, an anticipatory response of gastric acidification to regular feeding patterns would be beneficial for finely tuning digestion in fish. Another crucial factor involved in food consumption and breakdown is the secretion of digestive enzymes. In tilapia species, several enzymes involved in digestion have been described to date, such as pepsin-like, amylase, trypsin, alkaline phosphatase and esterase (Nagase 1964, Cockson and Bourne 1972, Moriarty 1973, Klaren et al. 1993, Li and Fan 1997, Tengjaroenkul et al. 2000). A few studies have investigated the existence of daily rhythmicity in the activity of digestive enzymes in Nile tilapia, along with their entrainment by feeding time. In unfed tilapia, Montoya-Mejía et al. (2016) found generally greater activity at nighttime than in the daytime. Thus, pepsin-like activity increased with time of day, whereas alkaline proteases, trypsin and amylase activity declined in the early hours of the morning and peaked at night. Lipase activity showed minimum levels at midday and chymotrypsin activity showed no daily rhythmicity. However, when fish were fed *ad libitum*, pepsin-like activity was greater than under fasting conditions, and its profile showed a secondary peak in the morning and a major peak at night. For alkaline proteases, trypsin, chymotrypsin and amylase, marked activity was observed for most of the day, although a decrease took place at midday. These authors concluded that in the absence of food stimuli, the activity of most digestive enzymes showed daily rhythmicity and were synchronized to the light-dark cycle. Secretion of enzymes was generally lower immediately before sunrise and started to increase at sunset to reach maximum levels at nighttime. When tilapias were fed *ad libitum,* higher nocturnal secretion persisted but, in this case, the average activity levels were higher than in the fasting fish. These results suggest that designing feeding schedules that overlap the times of day when enzyme activity is maximal (i.e. at nighttime) would improve nutrient utilization. However, other studies have shown that Nile tilapia performance is lower when fish are fed at night (Sousa et al. 2012). Therefore, further research on this topic is needed to clarify the potential link among rhythms in digestive enzyme activity, feeding schedules and feed efficiency. In a subsequent study on tilapia, Guerra-Santos et al. (2017) investigated the potential of different mealtimes to entrain the secretion of digestive enzymes. Fish were divided into two groups and fed at either ML or MD. After 30 days under these feeding conditions, alkaline protease in the midgut showed a daily rhythm of activity in both experimental groups, but the time of day at which activity was maximal (rhythm acrophase) differed depending on feeding time. Consequently, the fish fed at ML showed the acrophase almost 2 h later (*Zeitgeber* Time, ZT = 00:43 h) than those fed at MD (ZT = 23:04 h) (Figure 5). However, amylase and acid protease activities showed no significant daily rhythmicity in any feeding group in that study. According to the authors of that study, lack of rhythmicity in amylase activity might be related to either the long digestive process that takes place in tilapia and/or their common consumption of food containing large amounts of carbohydrates which might require constant high amylase secretion levels to be digested. In addition to digestive enzymes, that study also looked at plasma glucose and found a remarkable effect on the acrophase of the daily rhythm, which was observed at ZT = 23:11 h in tilapias fed at ML and at ZT = 11:33 h in the MD group, almost in the antiphase (Guerra-Santos et al. 2017). The daily rhythm in blood glucose might be linked with several metabolic processes, such as hormone release, glucose uptake in peripheral organs, insulin resistance rhythmicity and feeding habits (López-Olmeda et al. 2009, Kalsbeek et al. 2014, Jhaa et al. 2015). In tilapia, glucose levels return to basal concentrations 12 h after feeding, whereas this takes between 18 and 25 h for carnivorous species (Sang-Min et al. 2003, Montoya et al. 2010b), which also seems to be related to the omnivorous habits of tilapia, a species that would better cope with a diet rich in carbohydrates. The feeding-entrainment of digestive enzyme secretion and blood glucose levels in tilapia agrees with the results obtained in other fish species, and supports the hypothesis that the synchronization of digestive functions to a regular meal schedule is extended among vertebrates and confers evolutionary advantages by gaining maximum food use.

More recently, the influence of different temperature regimes on the expression of the genes coding for digestive factors has also been investigated in tilapia larvae. To this end, two experimental groups were formed: daily thermocycle (TC; 31°C: 25°C) and constant temperature (28°C). The photoperiod was set at 12 h L:12 h D and, in the TC group, the thermophase (high temperature phase, 31°C) coincided with the light phase, whereas the cryophase (low temperature, 25°C) coincided with the dark phase. The results of this study revealed a correlation between the

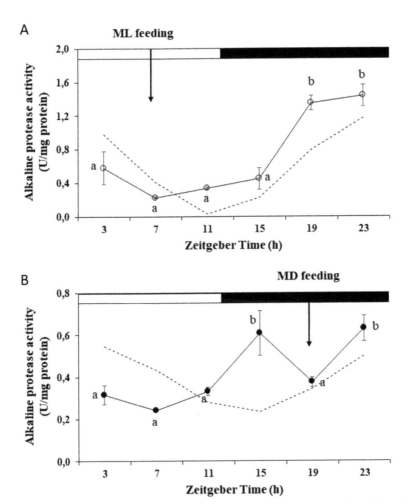

Fig. 5. Daily rhythm of alkaline protease in Nile tilapia fed in the middle of the light (A) and dark (B) phases. Values represent the mean±SEM. The white and black bars at the top of each graph indicate the light and dark periods, respectively. Different letters indicate statistically significant differences between times of day (ANOVA I, p < 0.05). Modified from Guerra-Santos et al. (2017).

existence of thermocycles and more robust rhythmicity in the expression of most of the investigated genes: *chymotrysinogen, trypsinogen, lipase, maltase, isomaltase, neuropeptide y* and *cholecystokinin*. The feeding schedule had a major effect on these rhythms, with acrophases found at around mealtimes (Espirito Santo et al. 2020). These data suggest that using thermocycles instead of constant temperatures might also improve digestive processes, which could be applied when designing tilapia larviculture protocols.

The photoperiod is also an important factor that can influence food utilization in Nile tilapia. In fingerlings, longer photoperiods, such as 18L:6D and 24L:0D, have resulted in better feed conversion rates and protein efficiency than short photoperiods

(6L:18D). However, fish subjected to 0L:24D and 12L:12D obtained similar values, which were also statistically comparable to those observed at 6L:18D, 18L:6D and 24L:0D. These results also correlated with higher growth rates in tilapia fingerlings under the 12L:12D, 18L:6D and 24L:0D photoperiods (Veras et al. 2013). These authors speculated that the best feed conversion rate and protein efficiency observed in tilapia under long photoperiods could be due to the longer intervals between the first and second feeding events, which would improve digestive processes and nutrients retention and, therefore, would increase fish growth. Previous research by Gross et al. (1965) supports this hypothesis as they showed that longer photoperiods can affect fish growth by both stimulating feed intake and improving the feed conversion ratio. A subsequent study looked at the effect of different light regimes on metabolic rates and energy loss in tilapia to find that fish better conserved energy under long photoperiods. For all the photoperiods tested in the study (3L:3D, 6L:6D; 12L:12D; 24L:24D), time-dependent variation appeared in the metabolic rate, which showed significant rhythmicity when fish were kept under a 12L:12D photoperiod, with higher oxygen consumption levels in the light phase. Feeding time also impacted the metabolic rate, with a post-prandial increase that was probably caused by the extra energy needed to transport food along the digestive tract and to carry out all digestive processes (Biswas and Takeuchi 2002).

To summarize, all these studies show that feeding regimes, thermocycles and photoperiods have a potent effect on digestive and metabolic rhythms in Nile tilapia. Accordingly, suitable protocols need to be designed in aquaculture facilities to optimize fish welfare and performance and, at the same time, to avoid unnecessary costs associated with aquafeed.

5. Rhythms of the endocrine system of tilapia

The pituitary is considered the master gland of the vertebrate endocrine system and the axis controlling it, the hypothalamus-pituitary axis, is a key regulator of vertebrate homeostasis (Löhr and Hammerschmidt 2011). In addition to this axis, other important endocrine tissues in vertebrates include the thyroid gland, gut, pancreas, liver, gonads and adrenal glands (interrenal cells in fish) (Kriegsfeld and Silver 2006, Haus 2007, Löhr and Hammerschmidt 2011, Isorna et al. 2017). Biological rhythms are pervasive to most factors produced by these glands and organs of the endocrine system of vertebrates, from fish to mammals (Haus 2007, Cowan et al. 2017, Isorna et al. 2017). These rhythms present different periodicities. Although daily rhythms are the most frequent, in some species several endocrine factors, such as the growth hormone (Gh) or glucocorticoids can display ultradian rhythms as they have pulsatile peaks of release in the daytime (Veldhuis and Bowers 2003, Tondsfelt and Chappell 2012). Seasonal rhythms are frequent in endocrine factors, especially those involved in reproduction (Cowan et al. 2017). In addition, not only hormones display biological rhythms in fish, but also other factors that play an essential role in endocrine physiology, such as hormone receptors and binding proteins (Cowan et al. 2017, Isorna et al. 2017).

In tilapia, daily rhythms have been reported in many hormones from several different endocrine axes and organs. For instance, Costa et al. (2016b) extensively analyzed the rhythmicity of the mRNA levels of factors from the somatotropic (growth) axis of Nile tilapia. This axis is the principal stimulator of growth in fish, mainly due to the actions of Gh and Insulin-like growth factor 1 (Igf-1) on cell hypertrophy and hyperplasia, especially in muscle (Reinecke 2010). Gh is produced in the pituitary gland and acts on many target tissues, with the liver and muscle being the most important (Montero et al. 2000). Besides stimulating growth, Gh induces the production of Igfs in target tissues, mainly in the liver which, in turn, also stimulates growth (Le Roith 2003). Both Gh and Igf exert their actions via their binding to the Gh and Igf receptors present in the cell membranes of target tissues (Canosa et al. 2007). In Nile tilapia, *igf1* expression levels display variations along the day, with the highest values located at around the beginning of the light phase, more precisely 3 h after lights on (Costa et al. 2016b). In their study, the authors did not find any daily rhythms in the other hormones from the somatotropic axis of tilapia, although such variations have been reported in other teleost species (Cowan et al. 2017). Despite lack of rhythms in most hormone transcripts, the authors found daily rhythms in most of their receptors for both Gh (*ghr1* and *ghr2*) and Igf-1 (*igf1ra* and *igf2r*) (Costa et al. 2016b). These rhythms occur in the two peripheral tissues analyzed in their study (liver and muscle), and suggest that daily rhythms in the somatotropic axis of tilapia, and possibly rhythms in growth, may be related to rhythms in receptors other than hormones.

These results also indicated an interesting issue: the response of peripheral tissues to Gh, which would lead to growth, may differ depending on the time of day. In order to test this hypothesis, Costa et al. (2016b) administered exogenous Gh intraperitoneally (i.p.) to Nile tilapia at two different times: in the daytime (3 h after lights on) and at nighttime (3 h after lights off). The physiological response in the target tissues differed depending on the time of Gh administration, with the highest response observed when Gh was injected at night. At this time, a significant increase took place in the *igf1* expression in the liver, which was not observed when Gh was injected in tilapia in the daytime. In addition, night Gh administration also stimulated the expression of Gh (*ghr1* and *ghr2*) and Igf-1 receptors (*igf1ra* and *igf2r*) in the liver (Costa et al. 2016b).

In addition to the somatotropic axis, cortisol rhythms have also been reported in Nile tilapia (Costa et al. 2019). Cortisol is a hormone produced by the interrenal cells of the head kidney in fish (Ellis et al. 2012). It is released after stress and is, hence, considered as one of the main indicators of stress in fish, especially acute stress (Ellis et al. 2012). Nevertheless, cortisol also plays important roles in other processes like energy homeostasis, behavior, reproduction and osmoregulation (Wendelar Bonga 1997, Mommsen et al. 1999, Cowan et al. 2017). Nile tilapia presents a daily plasma cortisol rhythm with maximum values at the beginning of the light phase (Costa et al. 2019). This plasma cortisol profile may be conserved between tilapia species as a similar daily rhythm, with a peak also located at around dawn was reported in *Oreochromis mossambicus* (Binuramesh and Michael 2011). The phase of cortisol rhythms is tightly related to the onset of the species' daily activity patterns. In diurnal animals, cortisol peaks at around dawn, whereas it peaks at around dusk in nocturnal animals (Kaalsbek et al. 2012). This is related to one cortisol function: this hormone is an arousal signal, used to prepare the organism for the activity phase by increasing alertness and stress (Cowan et al. 2017). Thus, the cortisol rhythm in tilapia coincides with what should be expected from a mostly diurnal animal.

Regarding the hormones produced by the gut, Espirito Santo et al. (2020) reported the existence of daily rhythms in the expression of *neuropeptide Y* (*npy*) and *cholecystokinin* (*cck*) on 13 dpf Nile tilapia larvae. The rhythms of both hormone transcripts were located in the first half of the light phase, which coincided with mealtime and indicated synchronization to feeding. These authors also reported an effect of temperature cycles (thermocycles) during development on the ontogeny of the rhythms of these hormones. Both *npy* and *cck* only displayed significant rhythms in the 13-dpf larvae reared under thermocycles, but not in the larvae raised under constant temperature conditions (Espirito Santo et al. 2020).

Melatonin is a hormone that is also known as the time-keeping molecule of vertebrates. Melatonin provides information on day/night and daylength to all peripheral tissues and cells, which contributes to the synchronization of biological rhythms (Cowan et al. 2017). The main contributor of melatonin to the bloodstream is the pineal gland and, thus, plasma melatonin profiles reflect production by this gland (Falcón et al. 2010). A conserved feature of plasma melatonin is that it is produced at nighttime, regardless of whether the animal is diurnal or nocturnal (Cowan et al. 2017). Hence, melatonin in Nile tilapia presents a daily rhythm with high levels throughout the dark phase, with average concentrations of 40-60 pg/ml, and low levels in the daytime, with average values of 10-15 pg/ml (Martinez-Chavez et al. 2008). Daily rhythms in plasma melatonin have also been reported in another tilapia species, *O. mossambicus* (Nikaido et al. 2009). As expected, the profile of melatonin rhythms in *O. mossambicus* was similar to Nile tilapia and the other vertebrate species, with higher melatonin levels at night. However, nocturnal melatonin levels were higher in *O. mossambicus* compared to Nile tilapia, with average levels ranging approximately from 100 to 200 pg/ml (Nikaido et al. 2009). Such differences in nocturnal melatonin levels seem to be species-dependent (Migaud et al. 2007), but probably do not translate into big differences in melatonin action because, for this hormone to play its main physiological role, differences between the day and night plasma concentrations are the most important.

5.1. Daily and seasonal rhythms of reproduction

Reproduction in vertebrates is controlled by the endocrine system through the brain-pituitary-gonad (BPG) axis (Levavi-Sivan et al. 2010, Zohar et al. 2010). In the hypothalamus, environmental signals drive the production of gonadotropin-releasing hormones (Gnrh) which, in turn, induce the production of two gonad-stimulating hormones, or gonadotropins (Gths), in the pituitary: follicle-stimulating hormone (Fsh) and luteinizing hormone (Lh) (Levavi-Sivan et al. 2010). In addition, other brain systems are involved in the control of Gth production, such as the dopaminergic system, kisspeptins and gonadotropin-inhibitory hormone (Gnih) (Zohar et al. 2010). Fsh and Lh are released to the bloodstream and their main target tissues are the gonads, where they stimulate the synthesis of sex steroids (Levavi-Sivan et al. 2010) (see Chapter 9 of the present book for further details on the BPG axis of tilapia).

In teleosts, rhythms on the BPG axis occur at all its levels. These rhythms are coordinated by the environmental and ecological oscillations that result from different superimposed environmental cycles (diurnal, annual, lunar). Ultimately, this synchronizes reproductive partners and sets the species' spawning window to coincide with the most favorable conditions by increasing both success and progeny survival (Cowan et al. 2017). Among the rhythms on the BPG axis of fish, seasonal rhythms have been classically studied the most, mainly for their implication in the seasonal reproduction of many species (Migaud et al. 2010, Oliveira and Sánchez-Vázquez 2010, Cowan et al. 2017). Besides, daily rhythms have also been reported for most of the factors from the BPG axis of fish (Oliveira and Sánchez-Vázquez 2010, Cowan et al. 2017).

In Nile tilapia, the existence of daily rhythms on the BPG axis has been reported by de Alba et al. (2019). In their study, these authors described daily variations in the mRNA expression of factors from the brain (the three *gnrh* isoforms, *kisspeptin2* and *gnih*) and from the pituitary (*fshβ* and *lhβ*) in both males and females reared under a 12:12 LD cycle (de Alba et al. 2019). At the gonadal level, Nile tilapia showed daily rhythms in the mRNA expression of some factors involved in the synthesis pathway of sex steroids. Interestingly, these factors include the genes for antimüllerian hormone (*amh*) and gonadal aromatase (*cyp19a1a*), which codify for the last step enzymes in the synthesis of sex steroids in males (testosterone) and females (17β-estradiol), respectively (Tokarz et al. 2015, de Alba et al. 2019). All in all, these rhythmic factors generate daily rhythms in plasma sex steroids in breeding Nile tilapia, with the highest values located at around the second half of the light phase (Figure 6).

Sex steroids are essential for gametogenesis and reproductive behaviors, and also for feedback to the BPG axis (Levavi-Sivan et al. 2010, Tokarz et al. 2015). In Nile tilapia, rhythms in sex steroids seem to lead to daily rhythms in some factors involved in oocyte quality and fertility, such as a protein from the chorion (Zp3b) and a protein involved in the calcium wave (Fyna) (de Alba et al. 2019). In the end, all these daily rhythms in the endocrine factors from the BPG axis are translated into the existence of daily rhythms in spawning, which have been reported in several fish species, including some tilapia species like Nile tilapia (Figure 6), Mozambique tilapia and *O. aureus* (Baroiller and Toguyeni 2004). In the tilapia reared under laboratory conditions and in a 12:12 LD cycle, spawning activity has been reported to occur in the last hours of the light phase (Figure 6) (Baroiller and Toguyeni 2004). This event would then be preceded by the peaks in sex steroids reported by de Alba et al. (2019), and reinforces the idea that the daily rhythms in the endocrine factors of the BPG axis are coordinated to ultimately time spawning to occur within a narrow daily window.

Finally, tilapia also displays seasonal (annual) reproduction rhythms. Regarding the BPG axis, annual variations in Fsh, Lh and sex steroids have been reported in Mozambique tilapia (Cornish 1998). These variations seem to generate a seasonal rhythm in the gonadosomatic index (GSI) in this species. With Nile tilapia, the existence of seasonal rhythms on the BPG axis remains unknown to date. However, seasonal rhythms of other breeding-related parameters are better characterized. In the wild, Nile tilapia normally breeds throughout the year, but can display a peak in greater activity during some seasons, thus presenting annual spawning rhythms (Baroiller and Toguyeni 2004). The environmental factors that can synchronize this rhythm are still

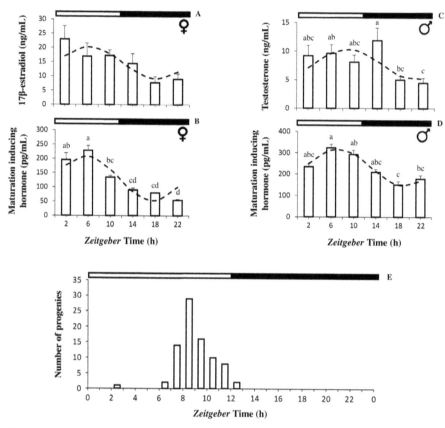

Fig. 6. Daily rhythms in plasma sex steroids of female (A, B) and male (C, D) tilapia, and daily rhythms of spawning (E). The following sex steroids were analyzed by ELISA during a 24-hour cycle: 17β-estradiol in females (A), testosterone in males (C), and maturation-inducing hormone (MIH) (also known as 17α, 20β-dihydroxy-4-pregnen-3-one, or 17α, 20β DHP) in both sexes (B, D). The data of the sex steroid concentrations are expressed as the mean±S.E.M. The spawning activity data are expressed as the number of progenies detected during each hour. The sinusoidal dashed line represents the adjustment to a sinusoidal rhythm. Different letters indicate statistically significant differences among the time points in the same graph (one-way ANOVA, $p < 0.05$). The white and black bars above each graph represent the light phase and the dark phase, respectively. Fish were maintained under a 12:12 LD cycle. Time scale (x-axis) is expressed as *Zeitgeber* Time (ZT), in which ZT0 h corresponds to lights on. Data were adapted from Baroiller and Toguyeni (2004), and from de Alba et al. (2019).

unclear. In an equatorial region like Lake Victoria, spawning activity coincides with the rainy season (Njiru et al. 2006). However, rainfall and greater spawning activity of Nile tilapia do not always correlate in tropical regions (Duponchelle et al. 1999). Other factors, such as seasonal variations in temperature, pH, dissolved oxygen and chlorophyll A levels, have been correlated with seasonal spawning rhythms and may influence them (Duponchelle et al. 1999, Novaes and Carvalho 2012). In addition, daylength (photoperiod) also affects the spawning of Nile tilapia, especially in subtropical and temperate regions, and under laboratory conditions (Duponchelle et al. 1999, Baroiller and Toguyeni 2004). Longer photoperiods seem to stimulate breeding, whereas short photoperiods reduce or inhibit it (see also Chapter 9 of the present book for further details). Such a correlation between the photoperiod and breeding has been reported for Nile tilapia even in tropical regions, where seasonal variations in daylength are minor (e.g. 11.5-12.7 h) (Duponchelle et al. 1999).

6. Concluding remarks

Biological rhythms are ubiquitous at all levels of Nile tilapia physiology and behavior. These rhythms are synchronized by environmental factors, of which the most important are photoperiod, water temperature and food availability. Hence, most biological processes in Nile tilapia are not constantly active, but are restricted and occur at specific times. This confers tilapia major adaptive advantages that allow it to time physiological processes to occur at the most favorable times of the day or year. All this enhances their success, minimizes energy expenditure and reduces the risk of mortality (e.g. due to predation).

The study of these rhythms is important from the point of view of basic knowledge on Nile tilapia biology, but can also prove to be a valuable tool to improve the aquaculture of this species. Indeed this statement can be applied to the aquaculture of any fish species. Despite their importance, research on biological rhythms in fish is still scarce, especially as regards processes of much importance in this industry, such as feeding, growth, reproduction, development and welfare. Hence, biological rhythms should be considered by both researchers and fish farmers.

Acknowledgments

The present research was partially funded by project CHRONOLIPOFISH (RTI2018-100678-A-I00), BLUESOLE (AGL2017-82582-C3-3-R) and CRONOFISH (RED2018-102487-T), granted to JFLO and FJSV by the Spanish Ministry of Science, Innovation and Universities, and co-funded by FEDER. JFLO and LMV were also funded by a "Ramón y Cajal" research fellowship (RYC-2016-20959 and RYC-2017-21835, respectively) granted by the Spanish MINECO.

References cited

Baroiller, J.F. and A. Toguyeni. 2004. The Tilapiini tribe: Environmental and social aspects of reproduction and growth. Fisheries and Aquaculture. Eolss Publishers, Oxford, UK.

Ben-Moshe, Z., G. Vatine, A. Shahar, A. Tovin, P. Mracek, N.S. Foulkes, et al. 2010. Multiple PAR and E4BP4 bZIP transcription factors in zebrafish: Diverse spatial and temporal expression patterns. Chronobiol. Int. 27: 1509–1531.

Betancor, M.B., E. McStay, M. Minghetti, H. Migaud, D.R. Tocher and A. Davie. 2014. Daily rhythms in expression of genes of hepatic lipid metabolism in Atlantic salmon (*Salmo salar* L.). PLoS ONE 9: e106739.

Betancor, M.B., M. Sprague, A. Ortega, F. De la Gándara, D.R. Tocher, R. Ruth, et al. 2020. Central and peripheral clocks in Atlantic bluefin tuna (*Thunnus thynnus*, L.): Daily rhythmicity of hepatic lipid metabolism and digestive genes. Aquaculture 523: 735220.

Binuramesh, C. and R.D. Michael. 2011. Diel variations in the selected serum immune parameters in *Oreochromis mossambicus*. Fish Shellfish Immun. 30: 824–829.

Biswas, A.K. and T. Takeuchi. 2002. Effect of different photoperiod cycles on metabolic rate and energy loss of fed and unfed adult tilapia *Oreochromis niloticus*: Part II. Fisheries Sci. 68: 543–553.

Boujard, T. and J.F. Letherland. 1992. Circadian rhythms and feeding time in fishes. Environ. Biol. Fish. 35: 109–131.

Cahill, G.M. 2002. Clock mechanisms in zebrafish. Cell Tissue Res. 309: 27–34.

Caldini, N.N., N.V. Pereira, V.T. Rebouças and M.V.C. Sá. 2013. Can a small change in tilapia's on-going feeding strategy impair its growth? Acta Sci. 35: 227–234.

Canosa, L.F., J.P. Chang and R.E. Peter. 2007. Neuroendocrine control of growth hormone in fish. Gen. Comp. Endocrinol. 151: 1–26.

Chemineau, P., B. Malpaux, J.P. Brillard and A. Fostier. 2007. Seasonality of reproduction and production in farm fishes, birds and mammals. Animal 1: 419–432.

Cockson, A. and D. Bourne. 1972. Enzymes in the digestive tract of two species of euryhaline fish. Comp. Biochem. Physiol. A 41: 715–718.

Cornish, D.A. 1998. Seasonal steroid hormone profiles in plasma and gonads of the tilapia, *Oreochromis mossambicus*. Water SA 24: 257–264.

Costa, L.S., I. Serrano, F.J. Sánchez-Vázquez and J.F. López-Olmeda. 2016a. Circadian rhythms of clock gene expression in Nile tilapia (*Oreochromis niloticus*) central and peripheral tissues: Influence of different lighting and feeding conditions. J. Comp. Physiol. B 186: 775–785.

Costa, L.S., P.V. Rosa, R. Fortes-Silva, F.J. Sánchez-Vázquez and J.F. López-Olmeda. 2016b. Daily rhythms of the expression of genes from the somatotropic axis: The influence on tilapia (*Oreochromis niloticus*) of feeding and growth hormone administration at different times. Comp. Biochem. Physiol. C 181–182: 27–34.

Costa, L.S., F.G. Araújo, R.R. Paulino, L.J. Pereira, E.J.D. Rodrigues, P.A.P. Ribeiro, et al. 2019. Daily rhythms of cortisol and glucose and the influence of the light/dark cycle on anaesthesia in Nile tilapia (*Oreochromis niloticus*): Does the timing of anaesthetic administration affect the stress response? Aquac. Res. 50: 2371–2379.

Cowan, M., C. Azpeleta and J.F. López-Olmeda. 2017. Rhythms in the endocrine system of fish: A review. J. Comp. Physiol. B 187: 1057–1089.

Cuesta, I.H., K. Lahiri, J.F. López-Olmeda, F. Loosli, N.S. Foulkes and D. Vallone. 2014. Differential maturation of rhythmic clock gene expression during early development in medaka (*Oryzias latipes*). Chronobiol. Int. 31: 468–478.

Da Silva Reis, Y., J.L.R. Leite, C.A.L. de Almeida, D.S.P. Pereira, L.V.O. Vidal, F.G. de Araujo, et al. 2019. New insights into tambaqui (*Colossoma macropomum*) feeding behavior and digestive physiology by the self-feeding approach: Effects on growth, dial patterns of food digestibility, amylase activity and gastrointestinal transit time. Aquaculture 498: 116–122.

Dabrowski, K. 1975. The point of no return in early fish life: An attempt to determine the minimal food requirement. Wiadomosci Ekologiczne 21: 277–293.

Davidson, A.J., S.A. Poole, S. Yamazaki and M. Menaker. 2003. Is the food-entrainable circadian oscillator in the digestive system? Genes Brain Behav. 2: 32–39.

de Alba, G., N.M.N. Mourad, J.F. Paredes, F.J. Sánchez-Vázquez and J.F. López-Olmeda. 2019. Daily rhythms in the reproductive axis of Nile tilapia (*Oreochromis niloticus*): Plasma steroids and gene expression in brain, pituitary, gonad and egg. Aquaculture 507: 313–321.

DeCoursey, P.J. 2004. The behavioral ecology and evolution of biological timing systems. pp. 27–65. *In*: J.C. Dunlap, J.J. Loros and P.J. DeCoursey [eds.]. Chronobiology: Biological Timekeeping. Sinauer Associates, Sunderland, USA.

de Mairan, J.J. 1729. Observation botanique. L´Histoire de l´Academie Royal Scientifique, pp. 47–48.

Dibner, C., U. Schibler and U. Albrecht. 2010. The mammalian circadian timing system: Organization and coordination of central and peripheral clocks. Annu. Rev. Physiol. 72: 517–549.

Dunlap, J.C. 1999. Molecular bases for circadian clocks. Cell 96: 271–290.

Duponchelle, F., P. Cecchi, D. Corbin, J. Nuñez and M. Legendre. 1999. Spawning season variations of female Nile tilapia, *Oreochromis niloticus*, from man-made lakes of Côte d'Ivoire. Env. Biol. Fish. 56: 375–387.

Edwards, L.N. 1988. Cellular and Molecular Basis of Biological Clocks. Springer-Verlag, Berlin.

Ellis, T., H.Y. Yildiz, J. López-Olmeda, M.T. Spedicato, L. Tort, Ø. Øverli, et al. 2012. Cortisol and finfish welfare. Fish Physiol. Biochem. 38: 163–188.

Ellison, A.R., T.M.U. Webster, O. Rey, C.G. de Leaniz, S. Consuegra, P. Orozco-Wengel, et al. 2018. Transcriptomic response to parasite infection in Nile tilapia (*Oreochromis niloticus*) depends on rearing density. BMC Genomics 19: 723.

Espirito Santo, A.H., G. De Alba, Y.S. Reis, L.S. Costa, F.J. Sánchez-Vázquez, R.K. Luz, et al. 2020. Effects of temperature regime on growth and daily rhythms of digestive factors in Nile tilapia (*Oreochromis niloticus*) larvae. Aquaculture 528: 735545.

Falcón, J., H. Migaud, J.A. Muñoz-Cueto and M. Carrillo. 2010. Current knowledge on the melatonin system in teleost fish. Gen. Comp. Endocrinol. 165: 469–482.

Gekakis, N., D. Staknis, H.B. Nguyen, F.C. Davis, L.D. Wilsbacner, D.P. King, et al. 1998. Role of the CLOCK protein in the mammalian circadian mechanism. Science 280: 1564–1569.

Getachew, T. 1989. Stomach pH, feeding rhythm and ingestion rate in *Oreochromis niloticus* L. (Pisces: Cichlidae) in Lake Awasa, Ethiopia. Hydrobiologia 174: 43–48.

Giebultowicz, J.M., R. Stanewsky, J.C. Hall and D.M. Hege. 2000. Transplanted Drosophila excretory tubules maintain circadian clock cycling out of phase with the host. Curr. Biol. 10: 107–110.

Goldman, B., E. Gwinner, F.J. Karsch, D. Saunders, I. Zucker and G.F. Ball. 2004. Circannual rhythms and photoperiodism. pp. 107–142. *In*: J.C. Dunlap, J.J. Loros and P.J. DeCoursey [eds.]. Chronobiology: Biological Timekeeping. Sinauer Associates, Sunderland, USA.

Gómez-Boronat, M., E. Isorna, A. Armirotti, M.J. Delgado, D. Piomelli and N. de Pedro. 2019. Diurnal profiles of N-acylethanolamines in goldfish brain and gastrointestinal tract: Possible role of feeding. Front. Neurosci. 13: 450.

Gross, W.L., E.W. Roelofs and P.O. Fromm. 1965. Influence of photoperiod on growth of green sunfish, *Lepomis cyanellus*. J. Fish. Res. Board Can. 22: 1379–1386.

Guerra-Santos, B., J.F. López-Olmeda, B.O. de Mattos, A.B. Baiao, D.S.P. Pereira, F.J. Sánchez-Vázquez, et al. 2017. Synchronization to light and mealtime of daily rhythms of locomotor activity, plasma glucose and digestive enzymes in the Nile tilapia (*Oreochromis niloticus*). Comp. Biochem. Physiol. A 204: 40–47.

Haus, E. 2007. Chronobiology in the endocrine system. Adv. Drug Deliv. Rev. 59: 985–1014.

Hogenesch, J.B., Y.Z. Gu, S. Jain and C.A. Bradfield. 1998. The basic helix-loop-helix-PAS orphan MOP3 forms transcriptionally active complexes with circadian and hypoxia factors. Proc. Nat. Acad. Sci. USA 95: 5474–5479.

Ibáñez, C. 2017. The 2017 Nobel prize in physiology or medicine – Advanced information: Discoveries of molecular mechanisms controlling the circadian rhythm. Nobelprize.org. Nobel Media AB 2014. http://www.nobelprize.org/nobel_prizes/medicine/laureates/2017/advanced.html [Accessed 10 January 2018].

Iigo, M., H. Kezuka, T. Suzuki, M. Tabata and K. Aida. 1994. Melatonin signal transduction in the goldfish *Carassius auratus*. Neurosci. Biobehav. R. 18: 563–569.

Isorna, E., N. de Pedro, A.I. Valenciano, A.L. Alonso-Gómez and M.J. Delgado. 2017. Interplay between the endocrine and circadian system in fishes. J. Endocrinol. 232: R141–R159.

Jhaa, P.K., E. Challet and A. Kalsbeek. 2015. Circadians rhythm in glucose and lipid metabolism in nocturnal and diurnal mammals. Mol. Cell. Endocrinol. 418: 74–88.

Jin, X., L.P. Shearman, D.R. Weaver, M.J. Zylka, G.J. de Vries and S.M. Reppert. 1999. A molecular mechanism regulating rhythmic output from the suprachiasmatic circadian clock. Cell 96: 57–68.

Johnson, C.H., J. Elliott, R. Foster, K. Honma and R. Kronauer. 2004. Fundamental properties of circadian rhythms. pp. 67–105. *In*: J.C. Dunlap, J.J. Loros and P.J. De Coursey [eds.]. Chronobiology: Biological Timekeeping. Sinauer Associates, Sunderland, USA.

Kalsbeek, A., R. van der Spek, J. Lei, E. Endert, R.M. Buijs and E. Fliers. 2012. Circadian rhythms in the hypothalamo-pituitary-adrenal (HPA) axis. Mol. Cell. Endocrinol. 349: 20–29.

Kalsbeek, A., S. Fleur and E. Fliers. 2014. Circadian control of glucose metabolism. Mol. Metab. 3: 372–383.

Klaren, P.H.M., G. Flik, R.A.C. Lock and S.E. WendelaarBonga. 1993. Ca^{2+} transport across intestinal brush border membranes of the cichlid teleost, *Oreochromis mossambicus*. J. Membrane Biol. 132: 157–166.

Kriegsfeld, L.J. and R. Silver. 2006. The regulation of neuroendocrine function: Timing is everything. Horm. Behav. 49: 557–574.

Kume, K., M.J. Zylka, S. Sriram, L.P. Shearman, D.R. Weaver, X. Jin, et al. 1999. mCRY1 and mCRY2 are essential components of the negative limb of the circadian clock feedback loop. Cell 98: 193–205.

Lazado, C.C., P.B. Pedersen, H.Q. Nguyen and I. Lund. 2017. Rhythmicity and plasticity of digestive physiology in a euryhaline teleost fish, permit (*Trachinotus falcatus*). Comp. Biochem. Physiol. A 212: 107–116.

Le Roith, D. 2003. The insulin-like growth factor system. Exp. Diab. Res. 4: 205–212.

Levavi-Sivan, B., J. Bogerd, E. Mañanos, A. Gómez and J.J. Lareyre. 2010. Perspectives on fish gonadotropins and their receptors. Gen. Comp. Endocrinol. 165: 412–437.

Li, S.N. and D.F. Fan. 1997. Activity of esterases from different tissues of freshwater fish and responses of their isoenzymes to inhibitors. J. Toxicol. Env. Health 51: 149–157.

Li, B.J., Z.X. Zhu, H. Qin, Z.N. Meng, H.R. Lin and J.H. Xia. 2020. Genome-wide characterization of alternative splicing events and their responses to cold stress in tilapia. Front. Genet. 11: 244.

Löhr, H. and M. Hammerschmidt. 2011. Zebrafish in endocrine systems: Recent advances and implications for human disease. Annu. Rev. Physiol. 73: 183–211.

López-Olmeda, J.F. 2017. Nonphotic entrainment in fish. Comp. Biochem. Physiol. A. 203: 133–143.

López-Olmeda, J.F., M. Egea-Álvarez and F.J. Sánchez-Vázquez. 2009. Glucose tolerance in fish: Is the daily feeding time important? Physiol. Behav. 96: 631–636.

López-Olmeda, J.F., E.V. Tartaglione, H.O. de la Iglesia and F.J. Sánchez-Vázquez. 2010. Feeding entrainment of food-anticipatory activity and per1 expression in the brain and liver of zebrafish under different lighting and feeding conditions. Chronobiol. Int. 27: 1380–1400.

López-Olmeda, J.F., I. López-García, M.J. Sánchez-Muros, B. Blanco-Vives, R. Aparicio and F.J. Sánchez-Vázquez. 2012. Daily rhythms of digestive physiology, metabolism and behaviour in the European eel (*Anguilla anguilla*). Aquacult. Int. 20: 1085–1096.

Madrid, J.A. 2006. Los relojes de la vida. Una introducción a la cronobiología. pp. 39–81. *In*: J.A. Madrid and M.A. Rol de Lama [eds.]. Cronobiología básica y clínica. Editec@red, Madrid, Spain.

Martín-Robles, A.J., D. Whitmore, F.J. Sánchez-Vázquez, C. Pendón and J.A. Muñoz-Cueto. 2012. Cloning, tissue expression pattern and daily rhythms of *Period1*, *Period2*, and *Clock* transcripts in the flatfish Senegalese sole, *Solea senegalensis*. J. Comp. Physiol. B 182: 673–685.

Martinez-Chavez, C.C., S. Al-Khamees, A. Campos-Mendoza, D.J. Penman and H. Migaud. 2008. Clock-controlled endogenous melatonin rhythms in Nile tilapia (*Oreochromis niloticus*) and African Catfish (*Clarias gariepinus*). Chronobiol. Int. 25: 31–49.

Migaud, H., A. Davie, C.C. Martinez-Chavez and S. Al-Khamees. 2007. Evidence for differential photic regulation of pineal melatonin synthesis in teleosts. J. Pineal Res. 43: 327–335.

Migaud, H., A. Davie and J.F. Taylor. 2010. Current knowledge on the photoneuroendocrine regulation of reproduction in temperate fish species. J. Fish Biol. 76: 27–68.

Mistlberger, R.E. 2009. Food-anticipatory circadian rhythms: Concepts and methods. Eur. J. Neurosci. 30: 1718–1729.

Mommsen, T.P., M.M. Vijayan and T.W. Moon. 1999. Cortisol in teleosts: Dynamics, mechanisms of action, and metabolic regulation. Rev. Fish Biol. Fish. 9: 211–268.

Montero, M., L. Yon, S. Kikuyama, S. Dufour and H. Vaudry. 2000. Molecular evolution of the growth hormone-releasing hormone/pituitary adenylate cyclase-activating polypeptide gene family. Functional implication in the regulation of growth hormone secretion. J. Mol. Endocrinol. 25: 157–168.

Montoya, A., J.F. López-Olmeda, M. Yúfera, M.J. Sánchez-Muros and F.J. Sánchez-Vázquez. 2010a. Feeding time synchronises daily rhythms of behaviour and digestive physiology in gilthead seabream (*Sparusaurata*). Aquaculture 306: 315–321.

Montoya, A., J.F. López-Olmeda, A.B.S. Garayzar and F.J. Sánchez-Vázquez. 2010b. Synchronization of daily rhythms of locomotor activity and plasma glucose, cortisol and thyroid hormones to feeding in Gilthead seabream (*Sparusaurata*) under a light-dark cycle. Physiol. Behav. 101: 101–107.

Montoya-Mejía, M., H. Rodríguez-González and H. Nolasco-Soria. 2016. Circadian cycle of digestive enzyme production at fasting and feeding conditions in Nile tilapia, *Oreochromis niloticus* (Actinopterygii: perciformes: cichlidae). Acta Ichthyol. Piscat. 46: 163–170.

Moriarty, D.J.W. 1973. The physiology of digestion of blue-green algae in the cichlid fish, *Tilapia nilotica*. J. Zool. 171: 25–39.

Mracek, P., C. Santoriello, M.L. Idda, C. Pagano, Z. Ben-Moshe, Y. Gothilf, et al. 2012. Regulation of *per* and *cry* genes reveals a central role for the D-box enhancer in light-dependent gene expression. PLoS ONE 7: e51278.

Nagase, G. 1964. Contribution to the physiology of digestion in Tilapia mossambica Peters: Digestive enzymes and the effects of diets on their activity. Zeitschriftfürvergleichende Physiologie 49: 270–284.

Nikaido, Y., S. Ueda and A. Takemura. 2009. Photic and circadian regulation of melatonin production in the Mozambique tilapia *Oreochromis mossambicus*. Comp. Biochem. Physiol. A 152: 77–82.

Njiru, M., J.E. Ojuok, J.B. Okeyo-Owuor, M. Muchiri, M.J. Ntiba and I.G. Cowx. 2006. Some biological aspects and life history strategies of Nile tilapia *Oreochromis niloticus* (L.) in Lake Victoria, Kenya. Afr. J. Ecol. 44: 30–37.

Novaes, J.L.C. and E.D. Carvalho. 2012. Reproduction, food dynamics and exploitation level of *Oreochromis niloticus* (Perciformes: Cichlidae) from artisanal fisheries in Barra Bonita Reservoir, Brazil. Rev. Biol. Trop. 60: 721–734.

Oliveira, C. and F.J. Sánchez-Vázquez. 2010. Reproduction rhythms in fish. pp. 185–215. *In*: Kulczykowska, E., W. Popek and B.G. Kapoor [eds.]. Biological Clock in Fish. Science Publishers, Enfield, USA.

Panda, S., M.P. Antoch, B.H. Miller, A.I. Su, A.B. Schook, M. Straume, et al. 2002. Coordinated transcription of key pathways in the mouse by the circadian clock. Cell 109: 307–320.

Pando, M.P. and P. Sassone-Corsi. 2002. Unraveling the mechanisms of the vertebrate circadian clock: Zebrafish may light the way. BioEssays 24: 419–426.

Paredes, J.F., L.M. Vera, I. Navarro, F.J. Martínez-López and F.J. Sánchez-Vázquez. 2014. Circadian rhythms of gene expression of lipid metabolism in Gilthead Sea bream liver: Synchronization to light and feeding time. Chronobiol. Int. 31: 613–626.

Paredes, J.F., J.F. López-Olmeda, F.J. Martínez-López and F.J. Sánchez-Vázquez. 2015. Daily rhythms of lipid metabolic gene expression in zebrafish liver: Response to light/dark and feeding cycles. Chronobiol. Int. 32: 1438–1448.

Plautz, J.D., M. Kaneko, J.C. Hall and S.A. Kay. 1997. Independent photoreceptive circadian clocks throughout Drosophila. Science 278: 1632–1635.

Preitner, N., F. Damiola, J. Zakany, D. Duboule, U. Albrecht and U. Schibler. 2002. The orphan nuclear receptor REV-ERBa controls circadian transcription within the positive limb of the mammalian circadian oscillator. Cell 110: 251–260.

Reebs, S.G. 2002. Plasticity of diel and circadian activity rhythms in fishes. Rev. Fish Biol. Fish. 12: 349–371.

Reinecke, M. 2010. Influences of the environment on the endocrine and paracrine fish growth hormone-insulin-like growth factor-I system. J. Fish Biol. 76: 1233–1254.

Rensing, L. and P. Ruoff. 2002. Temperature effect on entrainment, phase shifting, and amplitude of circadian clocks and its molecular bases. Chronobiol. Int. 19: 807–864.

Reppert, S.M. and D.R. Weaver. 2002. Coordination of circadian timing in mammals. Nature 418: 935–941.

Ripperger, J.A., L.P. Shearman, S.M. Reppert and U. Schibler. 2000. CLOCK, an essential pacemaker component, controls expression of the circadian transcription factor DBP. Gene Dev. 14: 679–689.

Sánchez, J.A. and F.J. Sánchez-Vázquez. 2009. Feeding entrainment of daily rhythms of locomotor activity and clock gene expression in zebrafish brain. Chronobiol. Int. 26: 1120–1135.

Sánchez, J.A., J.F. López-Olmeda, B. Blanco-Vives and F.J. Sánchez-Vázquez. 2009. Effects of feeding schedule on locomotor activity rhythms and stress response in sea bream. Physiol. Behav. 98: 125–129.

Sánchez-Vázquez, F.J., M. Iigo, J.A. Madrid and M. Tabata. 2000. Pinealectomy does not affect the entrainment to light nor the generation of the circadian demand-feeding rhythms of rainbow trout. Physiol. Behav. 69: 455–461.

Sang-Min, L., K. Kioung-Duck and P.L. Santosh. 2003. Utilization of glucose, maltose, dextrin and cellulose by juvenile flounder (*Paralichthys olivaceus*). Aquaculture 221: 427–438.

Shearman, L.P., S. Sriram, D.R. Weaver, E.S. Maywood, I. Chaves, B. Zheng, et al. 2000. Interacting molecular loops in the mammalian circadian clock. Science 288: 1013–1019.

Sousa, R.M.R., C.A. Agostinho, F.A. Oliveira, D. Argentim, P.K. Novelli and S.M.M. Agostinho. 2012. Productive performance of Nile tilapia (*Oreochromis niloticus*) fed at different frequencies and periods with automatic dispenser. Arq. Bras. Med. Vet. Zoo. 64: 192–197.

Tengjaroenkul, B., B.J. Smith, T. Caceci and S.A. Smith. 2000. Distribution of intestinal enzyme activities along the intestinal tract of cultured Nile tilapia, *Oreochromis niloticus* L. Aquaculture 182: 317–327.

Tessmar-Raible, K., F. Raible and E. Arboleda. 2011. Another place, another timer: Marine species and the rhythms of life. BioEssays 33: 165–172.

Tokarz, J., G. Möller, M. Hrabe De Angelis and J. Adamski. 2015. Steroids in teleost fishes: A functional point of view. Steroids 103: 123–144.

Tonsfeldt, K.J. and P.E. Chappell. 2012. Clocks on top: The role of the circadian clock in the hypothalamic and pituitary regulation of endocrine physiology. Mol. Cell. Endocrinol. 349: 3–12.

Ueda, H.R., W. Chen, A. Adachi, H. Wakamatsu, S. Hayashi, T. Takasugi, et al. 2002. A transcription factor response element for gene expression during circadian night. Nature 418: 534–539.

Vatine, G., D. Vallone, Y. Gothilf and N.S. Foulkes. 2011. It's time to swim! Zebrafish and the circadian clock. FEBS Lett. 585: 1485–1494.

Velarde, E., R. Haque, P.M. Iuvone, C. Azpeleta, A.L. Alonso-Gómez and M.J. Delgado. 2009. Circadian clock genes of goldfish, *Carassiusauratus*: cDNA cloning and rhythmic expression of *Period* and *Cryptochrome* transcripts in retina, liver, and gut. J. Biol. Rhythms 24: 104–113.

Veldhuis, J.D. and C.Y. Bowers. 2003. Human GH pulsatility: An ensemble property regulated by age and gender. J. Endocrinol. Invest. 26: 799–813.

Vera, L.M., N. De Pedro, E. Gómez-Milán, M.J. Delgado, M.J. Sánchez-Muros, J.A. Madrid, et al. 2007. Feeding entrainment of locomotor activity rhythms, digestive enzymes and neuroendocrine factors in goldfish. Physiol. Behav. 90: 518–524.

Vera, L.M., L. Cairns, F.J. Sánchez-Vázquez and H. Migaud. 2009. Circadian rhythms of locomotor activity in the Nile tilapia *Oreochromis niloticus*. Chronobiol. Int. 26: 666–681.

Vera, L.M., P. Negrini, C. Zagatti, E. Frigato, F.J. Sánchez-Vázquez and C. Bertolucci. 2013. Light and feeding entrainment of the molecular circadian clock in a marine teleost (*Sparusaurata*). Chronobiol. Int. 30: 649–661.

Veras, G.C., L.D. Solis Murgas, P.V. Rosa, M.G. Zangeronimo, M.S. Silva Ferreira and J.A. Solis-De Leon. 2013. Effect of photoperiod on locomotor activity, growth, feed efficiency and gonadal development of Nile tilapia. Rev. Bras. Zootecn. 42: 844–849.

Welsh, D.K., J.S. Takahashi and S.A. Kay. 2010. Suprachiasmatic nucleus: Cell autonomy and network properties. Annu. Rev. Physiol. 72: 551–577.

Wendelaar Bonga, S.E. 1997. The stress response in fish. Physiol. Rev. 77: 591–625.

Whitmore, D., N.S. Foulkes, U. Strähle and P. Sassone-Corsi. 1998. Zebrafish *Clock* rhythmic expression reveals independent peripheral circadian oscillators. Nature Neurosci. 1: 701–707.

Whitmore, D., N.S. Foulkes and P. Sassone-Corsi. 2000. Light acts directly on organs and cells in culture to set the vertebrate circadian clock. Nature 404: 87–91.

Wu, P., W. Chu, X. Liu, X. Guo and J. Zhang. 2018. The influence of short-term fasting on muscle growth and fiber hypotrophy regulated by the rhythmic expression of clock genes and myogenic factors in Nile tilapia. Mar. Biotechnol. 20: 750–768.

Wu, P., J. Cheng, L. Chen, J. Xiang, Y. Pan, Y. Zhang, et al. 2020. *Nr1d1* affects autophagy in the skeletal muscles of juvenile Nile tilapia by regulating the rhythmic expression of autophagy-related genes. Fish Physiol. Biochem. 46: 891–907.

Yadu, Y. and M. Shedpure. 2002. Pinealectomy does not modulate the characteristics of 24-h variation in air-gulping activity of *Clarias batrachus*. Biol. Rhythm Res. 33: 141–150.

Yúfera, M., F.J. Moyano, A. Astola, P. Pousao-Ferreira and G. Martínez-Rodríguez. 2012. Acidic digestion in a teleost: Postprandial and circadian pattern of gastric pH, pepsin activity, and pepsinogen and proton pump mRNAs expression. PLoS ONE 7: e33687.

Zhdanova, I.V. and S.G. Reebs. 2006. Circadian rhythms in fish. pp. 197–238. *In*: K.A. Sloman, R.W. Wilson and S. Balshine [eds.]. Behaviour and Physiology of Fish. Academic Press, Vol. 24.

Zohar, Y., J.A. Muñoz-Cueto, A. Elizur and O. Kah. 2010. Neuroendocrinology of reproduction in teleost fish. Gen. Comp. Endocrinol. 165: 438–455.

CHAPTER
13

Tilapia Production in Aquaponics

Morris Villarroel[1]*, Manuel Martín Mariscal-Lagarda[2] and Eugenio García Franco[3]

[1] CEIGRAM-Agricultural Production, ETSIAAB-UPM, Avenida Puerta de Hierro 2, 28040, Madrid, Spain

[2] Universidad Estatal de Sonora-Benito Juárez, Laboratorio de sistemas acuapónicos delprograma de IngenieroenHorticultura.Fraternidad y 5 de mayo, Villa Juárez, Sonora, México

[3] Aquacria C.A. Madrid, Spain

1. An introduction to aquaponics

Freshwater is a scarce resource, making competition for access to it one of the most important challenges in developing countries (FAO 2012). In addition, the effects of climate change, extreme meteorological phenomena and financial crises may hinder the ability to feed a growing human population. In a recent conference of the United Nations (Río+20) to develop the 2030 agenda for sustainable development, members aimed to create an institutional framework to support a green economy, mostly by improving management and efficiency throughout the food value chain. One of those proposed food production systems is aquaponics, which optimizes the use of available resources (water, soil, space and capital), by producing plants and aquatic animals in a sustainable manner, taking more advantage of the water used while reducing environmental contamination and the dependence on agrochemicals (Stevenson et al. 2010, Mariscal-Lagarda et al. 2012). In 2015, the European Parliament included aquaponics as one of the "10 technologies which could change our lives" (Van Woensel and Archer 2015).

1.1. What is aquaponics?

In general, aquaponics is defined as a system that combines the production of aquatic species in recirculation units with plants in hydroponic units (Rakocy et al. 2006, Enduta et al. 2011). According to Goddek et al. (2019), the agricultural production of Aztecs (México) known as *chinampas* may have been one of the first examples of a type of aquaponics, developed between 1150 and 1350 BC. There, plants were placed in floating rafts in shallow lakes with a muddy substrate, along with fish

*Corresponding author: morris.villarroel@upm.es

(Crossley 2004). Probably, there were even earlier attempts to develop aquaponics in rice fields in southern China, extending to south-eastern Asia along with the help of Chinese colonists from Yunnan. With technological advances in the 20th century, Lewis and Naegel developed the first practical and efficient aquaponics system using fish and plants in the 1970s. Later, Waten and Busch, and then Rakocy, made significant advances in the design and development of those systems, defining the term aquaponics as we understand it today (Palm et al. 2018). Recently, Goddek et al. (2019) argued that aquaponic systems can also be decoupled, that is, that the water in the system is not recirculated continuously among fish and plants but only when desirable. Thus, modern day aquaponics can be defined as a production system where water is recycled continuously or intermittently between a recirculation unit with fish and a hydroponic unit with plants.

1.2. Advantages and disadvantages of aquaponics

According to several authors, aquaponic systems offer advantages over traditional aquaculture and hydroponics since they optimize the use of water, decrease the need for nutrients (for the plants), decrease environmental contamination, increase the sources of income (economic return), and allow producers to use land not normally suitable for traditional agriculture (Flaherty and Halwart 2000, Rakocy et al. 2006, Enduta et al. 2011, Mariscal-Lagarda et al. 2012, Somerville et al. 2014, Henares et al. 2019).

However, some disadvantages include a greater risk for the transmission of disease or contaminants between plants and fish, and more complex system handling since fish and plants have different nutritional and environmental requirements, so that small errors can cause a complete collapse of the system (see Rakocy et al. 2006, Somerville et al. 2014). Normally, the plants require additional fertilizer since the fish cannot always provide all their essential nutrients via feed waste. Lastly, another disadvantage is that recirculation aquaculture systems (RAS) are highly dependent on electrical energy to run the pumps and assure a constant water flow.

1.3. Components of an aquaponic system

Aquaponic systems consist of three main parts, which house the three main organisms grown: the aquaculture or fish section, the filtration or bacteria section, and the hydroponic or plant section (Figure 1). The whole system is usually kept in a greenhouse, which provides protection from the elements but allows free access to sunlight. The aquaculture section consists of tanks that house the fish and provide aeration, feeders and recirculating water. The waste from the fish (both solid and liquid waste) are removed from the tank (in the water that leaves the tank) to a solids filter, followed by a biofilter where bacteria are kept at high densities for the process of nitrification (Rakocy et al. 2006). The hydroponic system is usually one of three types, which will be considered in more detail below (floating rafts, nutrient film technique, or substrate beds), with an irrigation system (Rakocy et al. 2006, Enduta et al. 2011, Somerville et al. 2014, Henares et al. 2019).

Normally, the fish are grown in circular tanks, although they can also be square shaped, rectangular or other geometric shapes. Independent of the shape of the tanks,

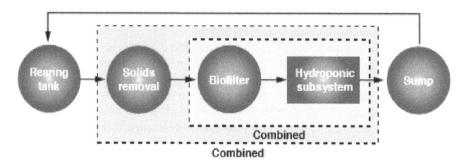

Fig. 1. Optimal arrangement of aquaponic system components (not to scale), based on Rakocy et al. (2006). Depending on the system, the hydroponic system can act as a biofilter as well (combined), or the solids removal can also be combined with the biofilter and hydroponic components (as, for example in the media-bed technique).

these receptacles are the main repository for water in the aquaponics system, and so their cost should be carefully considered before installation. For example, for circular tanks it is important to note that the greater their diameter, the higher the construction costs, but the lower the cost of water storage.

1.4. Basic working principles

The feed used to grow the fish is composed of several ingredients that provide different nutrients (macro and micronutrients). Among the main nutrients are nitrogen (N, from protein), phosphorous (P), potassium (K), calcium (Ca), magnesium (Mg), iron (Fe), manganese (Mg), zinc (Zn) and copper (Cu). The N and P waste are the most important nutrients for plant growth. According to Páez-Osuna et al. (2005), fish incorporate only about 55% of the N and 75% of the P in feed. Incorporation rates for the other nutrients have not been studied in detail and are often more difficult to calculate due to their low concentration in feeds. In any case, it is safe to assume that the nutrients that are not absorbed by the fish are dissolved in faeces or urine which dissolve in the water, representing a net economic loss and increased environmental contamination if not reused (Mariscal-Lagarda et al. 2012).

The main nitrogen source for plants comes from the feed given to the fish, via metabolic waste and from unconsumed feed. Most nitrogen is excreted directly into the water via the gills as unionized ammonia (NH_3^-), which is ionized to ammonium (NH_4^-) and oxidized via the action of the bacteria in the biofilter through nitrification, producing nitrite (NO_2^-) and nitrate (NO_3^-), via the intervention of *Nitrosomonas* and *Nitrobacter*, respectively. Both NH_3^- and NO_2^- are quite toxic for fish, even at low concentrations, although they can be used by plants as nutrients for growth, but NO_3^- are preferred. Although the toxicity of NH_3^- and NO_2^- depends on the species and age of fish, in general, their concentrations should be less than 0.1 mg/l (Van Wyk 1999, Nicovita 2002, Somerville et al. 2014). Nitrate is less toxic and can easily reach concentrations above 100-200 mg/l, although levels above 500 mg/l are thought to be toxic for tilapia. In aquaponics, the bacteria in the biofilter eliminate the NH_3^- and NO_2^- while the plants absorb the NO_3^-, which assures that all the nitrogen waste products from the feed are maintained at appropriate levels. Thus, in aquaponics, the biofilter and plants effectively treat the effluent from the fish, cleaning the water and generating environmental benefits. According to Boyd (2002) and Collins et al. (2005), the salinity of the freshwater used in aquaponics is typically low, around 1 to 10 g/l of salt and an electrical conductivity of less than 1500 μS/cm.

1.5. Types of hydroponic units

The word hydroponics comes from the Greek word *hydro* or water and *ponos* or cultivation or work. Hydroponics is used to grow plants without soil, which is very efficient (Samperio-Ruíz 2000, Rodríguez-Delfínetal.2004). Based on Love et al. (2015), three main systems are used for aquaponics: the method of floating rafts (60% of all companies surveyed), followed by substrate beds (19%) and the nutrient film technique (NFT) system (15%). All three systems are shown in Figure 2. Briefly, in the floating raft system the roots of the plants are submerged in the water either directly above or beside the fish, usually on a polystyrene sheet. Sometimes the plant stem can be placed in a sponge or inside a plastic cup with gravel within the sheet. This system works well for green leaf plants. Typically, the rafts float on long narrow tanks about 30 m long, 1.2 m wide and 20-30 cm deep, with a slope of about 1-2% towards the exit. The polystyrene sheets which hold the plants are about 2.44 m by 1.22 m and about 3-5 cm thick.

Fig. 2. Students showing the roots of lettuce in a floating raft system, under which tilapia are housed (A); a tank being prepared to house tilapia and a series of floating rafts on top (B); an NFT system with lettuce growing in pipes (C); a media bed system with plants ready to harvest (D). All photos taken of units at the Universidad Estatal de Sonora-Benito Juárez, by M. Mariscal-Lagarda.

With the NFT system, plant roots are placed in pockets in a tube or plastic bag with a narrow stream of water. This system is recommended for leafy greens (lettuce, spinach, etc.) and can take up less space than the other types since several levels/ floors of plants can be produced to increase the overall surface area.

Plants can also be placed in substrate (solid media) which serves mostly as a support and anchor (Rodríguez-Delfín et al. 2004). Almost any type of plant can be used in this type of system, but it is mostly for larger plants which require some sort of supporting structure. The substrate is normally placed in pots or cement blocks filled with plastic or a polyethylene liner. The substrate itself can be organic (coco fibre or rice hulls), or inorganic (sand, gravel or perlita). According to Rodríguez-Delfín et al. (2004), the best substrates have a high water retention, enough air circulation to provide air to the roots, adequate particle size to balance air and water, low salinity, are light-weight, easy to mix and clean, cost effective and have little effect on water pH. Any substrate bed needs its own irrigation system, valves to drain the bed, a sump to collect water from substrate and pumps to recirculate the water.

2. Aquaponics with tilapia

In terms of plants, according to Somerville et al. (2014), approximately 150 species have been used in aquaponics (vegetables, flowers, aromatic herbs and small trees), generally classified in terms of their nutritional requirements: (1) low requirements (herbs and green leafy vegetables, such as lettuce, chard, basil, mint, parsley, coriander, chives, pakchoi and watercress and some legumes such as peas and beans), (2) medium requirements (kale, cauliflower, broccoli and kohlrabi, beet, onion and carrot), and (3) high requirements (tomato, cucumber, peppers, squash, eggplant and strawberries).

In terms of scientific publications, the most common pairings of tilapia and plants involve lettuce, aromatic herbs (basil, coriander, peppermint), tomato and cucumber (McMurtry et al. 1997, Rakocy et al. 2006, Hanson et al. 2008, Yousef et al. 2008, Graber and Junge 2009, Halwart and Van Dam, 2010, Abdul-Rahman et al. 2011, Liang and Chien 2013, Limbu et al. 2017, Paulus et al. 2019, Thomas et al. 2019, Effendi and Pratiwi 2019). In their review of commercial practice in aquaponics, Love et al. (2015) found that tilapia were used in 69% of the companies and lettuce and basil in 70% of cases, with tomato, cucumber and peppers making up around 20%. Nonetheless, the most common combination is tilapia with lettuce or tilapia with basil, probably due to the characteristics and ease with which they can be grown.

Since aquaponics uses clean freshwater, a number of species of fish can be used. Worldwide, tilapia is the most common, including several species such as *Oreochromis niloticus, Oreochromis aureus* and *Oreochromis mossambicus,* and their hybrids such as red tilapia and white or "*Rocky Mountain*" tilapia. In this chapter, when we refer to tilapia we are referring to *Oreochromis niloticus*, unless specified otherwise. An international survey about commercial production systems in aquaponics by Love et al. (2015) found that, of the 257 producers surveyed, 69% used tilapia (*Oreochromis* sp.), while 43% used ornamental fish, 25% used catfish, and fewer proportions for other species (the sum of percentages adds up to over 100% since many producers grow more than one species).

There are several reasons why tilapia is favoured over other species for aquaponics. Those can be divided into three main advantages: market demand, good growth and tolerance to different levels of water quality. In terms of markets, tilapia is a well-known fish on an international level with a high commercial acceptance, mostly due to its neutral taste and firm flesh, making it quite versatile for a number of dishes and recipes (see Chapter 15 of this book for more details). Tilapia also comes in a number of colours, like grey, white and red, which can be more or less preferred by different consumers.

In terms of growth, new lines of tilapia are among the fastest growing fish in aquaculture, with a good conversion factor (feed/growth), reaching market size (600 g in 6–8 months, Somerville et al. 2014). Normally, in commercial aquaponics fish densities can range from 15 to 25 kg/m^3, although, according to Rakocy et al. (2006), some systems can withstand densities of up to 70 kg/m^3. Tilapia can be grown in small tanks and at high densities, although there may be problems of aggression if the tanks are too small or if animal density is too low (too few fish per tank) since they can develop territorial behaviour and become quite aggressive. Since tilapia are omnivorous, they have lower protein requirements than carnivorous fish, which means that less nitrogen waste is produced per kilogram of fish, or to put it another way, more fish can be grown per m^3 for any plant production.

Along those same lines, tilapia tolerate a wide range of environmental conditions (i.e. water quality) (Nicovita 2002). First of all, they are eurythermal, tolerating a wide range of temperatures, from about 14°C to 36°C. This is important when they are kept in greenhouses since the water will be generally quite warm, favouring the optimal range for growth (27-30 °C). Also, tilapia will adapt to seasonal and daily variations in temperature typical of greenhouses. Tilapia are also quite resistant to low oxygen levels, which are often below 5 mg/l, which helps to decrease the cost of aerating the water but also provides a safety margin in case of an interruption in the supply of electrical energy to the pumps and aerators. Tilapia are also resistant to a wide range of pH, from 6.5 to 9, which can be good to accommodate the requirements of the plants. Finally, tilapia are resistant to high levels of suspended solids, as well as nitrogen, phosphorous and potassium compounds (Rakocy et al. 2006, Timmons and Ebling 2010).

Despite recent advances in commercial aquaponics, most companies still rely on strong ties with academia to help make their business more profitable. In the next section, we will review some of those technical advances over the years, most of which have been published in research journals or presented at scientific conferences.

2.1. Scientific advances to help foster aquaponics with tilapia

As mentioned above, tilapia is one of the most popular fish for aquaponic systems, starting with the work by Rakocy et al. (2004), who grew Nile tilapia with basil (*Ocimum basilicum* 'Genovese') in a commercial scale system, using batch and staggered harvests to compare it with open capped basil production, evaluating the proportion of fish feed to basil growing area. The results indicated that the biomass obtained from tilapia was 4.37 tons per year and the average basil production for the batch, staggered harvest and open field system was 2.0, 1.8 and 0.6 kg/m^2, respectively. The feed ratio: plant area was 81.4 g/m^2/day for the batch system and 99.6 g/m^2/day for the open field system. For the batch production, there were nutritional deficiencies in basil; however, in the staggered crops these deficiencies were not observed due to the low demand of nutrients in the initial phase, which saved them up for the final stages of growth. From this perspective, the most sustainable tilapia-basil system requires staggered basil production.

In another study, Yousef et al. (2008) developed an aquaponic system with tilapia and lettuce (*Lactuca sativa*) to maximize water use in terms of fish feed and plants. The fish density used was 160 tilapia/m^3 and the plant density was 42 lettuce plants/m^2. During 186 days of growth, only 1.4% of the total water volume in the system needed to be topped up to compensate for losses due to evaporation and transpiration. The tilapia (average final weight of 285 g) had a water consumption of 320 l/kg fish and a production yield of 33.5 kg/m^3. For lettuce, the average final harvested weight was 251 g. In conclusion, for that study, a ratio of 56 g of fish feed/ m^2/day provided enough nutrients for the lettuce, while better quality plants were obtained at a plant density of 28 lettuce/m^2.

Kotzen and Appelbaum (2010) worked on an aquaponic system using brackish water with an electrical conductivity of 4500 μS/cm to evaluate the production of tilapia with various vegetables (celery, chard, parsley and chives) in two hydroponic unit types, deep water culture (DWC) and media bed. After a 101 day cycle, the tilapia reached an average weight of 313 g with a yield of 12.5 kg/m^3. All the plants showed good growth in both hydroponic systems, although it was necessary to supplement with iron chelate to fulfil the needs of the plants (although no mention is made of how that may have affected the fish).

Another aquaponic experiment grew red tilapia (*Orechromis* sp.) and water spinach (*Ipomoea aquatica*) to evaluate growth and production using two photoperiods (12-h and 24-h light per day) and three feeding frequencies (2, 4 and 6 times per day). Results showed that water loss was 3.3% over the four weeks of the experiment, while water quality remained constant and at adequate levels for fish and plants. The average weight gain was 43.9% and 169% for tilapia and water spinach, respectively. The 24-hour light photoperiod improved fish growth by 2.4% and 12% for plants, while increasing feeding frequency favoured water quality and fish and plant growth by 4.9% and 11%, respectively (Liang and Chien 2013).

Goada et al. (2015) evaluated the economic characteristics of two aquaponic systems with tilapia and various vegetables, including broccoli. They compared the NFT with DWC and soil culture. The results showed that both aquaponic systems achieved better average net incomes compared to soil culture. Also, NFT and DWC were able to cover the operating costs and achieve an economic surplus of 53% and 47%, respectively. The results confirmed the ability of these two systems to bear the burden of increased production costs or a fall in the prices of the fish and vegetables produced (risk reduction). Both systems can be considered as successful aquaponic models, in which, under the conditions evaluated, the payback period was 2.75 years on average.

According to Diem et al. (2017), aquaponic systems can be used to grow vegetable crops without additional fertilization, which helps producers by reducing production costs. In their experiment, they grew tilapia (*Oreochromis niloticus*) with the *Ipomoea aquatica*, *Lactuca sativa* and *Canna glauca*, evaluating three recirculation rates (50%, 200% and 400%) to measure the effects on water quality and growth of tilapia at two densities (122 and 220 fish/m³). The results showed that most water quality parameters were within adequate ranges at 400% recirculation, while the best growth and survival of tilapia was at 122 fish/m³, with a 400% turnover. At that same recirculation rate, the plants could effectively remove harmful concentrations of NH_4-N and NO_2-N. As for the plants, growth was best at a 50% recirculation rate, with *C. glauca* producing the best biomass (3045 g DW/m²/year) followed by *I. aquatica* (1400 g DW/m²/year).

An evaluation of the growth, development and quality of the eucalyptus plant (*Eucalyptus grandis* W.) was conducted by irrigating with different proportions of effluent from a tilapia RAS system, and commercial nutrient solutions: (i) only fish water effluent, (ii) only nutrient solution, (iii) 50% effluent + 50% nutrient solution and (iv) 75% effluent and 25% nutrient solution, all applied daily. The concentration of NPK nutrients in the leaves of the eucalyptus and the height, diameter and quality index were better with 50% effluent and 50% nutrient solution. The final growth of the eucalyptus plant was 21.1 cm in 110 days, which was 21.74% higher than the plants irrigated with only nutrient solution. Therefore, it was concluded that the irrigation of eucalyptus plants can be carried out with the water from tilapia effluent, supplementing it with external nutrients to maintain the quality of the eucalyptus plants (Paulus et al. 2019).

2.2. Growth and survival studies

Few studies have compared the growth or food conversion ratio in tilapia in aquaponics versus simple recirculation systems with no plants. Castillo-Castellanos et al. (2016) found that tilapia survival was high (over 97%) and specific growth (SGR) and food conversion (FCR) ratios were 4.95 and 0.99, respectively, in an aquaponic system with Carolina cucumber (*Cucumis sativus*) and Parris Island lettuce (*Lactuca sativa*). Knausand Palm (2017) found that tilapia had an SGR and FCR of 2.81 and 0.91, respectively, in aquaponic systems used to grow culinary herbs. Pinho et al. (2018) found that tilapia had a FCR of around 2.0 in systems used to grow scallion and parsley. These figures compare well with other species such as Koi carp (*Cyprinus carpio* var. *koi*), where Nuwasi et al. (2019) found that the FCR was between 4 and 5 for after 60 days in an aquaponic system used to grow Gotukola (*Centella asiatica*), a traditional medicinal plant in India. Although there is quite a large variation in types of systems used (e.g. size, recirculation rate) and a general lack of data with regard to growth and FCR per different grow out stages, the authors find it safe to assume that the growth and food conversion rates of tilapia in aquaponics systems may be maintained within acceptable limits for commercial production given the appropriate conditions, especially with regard to water quality.

2.3. Welfare considerations

Very few articles have considered the effect of aquaponics on fish welfare. One of the first papers to address this issue was Yavuzcan et al. (2017), who argue that water quality parameters are the main issue in terms of fish health and welfare in aquaponic systems. The main concerns are with suspended solids, and the possible effects of dissolved minerals on fish metabolism. Apart from those, other management or environmental concerns are similar to normal recirculation aquaculture systems with tilapia but without plants. Those stressors include things like high stocking density, feeding procedures, group size and ammonia levels. In a more recent article, Espinal and Matulic (2019) mentioned that regarding welfare of fish in aquaponics, care must be paid to three main components, namely, chemical stressors, physical stressors and perceived stressors (perceived by the fish), but they do not provide specific examples or studies on tilapia in aquaponic systems.

Regarding the physiological effects of aquaponics cultivation on tilapia, there is no data available in the literature. In other species, the only study available is by Baßmann et al. (2017) who found that in catfish, glucose and cortisol levels were also within normal ranges in coupled aquaponic systems. Palm et al. (2019) have argued that fish in aquaponics have a higher welfare since agonistic behaviour decreases, presumably as a result of chemicals (allelochemicals) released by the plants (e.g. phosphatases from basil) that calm them down or have some effects that have not been described in detail. No such studies have been carried out with tilapia, however.

According to EU Directive 98/58/CE, the welfare of all vertebrate animals, including fish, must be considered in any production system, and attempts should be made to avoid any possible suffering. Advances in the field of fish welfare now allow us to assess their physiological levels of stress in different systems and in terms of different handling procedures, as detailed in part in the new Handbook for Salmon welfare (Noble et al. 2018). The study of water quality in RAS systems and fish welfare is fairly recent, but one of the key parameters in aquaponic systems is the water pH, which will affect the concentration of free ammonia (NH_3^-) as well as other minerals. However, few studies have been carried out on the effect of commercial fertilizers on tilapia in aquaponics systems with similar growth rates and added feed (Diver and Rinehart 2006). In our lab, we have compared how the addition of fertilizer rich in nitrogen compounds affects fish stress in the production of Nile tilapia and on lettuce plants (*Lactuca sativa*). Four experimental units (four independent aquaponic systems) were used, two with fertilizer added and two without fertilizer (López-Minguito 2018). The effect caused by the fertilizer on the fish was measured by taking blood samples in which we analysed plasma cortisol, glucose and triglycerides. We did not find significant differences among treatments for fish stress during the trial but among the tilapia that were challenged after the trial ended (using the fish net test, Brydges et al. 2009); the tilapia in aquaponic units with fertilizer were better able to cope with the stress afterward (their recovery to normal levels of glucose and cortisol was faster). In addition, the growth of lettuce was significantly greater in the systems supplemented with fertilizer.

3. An example of an aquaponic system using tilapia and lettuce

Estimating the appropriate size of an aquaponic system is difficult since one needs to consider a number of variables such as the size of the fish, the temperature of the air and water, the content of protein in the feed, fish density, type of crop and the nutrient demands of the plants. All of this can vary with the time of production and we may have to make adjustments daily to make the production successful (Somerville et al. 2014).

An accepted method to size the system is to take into consideration the average daily quantity of feed that is added to the system and adjust to make sure it is balanced, in the sense proposed by Rakocy et al. (2006). In general, we can quantify the plant surface area in terms of the aquatic part. A golden rule is that for each 60 g of feed per day we can grow 1 m² of leafy greens (lettuce etc.), while for each 100 g of feed daily we can grow 1 m² of fruit plants (tomato, cucumber). Sommerville et al. (2014) reported that to maintain 1 m² of leafy greens, we need an average of 45 g of feed per day and 65 g of feed per day to provide for the nutrient demands of 1 m² of fruit plants (tomato, cucumber and pepper). To give an idea of the production potential of this technology, we can estimate the size of an aquaponic unit using some parameters in Rakocy et al. (2006) for a system growing tilapia and lettuce (see Table 1).

Table 1. Technical data to produce tilapia in an aquaponic system with lettuce (based on trials carried out by the authors)

Tank diameter (m)	6.00
Number of tanks with tilapia	1.00
Tank volume (m³)	28.27
Fish density (fish/m³)	60.00
Initial number of fish	1 696
Survival (%)	85.00
Final number of fish	1 422
Final fish weight (g)	500.40
Food conversion index	1.71
Average feed per day (g)	5 833.03

For this example we can use a circular tank that is 6 m in diameter, with a water and tank height of 1.0 and 1.2 m, respectively. Based on Table 1, we add 5,883.03 g of feed to the tank, which can produce 5883.03/60 g = 98 m² surface area for lettuce growth, taking into account that for each 60 g of feed we can "fertilize" 1 m². Next, we need to calculate the number of floating rafts needed and their size, as well as the number of polystyrene sheets (usually 1.22 wide, 2.44 long and 3-4 cm thick). We can then calculate how many rafts fit in one tank and the total number of lettuce per sheet or tank. Normally, in order to allow for growth, lettuce seedlings are spaced 25 cm apart. In time, normally we can turnover four cycles of lettuce production per cycle of tilapia produced.

4. Future prospects

Aquaponics has several advantages over other food production systems, which have been touched on in this chapter, all of which suggest that the future is bright.

Recent reviews (e.g. Goddek et al. 2019) give an idea of the breadth of possible uses for aquaponics, from industrial food production to urban farming and architectural greening of urban spaces, to education and applications in developing countries. Most probably, most of these future systems will depend on tilapia, for the reasons also mentioned previously, with companies possibly even developing specific genetic lines of tilapia suitable for aquaponics. On a technical level, future efforts will probably focus on improving the efficiency of aquaponics systems to make them more autonomous, sustainable and profitable. We can expect great advances in "precision aquaponics", using sensors and complex algorithms to control and predict problems related to water quality and fish and plant health. In the developing world, upcoming goals may focus on developing autonomous source of energy (e.g. solar or wind) to run aquaponic systems in environments not normally suitable for aquaculture or agriculture. In all of these cases, tilapia welfare will need to be considered and investigated further, otherwise profits will most probably be only short term and production will not be as efficient.

References cited

Abdul-Rahman, S., I.P. Saoud, M.K. Owaied, H. Holail, N. Farajalla, M. Haidar, et al. 2011. Improving water use efficiency in semi-arid regions through integrated aquaculture/agriculture. J. Appl. Aquacult. 23: 212–230.

Baßmann, B., M. Brenner and H.W. Palm. 2017. Stress and welfare of African catfish (*Clarias gariepinus* Burchell, 1822) in a coupled aquaponic system. Water 9(7): 504.

Boyd, C.E. 2002. Standardized terminology for low-salinity shrimp culture. Global Aquacult. Adv. 7: 58–59.

Brydges, N.M., P. Boulcott, T. Ellis and V.A. Braithwaite. 2009. Quantifying stress responses induced by different handling methods in three species of fish. Appl. Anim. Behav. Sci. 116(2–4): 295–301.

Castillo-Castellanos, D., I. Zavala-Leal, J.M.J. Ruiz-Velazco, A. Radilla-García, J.T. Nieto-Navarro, C.A. Romero-Bañuelos, et al. 2016. Implementation of an experimental nutrient film technique-type aquaponic system. Aquacult. Int. 24(2): 637–646.

Collins, A., B. Russell, A. Walls and T. Hoang. 2005. Inland Prawn Farming: Studies into the Potential for Inland Marine Prawn Farming in Queensland. Department of Primary Industries and Fisheries. Queensland, Australia.

Crossley, P.L. 2004. Sub-irrigation in wetland agriculture. Agric. Hum. Values 21(2/3): 191–205.

Diem, T.N., D. Konnerup and H. Brix. 2017. Effects of recirculation rates on water quality and *Oreochromis niloticus* growth in aquaponic systems. Aquacult. Eng. 78: 95–104.

Diver, S. and L. Rinehart. 2006. Aquaponics—Integration of Hydroponics with Aquaculture. Attra, USA.

Effendi, H. and N. Pratiwi. 2019. The growth and survival rate of Nile tilapia, *Oreochromis niloticus* (Linnaeus, 1758) in the aquaponic system with different vetiver (*Vetiveria zizanioides* L. Nash) plant density. J. Iktiologi Indonesia, 19(1): 157–166.

Enduta, A., N. Jusoh, N. Ali and W.B. Wan Nik. 2011. Nutrient removal from aquaculture wastewater by vegetable production in aquaponics recirculation system. Desalin. Water Treat. 32: 422–430.

Espinal, C.A. and D. Matulić. 2019. Recirculating aquaculture technologies. *In*: S. Goddek, A. Joyce, B. Kotzen and M. Dos-Santos (eds.). Aquaponics Food Production System: Combined Aquaculture and Hydroponic Production Technologies for the Future. Springer Open. Switzerland.

FAO. 2012. El estadomundial de la pesca y la acuicultura, Departamento de pesca y acuicultura, Rome.

Flaherty, M., B. Szuster and P. Miller. 2000. Low salinity inland shrimp farming in Thailand. Ambio 29(3): 174–179.

Goada, A.M., M.A. Essa, M.S. Hassaan and Z. Sharawy. 2015. Bio-economic features for aquaponic systems in Egypt. Turk. J. Fish. Aquat. Sc. 15: 525–532.

Goddek, S., A. Joyce, B. Kotzen and M. Dos-Santos. 2019. Aquaponics and global food challenges. pp. 15–29. *In*: S. Goddek, A. Joyce, B. Kotzen and M. Dos-Santos (eds.). Aquaponics Food Production System: Combined Aquaculture and Hydroponic Production Technologies for the Future. Springer Open. Switzerland.

Graber, A. and R. Junge. 2009. Aquaponic systems: Nutrient recycling from fish wastewater by vegetable production. Desalin. 246: 147–156.

Halwart, A.A. and A.A. Van Dam. 2010. Integración de sistemas de irrigación y acuiculturaen África occidental, conceptos, prácticas y potencial. Organización de las Naciones Unidas para la Agricultura y la Alimentación. Centro Mundial de Pesca, Rome 193 p.

Hanson, A., J. Yabes Jr. and L.P. Primavera. 2008. Cultivation of lemon basil, (*Ocimum americanum*), in two different hydroponic configurations supplemented with various concentrations of tilapia aquaculture green water. BIOS 79(3): 92–102.

Henares, M.N., M.V. Medeiros and A.F. Camargo. 2019. Overview of strategies that contribute to the environmental sustainability of pond aquaculture: Rearing systems, residue treatment, and environmental assessment tools. Rev. Aquacult. 12(1): 453–470.

Knaus, U. and H.W. Palm. 2017. Effects of fish biology on ebb and flow aquaponical cultured herbs in northern Germany (Mecklenburg Western Pomerania). Aquaculture 466: 51–63.

Kotzen, B. and S. Appelbaum. 2010. An investigation of aquaponics using brackish water resources in the Negev Desert. J. Appl. Aquacult. 22: 297–320.

Liang, J.Y. and Y.H. Chien. 2013. Effects of feeding frequency and photoperiod on water quality and crop production in a tilapia-water spinach raft aquaponics system. Int. Biodeter. Biodegr. 85: 693–700.

Limbu, S.M., A.P. Shoko, H.A. Lamtane, M.A. Kishe-Machumu, M.C. Joram, A.S. Mbonde, et al. 2017. Fish polyculture system integrated with vegetable farming improves yield and economic benefits of small-scale farmers. Aquacult. Res. 48(7): 3631–3644.

López-Minguito, A. 2018. Effects of using fertilizer for lettuce (*Lactuca sativa*) on the welfare of Nile tilapia (*Oreochromis niloticus*) in aquaponic units. Master's thesis, Technical University of Madrid.

Love, D.C., J.P. Fry, X. Li, E.S. Hill, L. Genello, K. Semmens, et al. 2015. Commercial aquaponics production and profitability: Findings from an international survey. Aquaculture 435: 67–74.

Mariscal-Lagarda, M.M., F. Páez-Osuna, J.L. Esquer-Méndez, I. Guerrero-Monroy, A. Romo del Vivar and R. Félix-Gastelum. 2012. Integrated culture of white shrimp

(*Litopenaeus vannamei*) and tomato (*Lycopersicon esculentum* Mill) with low salinity groundwater: Management and production. Aquaculture 366: 76–84.

McMurtry, M.R., D.C. Sanders, R.G. Hodson, B.C. Haning and P.C. Amand. 1997. Efficiency of water use of an integrated fish/vegetable co-culture system. J. World Aquacult. Soc. 28: 420–428.

Nicovita. 2002. Manual de Crianza de Tilapia. pp. 49. Lima, Peru.

Noble, C., K. Gismervik, M.H. Iversen, J. Kolarevic, J. Nilsson, L.H. Stien, et al. 2018. Welfare indicators for farmed Atlantic salmon: Tools for assessing fish welfare. Handbook, Norway.

Nuwansi, K.K., A.K. Verma, G. Rathore, C. Prakash, M.H. Chandrakant and G.P. Prabhath. 2019. Utilization of phytoremediated aquaculture wastewater for production of koi carp (*Cyprinus carpio* var. koi) and gotukola (*Centella asiatica*) in an aquaponics. Aquaculture 507: 361–369.

Páez-Osuna, F. and A.C. Ruiz-Fernández. 2005. Environmental load of nitrogen and phosphorus from extensive, semi-intensive, and intensive shrimp faros in the Gulf of California ecoregion. Bull. Environ. Contam. Toxicol. 74: 681–688.

Palm, H.W., U. Knaus, S. Appelbaum, S. Goddek, S.M. Strauch, T. Vermeulen, et al. 2018. Towards commercial aquaponics: A review of systems, design, scales and nomenclature. Aquacult. Int. 26(3): 813–842.

Palm, H.W., U. Knaus, S. Appelbaum, S.M. Strauch and B. Kotzen. 2019. Coupled aquaponics systems. *In*: S. Goddek, A. Joyce, B. Kotzen and M. Dos-Santos (eds.). Aquaponics Food Production System: Combined Aquaculture and Hydroponic Production Technologies for the Future. Springer Open. Switzerland.

Paulus, D., I.C. Zoorzi, F. Rankrape and G.A. Nava. 2019. Wastewater from fish farms for producing *Eucalyptus grandis* seedlings. J. Exp. Agric. Int. 36(2): 1–11.

Pinho, S.M., G.L. de Mello, K.M. Fitzsimmons and M.G. Emerenciano. 2018. Integrated production of fish (pacu *Piaractus mesopotamicus* and red tilapia *Oreochromis* sp.) with two varieties of garnish (scallion and parsley) in aquaponics system. Aquacult. Int. 26: 99–112.

Rakocy, J., R.C. Shultz, D.S. Bailey and E.S. Thoman. 2004. Aquaponic production of tilapia and basil: Comparing a batch and staggered cropping system. Acta Hort. 648: 63–69.

Rakocy, J.E., M.P. Masser and T.M. Losordo. 2006. Recirculating aquaculture tank production systems: Aquaponics—integrating fish and plant culture. SRAC Publ. – South. Reg. Aquac. Cent. 16.

Rodríguez-Delfín, A., M. Chang-La Rosa, M. Hoyos-Rojas and F. Falcón-Gutiérrez. 2004. Manual Práctico de Hidroponía. (ed.) Universidad Nacional Agraria La Molina. Lima, Perú.

Samperio-Ruiz, G. (2000). Hidroponía Básica. Ed. DIANA, México, D.F. 157 p.

Stevenson, K.T., K.M. Fitzsimmons, P.A. Clay, L. Alessa and A. Kliskey. 2010. Integration of aquaculture and arid lands agriculture for water reuse and reduced fertilizer dependency. Exp. Agric. 46: 173–190.

Somerville, C., M. Cohen, E. Pantanella, A. Stankus and A. Lovatelli. 2014. Small-scale aquaponic food production: Integrated fish and plant farming. Fisheries Aquaculture Technical Paper 589. FAO.

Thomas, R.M., A.K. Verma, C. Prakash, H. Krishna, S. Prakash and A. Kumar. 2019. Utilization of inland saline underground water for bio-integration of Nile tilapia (*Oreochromis niloticus*) and spinach (*Spinacia oleracea*). Agric. Water Manag. 222: 154–160.

Timmons, M. and J. Ebling. 2010. Recirculating Aquaculture, 2nd ed. Ithaca NY, USA.

Van Woensel, L. and G. Archer. 2015. Ten technologies which could change our lives:

Potential impacts and policy implications. European Parliament Report from the Scientific Foresight STOA Unit.

Van Wyk, P.M. 1999. Principles of recirculating system design. *In*: J. Scarpa, P. Van Wyk, J. Scarpaeds, eds. Farming Marine Shrimp in Recirculating Freshwater Systems. Department of Agriculture and Consumer Services. Florida, USA.

Yavuzcan, H., L. Robaina, J. Pirhonen, E. Mente, D. Domínguez and G. Parisi. 2017. Fish welfare in aquaponic systems: Its relation to water quality with an emphasis on feed and faeces—A review. Water 9(1): 13.

Yousef, S.A., A. Aftab and S.B. Mohamed. 2008. Food production and water conservation in a recirculating aquaponic system in Saudi Arabia at different ratios of fish feed to plants. J. World Aquacult. Soc. 39(4): 510–520.

Biofloc Technology (BFT) in Tilapia Culture

Maurício Gustavo Coelho Emerenciano[1]*, Kevin Fitzsimmons[2], Artur Nishioka Rombenso[1], Anselmo Miranda-Baeza[3], Gabriel Bernardes Martins[4], Rafael Lazzari[5], Yenitze Elizabeth Fimbres-Acedo[6] and Sara Mello Pinho[7,8]

[1] CSIRO Agriculture and Food, Livestock & Aquaculture Program, Bribie Island Research Centre, Woorim, QLD, Australia
[2] Department of Soil Water and Environmental Science, University of Arizona, Tucson, AZ, USA
[3] Sonora State University (UES), Navojoa Unit, Navojoa, SO, Mexico
[4] Federal University of Pampa, Uruguaiana, RS, Brazil
[5] Animal Science and Biological Sciences Department, Federal University of Santa Maria (UFSM), Palmeira das Missões, RS, Brazil
[6] The Northwestern Center for Biological Research (CIBNOR), La Paz, BCS, Mexico
[7] São Paulo State University (UNESP), Aquaculture Center of Unesp (CAUNESP), Jaboticabal, SP, Brazil
[8] Mathematical and Statistical Methods (Biometris), Wageningen University, Wageningen, The Netherlands

1. History and basic aspects of BFT

The biofloc technology (BFT) was proposed as an alternative to the conventional extensive and semi-extensive aquaculture production systems. In terms of expansion, conventional systems face serious challenges such as competition for land and water and environmental regulations related to discharge of effluents which contain excess organic matter, nitrogenous compounds and other toxic metabolites that might compromise the adjacent culture areas and boost the spread of disease (Browdy et al. 2012). Currently, BFT is used in the cultivation of commercial species such as penaeid shrimp and tilapia (Figure 1), and more recently, as a management tool during early cultivation during nursery phases. In recent times, as the aquaculture industry has faced diverse issues, in terms of diseases and reduced production yields, BFT was considered (and still is) the new "blue revolution" (Stokstad 2010). Such system originated in the 1970s for shrimp at the French Research Institute for Exploitation of the Sea (IFREMER), located in Tahiti, and in partnership with private companies

*Corresponding author: mauricio.emerenciano@csiro.au

from the U. S. (Emerenciano et al. 2013, Anjalee-Devi and Madhusoodana-Kurup 2015). At this time, BFT was called a "moulinette" system (Emerenciano et al. 2012) due to the vortex created as a result of constant water aeration and movement. It was later expanded to commercial shrimp farms in both Tahiti and the U.S.

In the 1980s Steve Serfling, at his tilapia farms in Southern California and the Jordan River Valley, Jordan, developed an Organic Detrital Algal Soup System, (ODASS), which is called BFT today. At roughly the same time, researchers in Israel and China began experimenting with similar systems using heavy aeration and water motion to encourage vast production of bacterial and algae biomass to process fish or shrimp wastes *in situ* (Avnimelech 2015*).* In the 1990s, the Waddell Mariculture Center, in the United States of America, and at the Technion-Israel Institute of Technology initiated deep scientific studies and commercial pilot-scale trials. In the mid-2000s, the Texas A&M University (Corpus Christi campus, USA), and the Federal University of Rio Grande (FURG, Brazil), two major research centers in North and South America, respectively, began several studies that formed a baseline to the development of BFT technology. Thanks to the diverse research projects and training of human resources, various professionals have spread BFT knowledge, implemented and managed commercial farms regionally and also globally.

Since the early 2000s, there has been a significant increase in number of scientific publications on BFT worldwide. According to Scopus database, this number has increased from less than 20 in 2009 to more than 150 publications in 2018 (Scopus 2019). For Google Scholar, this number increased to more than 4,900 publications in March 2021 (Google Scholar 2021). The majority of such studies were carried out in Mexico, Brazil, USA, China, and India helping to spread and strengthen the technology, as well as boost the industry. The wide range of courses and lectures offered in both scientific and commercial events for the academia and farmers were also important factors for such progress.

Fig. 1. Biofloc technology (BFT) applied to tilapia culture in Mexico (left) and Israel (right). Source: Maurício G. C. Emerenciano and courtesy of Y. Avnimelech.

But how does BFT work? Microorganisms play a key role in BFT systems (Martínez-Córdova et al. 2015). Similar to an activated sludge water treatment system, fecal and nitrogenous wastes are oxidized to small organic compounds, CO_2 and NO_3 which are then assimilated by bacteria and algae. The bacteria and algae tend to form flocs. These flocs (Figure 2) are maintained in the water column by limited or zero water exchange and vigorous water motion/aeration (Emerenciano et al. 2013), making the floc available for consumption by detritivores like shrimp and tilapia. In addition, a high carbon to nitrogen ratio (C:N) is maintained since nitrogenous by-products can be easily taken up by heterotrophic bacteria (Avnimelech 1999). In the beginning of the culture cycles, these high ratio are required to guarantee optimum heterotrophic bacteria growth, while the chemoautotrophic bacterial community (i.e. nitrifying bacteria) stabilizes after approximately ~25-50 days (Avnimelech 2015). In addition, a minimum fish/shrimp biomass per cubic meter is required (>300 g/ m^3) for a proper system to develop and the establishment of the bacteria population (Emerenciano et al. 2012). In the late stages, the nitrifying community might be responsible for 2/3rds of the ammonia assimilation (Emerenciano et al. 2017); thus, the addition of external carbon should be reduced or even eliminated, and alkalinity consumed by the microorganisms must be continuously replaced by different carbonates/bicarbonates sources (Furtado et al. 2015).

The water quality stability will depend on the dynamic interaction among communities of bacteria, microalgae, fungi, protozoans, nematodes, rotifers etc. that will occur naturally (Martínez-Córdova et al. 2016), and suitable levels of various parameters such as dissolved oxygen, pH, alkalinity and suspended solids (Emerenciano et al. 2017). Besides the oxygen levels, excess of particulate organic matter and toxic nitrogen compounds (pH related) are the major concern in the BFT system. In this context, three pathways occur for the removal of ammonia nitrogen: at a lesser rate (i) photoautotrophic removal by algae, and at a higher rate (ii) heterotrophic bacterial conversion of ammonia nitrogen directly to microbial biomass; and (iii) chemoautotrophic bacterial conversion from ammonia to nitrate (Martínez-Córdova et al. 2015).

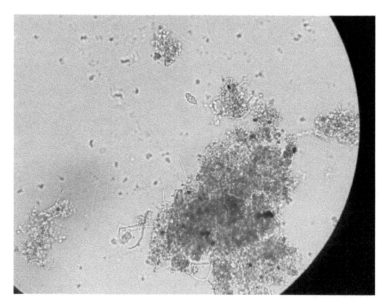

Fig. 2. Microbial aggregates from a biofloc technology (BFT) commercial production pond: diversity and abundance helping to maintain the water quality, nutrition and health of fish. Source: Maurício G. C. Emerenciano.

In BFT, the nutrients are continuously recycled and reused in the culture medium as a result of the *in situ* microorganism production and the minimum or zero water exchange (Avnimelech 2015). The microbial aggregates (bioflocs) are a natural protein-lipid rich source of food that become available 24 hours per day due to a complex interaction between organic matter, physical substrate and large range of microorganisms (Ray et al. 2010). The natural productivity in the form of microorganisms production plays three major roles in the tanks, raceways or lined ponds: (i) in the maintenance of water quality, by the uptake of nitrogen compounds generating *in situ* microbial protein; (ii) in nutrition, increasing culture feasibility by reducing feed conversion ratios and a decrease in feed costs; and (iii) competition with pathogens (Emerenciano et al. 2017, Walker et al. 2020).

It is important to note that application of biofloc technology has focused on primarily omnivorous aquatic organisms. Assessments of candidate species for BFT should include their adaptability to intensive farming conditions, the phase of their production cycle, as well as (i) tolerance to low-medium levels of ammonia nitrogen, nitrite, and suspended solids (Samocha et al. 2007, Baloi et al. 2013, Schveitzer et al. 2013), (ii) adequate morphological structure that will enable the cultivated species to graze the bioflocs properly (Kim et al. 2015), (iii) capacity to digest and assimilate the microbial aggregates (Azim et al. 2003, Avnimelech 2006) and (iv) good market value. In this sense, tilapia has proven to be a suitable species meeting the basic criteria to be raised in BFT conditions.

2. BFT in tilapia: Management approaches

In terms of general management procedures when BFT is applied to tilapia culture, it is important to address some key points such as (i) reliable source of water without pathogens and contaminants; (ii) adequate infrastructure with lined ponds, suspended tanks or race-ways normally covered by shade mesh (control of excessive phytoplankton blooms) and by greenhouses in cold zones; (iii) provide adequate dissolved oxygen levels (>5.0 mg/L) supporting the dominance of aerobic bacteria over phytoplankton; (iv) constant water motion to maintain all the bioflocs in suspension. In addition, (v) maintenance of a proper ratio of carbon to nitrogen in the system; (vi) proper feeding management and (vii) continuous water quality monitoring are also critical points to guarantee reliable production in BFT.

2.1. Research and field experiences on tilapia nursery phase

In the traditional culture systems worldwide, tilapia is produced in earthen ponds, and to a lesser degree, in cages. In earthen ponds, a single fish stocking is commonly used (0.5-1.0 g until final weight). Thereby, such situation enables an inaccurate control of growth, survival, feeding conversion rate and size homogeneity. This is at odds with the sustainability of aquaculture which should consider the responsible use of the natural resources such as water, land and feed.

These restrictions drive aquaculture farms to maximize biomass production in small areas with the minimum amount of water. The BFT systems meet these requirements, but at the same time they demand constant supply of energy to maintain the oxygen supply for fish and associated microorganisms as well as to avoid the precipitation of bioflocs. Considering the previous information, in BFT farms the use of a nursey phase can significantly reduce the water volume, and the aeration cost (García-Ríos et al. 2019). If the environmental conditions and the water quality are controlled, it is possible to reach a high net yield of fingerlings, avoiding stress and keeping the productive parameters. Therefore, for the intensive biofloc system, the use of small tanks permits the efficient use of aeration. The maintenance of dissolved oxygen is critical because the respiration rate varies according to the age and the body biomass; Sparks et al. (2003) indicated that the metabolism and oxygen consumption for Nile tilapia juveniles can be 2.5 times more than adult fish.

The cost of commercial feed for tilapia fingerlings is higher than the juvenile or adults since it contains high levels of crude protein (40-50%; El-Sayed 2007). The advantage of using a nursery phase is the control over feed consumption and consequently, a diminution of the feed conversion rate. In commercial farms, this strategy could reduce the overall production cost. The diseases and survival control is more efficient because it allows to sample more organisms by having them more concentrated. Some farmers believe that density could adversely affect size distribution and growth; however, commercial experiences in BFT farms in the NW of Mexico indicated that the fish's size homogeneity depends more on the genetic characteristics than the density of the fingerlings (A. Miranda-Baeza, personal communication). Under this consideration, it is recommended to evaluate different fingerlings hatcheries and compare their productive performance.

The nursery of Nile tilapia in biofloc technology brings some advantages such as increased control over growth and survival, improved disease control, reduction in water volume and aeration costs and more efficient use of commercial feeds. The strategies are diverse. Under commercial conditions, the use of two phases of nursery allows the efficient use of facilities. Experiences in small farms have shown that integrating two phases of nursery and two phases of final growth increases the productivity and reduces operating costs. In the nursery 1 phase, fry can be cultured from 0.25-0.5 g/ind. up to 15 g/ind. for a period of 45 d, reaching a biomass of 5-8 kg/m^2. The nursery 2 phase continues with organisms from 10-15 g to 200-250 g. The cultivation continues with one or two additional phases (Figure 3) depending on the final juvenile market size (A. Miranda-Baeza, personal communication; G.B. Martins, personal communication). The farms must evaluate the best strategy, since the one nursery phase requires less management (less workforce), but two phases seem to be more productive and improve land-use effectiveness.

Nursery facilities depend on the size of the farm as well as the environmental conditions as some companies require the use of greenhouses (Figure 4A) to control temperature, but in other cases this is not necessary (Figures 4B and 4C). In cold regions, biofloc systems paired with a greenhouse is an alternative to temperature maintenance, Crab et al. (2009) demonstrated good results for Nile tilapia grow-out with biofloc during the winter, keeping the temperature 0.4-4.9°C higher than the influent water. The same thermal control was obtained for Nile tilapia nursery during 2018 in Porto Alegre city, Brazil in the winter period (June to September) (G.B. Martins, personal communication). This strategy allows the possibility of performing an additional productive cycle in subtropical or temperate regions.

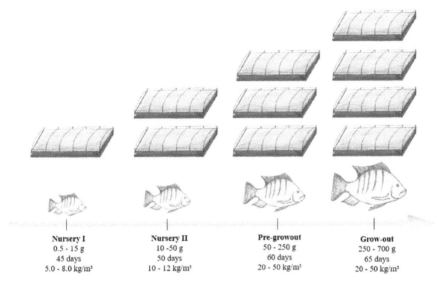

Fig. 3. General scheme of Nile tilapia culture in BFT using multiple phases.

Fig. 4. Facilities for Nile tilapia nursery in biofloc, Greenhouse used in Porto Alegre city, south of Brazil (**4A**) (Souce: G.B. Martins); and cylindrical tanks (120 m³) used in Sonora, Mexico (**4B**) (Source: A. Miranda-Baeza) and Acuícola Garza, Mérida city, México (**4C**) (Source: G.B. Martins)

2.2. Research on water quality, water types, C:N ratio and carbon sources

Some ranges obtained from studies performed with tilapia in BFT appear in Table 1. For the maintenance of biofloc, high quality water is fundamental. The biofloc concept determines that microbiological aggregates should remain suspended in the water column, and this occurs because of the movement caused by aeration. The aeration is necessary to maintain the oxygen above adequate level, as well as to maintain the biofloc in suspension. This is most commonly achieved by air blowers, but paddlewheels and pumping have also been used (for aeration in BFT, see the review by Piñeros-Roldan et al. 2020).

The diverse microorganisms that thrive within biofloc systems also have significant oxygen consumption requirements. Consequently, to estimate the total

Table 1. Water quality parameters registered in Tilapia cultivated in BFT systems

Parameter	Considerations	Range	Reference
pH	Juveniles (3.0–40.0 g)	6.5-7.5	Martins et al. (2019)
Total alkalinity (mg CaCO$_3$/L)	-	> 100	Emerenciano et al. (2017)
Total hardness (mg CaCO$_3$/L)	-	until 500	Martins (2016)
	Juvenile 79-163 g, in biofloc reused water	215	Gallardo-Colli et al. (2019)
Total suspended solids (mg/L)	-	< 500	Emerenciano et al. (2017)
Floc volume (mL/L)	Juveniles red tilapia (16-64 g), 35‰	354-537	Lopez-Elías et al. (2015)
	Fingerlings	5.0-20.0	Emerenciano et al. (2017)
	Juveniles and adults	20.0-50.0	
Chlorophyll a (µg/L)	Juveniles red tilapia (5.0-24.0 g), 35‰	226-780	Miranda-Baeza et al. (2017)
Salinity (‰)	Juveniles (0.6-20.0 g)	4.0-8.0	Alvarenga et al. (2018)
	Juveniles (66-98 g)	3.0	Lenz et al. (2017)
	Juvenile-adult GIFT (15.62-230 g)	0.0, 10.0 and 20.0	Luo et al. (2017)
	Juveniles red tilapia (5-24 g)	35.0	Miranda-Baeza et al. (2017)
Total ammonia nitrogen (TAN; mg/L)	Juvenile-adult, GIFT (15.62-230 g)	Interval 0-5 (peaks of 20-50)	Luo et al. (2017)
	Juveniles red tilapia (5.0-24.0 g)	Interval 0-1.5 (peaks of 3-5)	Miranda-Baeza et al. (2017)
	Fingerlings-juvenile, GIFT (3.0-20.0 g)	Interval 0.11-2.33	Brol et al. (2017)

Parameter	Description	Value	Reference
Nitrite (NO$_2$-N; mg/L)	Juvenile (79-163 g), in biofloc reused water	0.82	Gallardo-Colli et al. (2019)
	Juvenile-adult, GIFT (15.62-230 g)	Interval 0.0 -5.0 (peaks of 12.5-27.5)	Luo et al. (2017)
	Juveniles red tilapia (5.0 – 24.0 g)	Interval 0.0-1.5	Miranda-Baeza et al. (2017)
	Fingerlings-juveniles, GIFT (3.0 – 20.0 g)	Interval 0.01-0.4 0	Brol et al. (2016)
	Juvenile 79-163 g, in biofloc reused water	1.0	Gallardo-Colli et al. (2019)
Nitrate (NO$_3$-N; mg/L)	Juveniles-adults, GIFT (15.62-230 g)	Interval 0.0-10.0 (peaks of 40.0-75.0)	Luo et al. (2017)
	Juveniles red tilapia (5 -24 g)	Interval 5.0-20.0 (peaks of 60.0-70.0)	Miranda-Baeza et al. (2017)
	Fingerlings-juveniles, GIFT (3.0 – 20.0 g)	Interval 0.09-1.76	Brol et al. (2016)
	Juvenile 79-163 g, in biofloc reused water	24.42	Gallardo-Colli et al. (2019)
Bacteria ratio (pathogenic to degrading bacteria ratio)	Nursery 3-13 g (Nile tilapia), freshwater	<20%	Tubin (2017)
	Nursery, 4 g (Nile tilapia), freshwater	<25% (ideal <20%)	Monroy-Dosta et al. (2013)

oxygen demand both communities must be considered, bioflocs microbiota and tilapia biomass. In the BFT systems, to avoid the excessive consumption of oxygen, it is necessary to control the total solid suspension. There are two strategies by which these solids can be removed from the system: (i) water exchange (10-30%, commercial farms in Mexico; A. Miranda-Baeza, personal communication); (ii) using a clarification tank. It's important to mention when using such device that it is not unusual that biofloc remains in the water column and do not settle. This is related to the water chemical or microbiological characteristics. In some cases, the low alkalinity, total hardness or inappropriate pH are responsible for the low settling capability. However, it can also be associated with the microbiological composition of the biofloc, mainly due to the low concentration of bacterial exopolysaccharides or excessive presence of filamentous bacteria.

The water sources used for tilapia in BFT are similar to other systems: groundwater, rainwater or superficial water (lakes, rivers). The majority of tilapia production is cultivated in freshwater, but this organism could be cultured in brackish or seawater, obtained from estuaries or the sea (López-Elías et al. 2015, Luo et al. 2017). Depending on the water source, the chemical and biological composition vary. Generally, groundwater has limited contact with the soil and atmosphere, which results in the low concentration (or absence) of beneficial bacteria to be added to the system (specially nitrifying bacteria). In general, freshwater can have substantial variability in the concentrations of the ions Ca^{+2}, HCO_3^-, CO_3^- and Fe^+ as well as nutrients (mainly nitrogenous compounds), while the ion concentrations in seawater are more stable. Depending on the source, in some cases to increase the bacterial populations or to improve the diversity into the system, it is necessary to add microbial consortia (probiotics; bacteria and yeast), mainly during the first weeks. Previous evaluations must be developed to choose external consortia (commercial or native).

The amount of carbon (C:N ratio) and the source have an important influence on the abundance of heterotrophic bacteria as well as on the nutritional quality of biofloc (Crab et al. 2010a). The C:N ratio allows for the control of the type of biofloc in the cultivation system. Three major types of biofloc have been defined: photoautotrophic (dominated by microalgae), chemoautotrophic (dominated by both chemoautotrophic and heterotrophic bacteria with low microalgae prevalence) and heterotrophic (dominance of heterotrophic bacteria with low microalgae prevalence). In the first case, it seems that algae could be productive in their autotrophic or their heterotrophic aspect (for those with the capacity) but as the systems are almost always in a heavily aerated stage, the heterotrophic capability rarely comes to contribute. One way to promote the type of biofloc is the external addition of organic carbon, or inorganic nitrogen to a lesser degree. If an autotrophic biofloc is desired, a low C:N ratio (e.g. 6-8:1) is applied. If the aim is to produce a chemoautotrophic type biofloc (C:N ratio 9-12:1), it is necessary to add a moderate amount of external C, mainly in the first weeks of each phase. Control over light intensity is also required. Finally, if a heterotrophic medium is desired, the addition of significant external C source is necessary (C:N ratio; 12-20:1) and control over light intensity is suggested (Emerenciano et al. 2017).

To determine the C:N ratio, it is necessary to consider the protein content in the pelletized food. In his manual, Avnimelech (2015) indicates that in practice, the pelleted food contains approximately 50% of organic C. The dietary protein content allows for the determination of the amount of N. Some simple calculations with the respective mass of inorganic N or organic C determine which must be added to obtain a desired C:N ratio. In addition, the fish absorption of feed and feed dry matter also have to be taken into account (Emerenciano et al. 2017). There are several publications that explain in a simple way the calculation of this relationship (Avnimelech 1999, Emerenciano et al. 2017).

It is important to consider the presence of photoautotrophic organisms in BFT cultivation. Microalgae are an important source of protein, lipids, antioxidants and vitamins. These are of great importance to complement animal nutrition, since they are included in their diet in the natural environment (El Sayed 2007). In commercial farms, moderated C:N ratio is used. These may be between 9:1 to 15:1 and are suitable for promoting a "multitrophic" biofloc (which in terms of ecology means a mixture of different microorganisms' trophic levels in the medium; Ward 2019). Multitrophic conditions have been found to maintain system stability, especially with regards to nitrogenous compounds. In general, chemoautotrophic (nitrifying) bacteria prevail in low C:N conditions; however, at high C:N levels, the growth of heterotrophic bacteria is promoted as they compete with autotrophic bacteria for limited resources (Ren et al. 2019). The high C:N ratio leads to a great demand for external carbon sources and oxygen, which directly affect the production costs. The use of commercial feed with low crude protein, e.g. varying from 25 to 30%, yields C:N ratios higher than 10 (~13:1 to 11:1, respectively), and with these levels, it is not necessary to add an external C source.

The control of the C:N ratio is not sufficient to maintain the system in the expected condition (heterotrophic, chemoautotrophic or heterotrophic) as the biofloc cultures are highly dynamic. During the culture period, the water color could vary from green, soft green, soft brown and brown (Figure 5). The tonality is related with the biofloc composition and prevalence of different microorganisms' type (phytoplankton, bacteria and other heterotrophic microorganisms such as protozoa and fungi; Martínez-Córdova et al. 2015).

Environmental factors such as water temperature, precipitation, and light incidence (cloudy days and photoperiod duration) can significantly affect the abundance of microalgae; therefore, it is necessary to take the appropriate actions to regulate the presence of microalgae or bacteria in the biofloc. The periodical determination of the microorganism's diversity is highly recommended to detect undesirable microorganisms. In aquaculture, special attention is driven towards cyanobacteria (Miranda-Baeza et al. 2017, Pacheco-Veja et al. 2018). Throughout evolution, cyanobacteria have developed various ecophysiological adaptive strategies in response to adverse environmental conditions (López-Rodas et al. 2006). As a result, it can be found in places with extreme temperature, humidity, high solar radiation, diverse pH and high nutrient concentrations.

Cyanobacteria are opportunistic microorganisms and the conditions in biofloc systems can promote an explosive growth in their populations; Miranda-Baeza et

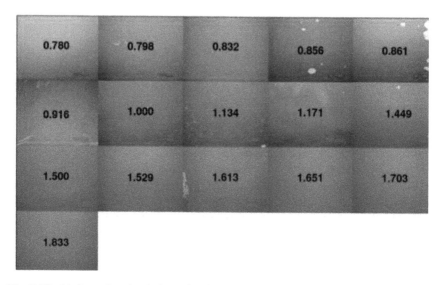

Fig. 5. The biofloc coloration index, related to the algal and bacterial prevalence according to the input of food. The transition between algal and bacterial systems occurs at a feed loading of 300 to 500 kg/ha per day, indicated by an MCCI between 1 and 1.2 (Source: Hargreaves 2013, and courtesy of D.E. Brune and K. Kirk).

al. (2017) demonstrated that a small inoculum (in heterotrophic conditions) could promote the prevalence of *Oscillatoria* sp. and cause adverse effects on the productive parameters of tilapia juveniles, which could be associated with production of toxins. The cyanotoxins are a diverse group of toxins with complex chemical structure and varied toxicological effects. In *Oscillatoria*, cyclic peptides and toxic alkaloids have been identified (Sivonen and Jones 1999) as other members of cyanobacteria group. Previous reports of tilapia cultured in biofloc have reported the presence of cyanobacteria but did not mention adverse effects (Monroy-Dosta et al. 2013, Emerenciano et al. 2017).

Several studies indicate that once a mature biofloc community is established and well managed, the nitrogen toxic compounds (TAN and NO_2-N) can be effectively controlled by either heterotrophic assimilation or chemoautotrophic nitrification. In the mature stage of BFT systems, concentrations of these compounds are maintained at acceptable ranges for the cultured organisms.

In terms of carbon sources, molasses is the most commonly used external carbon source for biofloc production, although some vegetable flours with high carbon content, such as cornmeal (García-Ríos et al. 2019), wheat flour (Mirzakhani et al. 2019), rice flour (Kumar et al. 2017), tapioca (Vermaet al. 2016), and cassava starch (Silva et al.2017), have also been employed. In addition, other compounds (polymers and polyesters) have been tested such as poly-β-hydroxybutyric and polycaprolactone (Luo et al. 2017), or industrial wastes, such as brewery by-products (Zhang et al. 2007).

When deciding on the carbon source for use in commercial farms, implementation of some comparative tests will help to determine the one that demonstrates the best performance as well as being both accessible and low cost. It is important to remark that molasses (sugar) is a monosaccharide, while the grain vegetable flours are polysaccharides meaning the molecular differences could affect the availability of organic C. In one experiment, García-Ríos et al. (2019) compared sugar, corn meal and wheat meal; they reported that sugar had a better productive performance in tilapia fingerlings production. Additionally, it was concluded that the organic carbon source modified the bioflocs composition and even the body composition of the cultivated organisms. Nevertheless, more studies must be developed to evaluate physiological parameters in tilapia, since Kumar et al. (2017) indicated that rice flour significantly improved the immune response in shrimp. In another study, Verma et al. (2016) compared different carbon sources in *Labeo rohita* culture, where they found that fishes reared in tapioca based biofloc showed significantly higher serum protein, serum albumin, total immunoglobulin, respiratory burst activity, myeloperoxidase activity and lower serum glucose and serum cortisol when compared to wheat, corn, and sugar bagasse based biofloc systems.

2.3. Research on grow-out and different stocking densities

Information on the productive parameters of tilapia culture at the growth out phase is scarce. For this manuscript, we included some reports (or personal experiences) with final weight greater than 200 g (Table 2).

2.4. BFT as a tool to control diseases and broodstock application

Biofloc has recently been projected as a possible novel strategy for disease and parasite management (Figure 6) with the natural probiotic effect in contrast to conventional approaches such as antibiotic, antifungal, and external probiotic and prebiotic application (Emerenciano et al. 2013). The natural and external microbiota

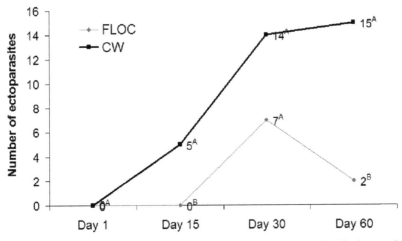

Fig. 6. Number of total ectoparasites in gills and ectoderm's mucous of fry tilapia reared under BFT limited water-exchange condition (FLOC) and conventional water-exchange system (CW) after 60 days (Emerenciano et al. 2009).

Table 2. Main productive parameter of tilapia cultured in BFT

Initial weight (g)	Final weight (g)	Net yield (kg/m³)	Final density (ind/m³)	Time (days)	Survival (%)	FCR	Reference (conditions)
38.4	251.5-280.8	16.2-18.0	70.4*	180	84.9-94.7	1.0-1.1	Pérez-Fuentes et al. (2016); experimental
15.6	230.3	35.8	155.4*	150	99.0	1.39	Luo et al. (2017); experimental
99.6	263.3	16.3	61.9*	56	99.8	1.19	Liu et al. (2018); experimental
72.6	247.6	12.1	48.9*	145	98.6	1.75	Lima et al. (2018); experimental
94.2	280.4	8.4	30.0	90	100	1.4	Souza et al. (2019); experimental
100	200.5	21.2	69.7	56	99.6	-	Pasco et al. (2018); experimental
1.0	450.0	23	51.1	150	90.0	1.08	G.B. Martins (personal communication, 2018); Brazilian commercial farm, two phases
0.60	488.2	14.0	26.7	211	84.0	1.29	A. Miranda-Baeza (personal communication, 2018); AQUAPRIM commercial farm; three phases
0.25	525.0	13.9	26.5	204	86.0	1.30	A. Miranda-Baeza (personal communication, 2019); AQUAPRIM commercial farm; three phases

* Estimates based on the author's data.

are in permanent contact with the animals. In this sense, Verschuere et al. (2000) suggested a new term of probiotic for aquaculture, which includes their effect over the microbial community of the water, and over the quality of the farming environment.

Beneficial bacterial communities can naturally develop in cultivation systems. The density of heterotrophic bacteria is a function of the culture age, as well as accumulated biomass in the system (Arias-Moscoso et al. 2018). The presence of beneficial bacteria that thrive on biofloc can act as a biocontrol by competitive exclusion over the pathogenic microbes, as has been documented with pathogenic species *Vibrio* (Crab et al. 2010b). Additionally, the bacteria and microalgae that live in biofloc produce particular extracellular polymeric substances (exopolysaccharides) with chemical characteristics related to the formation and properties of microbial aggregates (floc) (Arias-Moscoso et al. 2016). Some exopolysaccharides have antioxidant or antibacterial properties that promote the health of organisms. Furthermore, recently Durigon et al. (2019) observed that biofloc system seems to be a preventive tool against ectoparasite spread.

The stress associated to the culture conditions can cause immunosuppression and eventually diseases. Liu et al. (2018) demonstrated that BFT can promote an anti-crowding effect reducing stress on Nile tilapia juveniles when cultured in biofloc and also enhance the activity of digestive enzymes (lipase, trypsin, and amylase). To avoid stress, the maintenance of water movement is crucial. The recommended velocity for water movement generated by aeration system (especially paddlewheels and nozzles) is up to 15-20 cm/s for fingerlings and early juvenile stages and up to 30-35 cm/s for fish which are larger than 50 g. Excess energy consumption due to swimming can cause immune system depression with less disease resistance and eventually mortalities.

In terms of biofloc application to tilapia broodstock, diverse benefits have been demonstrated (Ekasari et al. 2013, Alvarenga et al. 2017) with similar benefits occurring for penaeid shrimp (Emerenciano et al. 2015) and African catfish (Ekasari et al. 2016). Ekasari et al. (2013) suggested that the application of BFT effectively enhanced tilapia reproductive performance providing broodstock with higher body weight, gonad size and fecundity. In addition, the authors observed higher blood glucose levels in broodstock raised in BFT as compared to control groups (improving the capacity of nutrient transfer into eggs). As a result, higher number of fry produced was observed suggesting the BFT as a way to increase tilapia fingerlings production. Alvarenga et al. (2017) suggested that BFT can alter the energy mobilization in the post-spawning period. The authors did not detect a reduction in the reproductive cycle length in Nile tilapia reared in BFT and highlighted no evidence of the negative effects of BFT on Nile tilapia reproduction, concluding that BFT might be used for breeder stocking of this species.

2.5. Practical tips

- The multiple nursery phase for juvenile Nile tilapia production is effective and beneficial because of the better control of survival, feeding, and disease monitoring and control, resulting in less manpower and increasing the total annual productivity in grow-out.

- Fertilization strategies – establishment of a parameter for the addition of organic carbon according to the ammonia concentration in the system (add organic carbon, for example, when TAN > 1.0 mg/L).
- The C:N ratio of 10-15:1 is often applied basically in the first weeks of culture or until stabilization of nitrification process.
- Organic carbon source – molasses and grain flours (e.g. rice, corn, tapioca and wheat bran) are also often used due to costs and availability. Alternative sources such as white/brown sugar, glycerol or by-products such as brewery by-products can also be used.
- The use of groundwater or rainwater requires precaution due to the possible inconsistencies around biofloc formation. So, it may be necessary to use biofloc inoculum, probiotics and/or yeasts.
- Excessive color – green or brown – should be avoided. For this, the luminosity and total solid suspension control is necessary.
- The total solid suspension should be controlled to avoid high oxygen consumption, stress on fish, increases in organic matter, and thus, pathogenic organisms. In this sense, avoid the excess of feed and when necessary use the sedimentation tank or apply the water exchange.
- Water bacteria monitoring is an important tool. The count and proportion of total or degrading bacteria to potential pathogenic bacteria is crucial. Reference values might be created for each farm.

3. BFT in tilapia: Nutrition and feeding

Biofloc is a well-known source of nutrients such as protein, lipids, minerals and vitamins (Ju et al. 2008, Emerenciano et al. 2013), which are available 24 hours per day contributing to: (i) the decrease of feed use due to biofloc consumption; (ii) decrease of dietary protein content; and (iii) increase of replacement percentage of premium feedstuff (e.g. fish meal and fish oil) by alternative/low-cost ones.

In terms of feeding management, some tools such as feed deprivation (Correa et al. 2019) have also been demonstrated to reduce labor and feeding costs. The authors observed that feed deprivation of up to three days does not influence fish weight, but considerably improves feed conversion. Clearly, a compensatory effect of the biofloc is perceived for tilapia, improving the feed efficiency, reducing the amount of feed required. According to the authors, with three days of deprivation there was a saving of over 40% in feed cost. These results are important because they were obtained with juvenile tilapia (4 to 45 g). At this stage, it is important that the fish have good growth rates, as it influences the final phase of rearing and fish uniformity and yield. In addition, a biofloc study conducted by Cavalcante et al. (2017) showed that 15% of deprivation didn't impact fish performance and promoted a reduction of ammonia and nitrite. In this sense, such strategy seems to be a promising tool in BFT tilapia culture.

A concern regarding the nutrition of tilapia in BFT is the flesh quality. It has already been shown that dietary protein in BFT can impact the concentration of off-flavor substances, altering the amount of geosmin in water (Green et al. 2019) and thus, the flesh flavor. Recent research has also demonstrated the effect of bioflocs on gut microbiota (Li et al. 2018), and on health and enzymatic activity (Durigon et al. 2019). Many studies on nutritional factors are still needed aiming to elucidate the immunological, physiological and biochemical responses of tilapia.

Future research and commercial application will involve the development of specific BFT feeds. In this sense, the necessity of integrating nutritional and production systems knowledge is clear; without it, feed efficiency in BFT nutrition might not be achieved. The development of BFT specific formulations is a reality in penaeid shrimp with some aquafeed companies already commercializing this kind of product. On the other hand, some limitations might occur due to variation in BFT management and biofloc composition.

3.1. Research on protein, lipid, and energy requirements and sources

The nutritional requirements for tilapia are well-known and largely investigated in terms of protein and lipid demands, which are essential along with other amino acids, and essential fatty acids requirements. The protein to energy ratio has been defined (Furuya et al. 2005, Gonçalves et al. 2009), and several complementary and alternative ingredients tested ranging from animal to vegetable-origin ingredients submitted to different pre-processing methods (NRC 2011, Ng and Romano 2013) (see also Chapter 4 of the present book for further details). Generally, tilapia grow-out feeds are recommended to contain 20-30% crude protein and 6-12% total lipid with a protein to energy ratio of 19.7 mg CP kJ/g (Mozambique tilapia), 15.3-16.4 mg CP kJ/g (Nile tilapia) and 27.3 mg kJ/g (hybrid tilapia) (El-Dahhar and Lovell 1995, Chou and Shiau 1996, Gao et al. 2011, Sweilum et al. 2005, Abdel-Tawwab et al. 2010). Slight differences in formulations might be applied based on the tilapia species, genetic strains, production systems, and feed management.

In BFT, the floc particles are a combination of organic matter, physical substrate and microorganisms (e.g. phytoplankton, bacteria, rotifers, flagellates, ciliates protozoa, nematodes, and copepods), providing an additional feed source for a series of aquatic species, especially medium to low trophic level species. Although the biofloc composition might be somewhat variable, its nutritional profile is satisfactory and often strategic. Biofloc is a natural feed source (i.e. non-processed); therefore, the chemical forms of its nutrients might also be beneficial with respect to its digestibility as well as availability. Generally, bioflocs contain essential nutrients (e.g. amino acids, fatty acids), vitamins and minerals, polymers (PHB and PHA), exopolysaccharides, organic acids, and immunostimulants, among others (Emerenciano et al. 2013; Table 3). Consequently, floc brings a complex mix of key natural nutrients, some in readily available chemical forms that contribute not only to growth but also to the immune system and overall health of the organisms. As tilapia consume biofloc at certain ratios (Avnimelech and Kochba 2009), some questions arise around the optimal feed management and formulation for tilapia culture in BFT. Are the traditional nutrition concepts applicable to BFT? Are standard formulations optimal for BFT? What would be the optimal feeding practices and management in BFT?

Table 3. Composition of fish meal and biofloc biomass

	Fish meal[1]	Biofloc biomass[2]
Crude protein (%)	65-72	12-49
Total lipid (%)	5-8	0.3-12
Fibre (%)	<2	0.8-16
Starch/carbohydrate (%)	<1	18-36
Ash	3-12	13-52
Methionine	1.75	0-1.4
n-3 fatty acids	~2	0.4-4

[1]Modified from Gatlin et al. (2007).
[2]Modified from Emerenciano et al. (2013).

Continuous feed availability in BFT is an advantage because the fish can feed at its own convenience and necessity. It is known that fish are able to self-select feeds based on their preferences and physiological demands (Norambuena et al. 2012). Avnimelech (2015) suggested that 20-30% of tilapia nutrition requirements could be met by the continuous consumption of biofloc. An ideal feed:biofloc intake ratio should be reached to achieve efficiency. In other words, industrially compounded formulated feeds are not the only source of nutrition in BFT and they should be adjusted. In this sense, the major challenge for the feed industry and research will be to formulate balanced and efficient diets for BFT system, preferably at a lower cost, considering the wide variation of microorganisms.

Recent research on tilapia nutrition in BFT is presented in Table 4. As would be expected, most research has been focusing on protein nutrition with most studies investigating the optimal level of dietary protein in BFT (Azim and Little 2008, Amany et al. 2019, Mansour and Esteban 2017, da Silva et al. 2018, Green et al. 2019). Overall, tilapia fed feeds containing equivalent dietary protein display higher growth and improved production performance in BFT compared to those raised in clear water. In BFT, dietary protein was successfully reduced by 4-11%, from 30% to 26% CP and 33% to 22% CP without impairing growth and production performance (Mansour and Esteban 2017, da Silva et al. 2018, Durigon et al. 2020). In terms of digestible protein (DP), Green et al. (2019) tested in 32 g hybrid tilapia juveniles (*Oreochromis aureus*×*O. niloticus*) diets formulated to contain 22.5%, 27.7%, and 32.3% DP and 6% lipid. Results suggested that by using the "ideal protein concept" and limiting amino acids' (Lys, Met, Thr, Ile) supplementation, digestible protein can be reduced from 32.3% to 27.7% without adversely affecting hybrid tilapia performance. Similarly, in a recent study Durigon et al. (2020) recommended 26% DP and 3150 kcal/kg of DE for 1g Nile tilapia under BFT in brackish water. Additionally, welfare indicators such as ectoparasite incidence, fin condition, plasma biochemistry and gill histology have also been evaluated and demonstrated no differences between BFT and clear water (Durigon et al. 2019, Green et al. 2019). In some instances, BFT was reported to enhance overall health of tilapia illustrated by immune and antioxidant status when compared to clear water, even when the dietary protein content was reduced by 10% (Mansour and Esteban 2017).

Fish meal replacement in tilapia feeds in BFT has also been investigated, being partially replaced with vegetable mix (30%) and brewer's yeast (60%) (López-Elías et al. 2015, Nhi et al. 2018). Interestingly, a recent study replaced 25% of soybean meal with floc meal without growth and production performance impairments (Mabroke et al. 2019). This is an upcoming tendency in aquaculture nutrition but has already been largely investigated in the poultry nutrition space so far.

Only one study evaluating different lipid sources (fish oil, soybean oil, lard oil) in tilapia feeds in BFT has been published to the author's knowledge. As expected, no differences in growth or production performance were noticed given the essential fatty acids were present at recommended levels in all dietary treatments and the fish oil-based feed yielded fillets with higher levels of omega-3 fatty acids compared to the other treatments (Inkam et al. 2018). Due to the higher stocking densities, in comparison to conventional earthen ponds, research on mineral and vitamins dietary content should also be investigated in BFT.

3.2. Research on alternative feedstuff

Tilapia as omnivores naturally include large amounts of plant material, algae, and decaying organic matter in their diets. In other production systems, such as RAS and flow-through, studies on alternative feedstuff have been increasing since 2012. Hussein et al. (2013) described the use of several algae meals to replace plant proteins in tilapia diets. Badwy et al. (2008) described the use of *Chlorella* and *Scenedesmus* meals as ingredients in tilapia diets. As microalgae production increases for use in biofuels, cosmetics, and pharmaceuticals, the co-products from this algae production are likely to be used as ingredients in aquatic feeds, tilapia feed formulations in particular. Yeast and other single cell proteins have also been used in tilapia diets (Ogunji 2004, Bob-Manuel et al. 2011, Al-Hafedh and Alam 2013) with generally positive results. While not as efficient as fish meal, the single cell proteins often provide all the essential amino acids and vitamins necessary for the tilapia and can be used to substitute for large fractions of the fish meal in the diet. Gamboa-Delgado and Márquez-Reyes (2018) and Simon et al. (2019) reviewed the potential and results for some of the newest bacteriabiomass to be used for feeding tilapia. Large volumes of these new single cell proteins are coming into the markets backed by huge investment in bioreactor technology from multi-national companies. In terms of insect meals, black soldier fly larvae (Bondari and Sheppard 1987, Webster et al. 2016, Muin et al. 2017, Devic et al. 2018), beetles, mealworm, cockroach and cricket (Jabir et al. 2012, Fontes et al. 2019) have been tested in conventional systems for tilapia with generally positive results in performance and fish flesh quality.

Specifically in BFT, nutritional demands for tilapia seem to be modified based on the biofloc diversity, composition and intake ratio enabling greater flexibility in formulations and inclusion of nonconventional ingredients. Azim and Little (2008) and López-Elías et al. (2015) also reported the composition and proximate analyses of bioflocs in tilapia farming systems. These studies documented the nutritional benefits to be derived from the consumption of the biofloc material by tilapia. The insect meals such as mealworm and cockroach have also been evaluated in BFT at different inclusion levels (0, 5, 10, 15 or 20%) and positive results were obtained up

Table 4. Summary of tilapia nutrition studies in BFT

Nutrition topic	Species	Initial weight (g)	Experimental design	Main parameters	Feed manufacture	Duration (days)	Reference
Protein							
	Nile tilapia	80-120	24 and 35% CP	Performance, floc composition and water quality	Press pelleting	84	Azim and Little (2008)
	Nile tilapia	48	20 and 30% CP and carbon source (rice bran and wheat milling by-product)	Growth, immune and antioxidant status	Not described	70	Mansour and Esteban (2017)
	Nile tilapia	10/50	17, 21, 25, 29 and 33% CP	Performance and water quality	Extrusion	61 / 98	da Silva et al. (2018)
	Nile tilapia	13-20	25, 30, 35% protein and restriction (80%) 30 and 35% CP	Performance and water quality	Extrusion	388	Amany et al. (2019)
	Hybrid tilapia	32	22.5, 27.7, and 32.3% DP	Performance, flesh quality and water quality	Extrusion	122	Green et al. (2019)
	Nile tilapia	1	Digestible protein (22, 26, 30%) and digestible energy (3000, 3150, and 3300 kcal DE/kg)	Performance and water quality	Press pelleting	42	Durigon et al. (2020)
	Nile tilapia	1	Digestible protein (22, 26, 30%) and digestible energy (3000, 3150, and 3300 kcal DE/kg)	Digestive enzymes and parasitology	Press pelleting	42	Durigon et al. (2019)

Category	Species		Treatment		Evaluation		Reference
	Hybrid red tilapia	16	FM replacement with vegetable mix (0, 33, 67 or 100%)	Not described	Performance and floc composition	42	Lopez-Elias et al. (2015)
	Nile tilapia	29	FM replacement 0, 30, 60 and 100% with brewer's yeast	Press pelleting	Performance and body composition	90	Nhi et al. (2018)
	Nile tilapia	12	Soybean meal replacement with biofloc meal (0.25 or 50%)	Not described	Performance, floc composition and water quality	56	Mabroke et al. (2019)
Lipid	Nile tilapia	30	FO, FO:SO, LO, LO:SO, FO:LO	ND	Performance and flesh quality	56	Inkam et al. (2018)
Carbohydrates	Nile tilapia	3	Pizzeria by-product meal (0, 20, 40, 60, 80 and 100%)	P	Performance, water quality and economic analysis	38	Sousa et al. (2019)
Feed additives	Nile tilapia	9	0 (clear water), 0 (BFT), 5, 10 and 20 g/kg orange peels derived pectin (OPDP)	E	Immune response, disease resistance and growth performance	56	Doan et al. (2018)
	Nile tilapia	6	0, 10 g/kg OPDP, 10^8 CFU and 10 g/kg OPDP+10^8 CFU	E	Immune response, disease resistance and growth performance	84	Doan et al. (2019)
	Nile tilapia	8	Vegetable choline 0, 400, 800, 1200 mg/kg	P	Performance, energy metabolism and antioxidant status	40	Sousa et al. (2020)

(Contd.)

Table 4. (*Contd.*)

Nutrition topic	Species	Initial weight (g)	Experimental design	Main parameters	Feed manufacture	Duration (days)	Reference
Insect meal							
	Nile tilapia	3	Cockroach meal (0, 5, 10, 15 and 20%)	Performance and water quality	P	37	Tubin (2017)
	Nile tilapia	2	*Tenebrio molitor* meal (0, 10, 15 and 20%)	Performance and water quality	P	42	Tubin et al. (2020)
Feed management							
	Nile tilapia	5	Feed deprivation 7F, 6F/1D, 5F/2D, 4F/3D	Performance and water quality	P	56	Correa et al. (2019)

*CP = Crude protein, DP = Digestible protein, DE = Digestible energy, FM = Fish meal, FO = Fish oil, SO = Soybean oil, LO = Lard oil

10% of inclusion without growth and production performance impairments (Tubin 2017, Tubin et al. 2020). Pizzeria by-product, a complimentary carbohydrate source, was suitable at 20% in juvenile Nile tilapia feeds used in BFT (Sousa et al. 2019). The authors also observed that in the early stages, up to 40% utilization of Pizzeria by-product may be an efficient economic strategy for fingerling farmers.

Feed additives focusing on the immune system and disease resistance have also been evaluated as BFT is known to enhance the organism's immune system. Orange peels derived pectin (OPDP) was added at 0, 5, 10 and 20 g/kg of feed (Doan et al. 2018). Juvenile tilapia fed 10 g/kg of OPDP displayed the highest resistance against *Streptococcus agalactiae* and improved growth and production performance compared to the other dietary treatments. In a following study, the same authors investigated the interaction of OPDP with *Lactobacillus plantarum* suggesting that 10 g/kg of OPDP combined with 10^8 CFU/g of *L. plantarum* enhanced growth, skin mucus, immunity parameters and resistance against *S. agalactiae* in comparison to the rest of the treatments (Doan et al. 2019). Phosphatidyl choline, vegetable choline, was investigated in juvenile Nile tilapia at 0, 400, 800 or 1200 mg/kg of feed (Sousa et al. 2020). Although no effects on production performance were noticed, vegetable choline stimulated energy metabolism and enhanced the antioxidant status.

3.3. Practical tips

- As BFT enables feed reductions (up to 20-30%) and savings in dietary protein levels, each farm needs to adapt and create its own feed management protocol. The adjustments will be done and refined over time according to the farm genetic/ strain, climate conditions and water quality parameters (temperature, DO, level of solids, etc.).
- BFT enables flexibility in tilapia feed formulation, in particular the addition of low-cost nonconventional ingredients. Optimized BFT management and consistency in biofloc composition is key to develop an optimal BFT specific formulation (i.e. reduced crude protein content and the inclusion of a range of alternative ingredients).
- Feed management is key in any animal production system. BFT is no different where it may even prove to be more important to achieve the expected outcomes. Avoid overfeeding and diets with excess crude protein during early stages (0.1-20 g) as TAN and NO_2 issues may occurs.
- In intensive systems with limited water exchange, the feed quality criteria should overcome the price criteria.
- Feed deprivation is suitable but needs to be further investigated and validated in commercial operations.

4. Integrating biofloc and plant production

4.1. Update review

The integration of aquatic organism productions (aquaculture) with soil-less plant culture (hydroponics) is technically viable by aquaponic systems. Aquaponics is a quasi-closed food production system where aquaculture effluent is used to nourish plants (Yep and Zheng 2019). This is in fact an ancient technique with our ancestors producing fish and vegetables in integrated and rustic systems approximately 2000 years ago before the term "aquaponics" was known (Palm et al. 2018). However, current systems have advanced a lot, thanks to the adaptation of engineering and technological concepts. The development of these was driven by the need to increase the efficiency of agri-aquaculture productions, i.e. to reduce water and land dependence and minimize effluent discharge (Goddek et al. 2019). In this modern design, the plants are cultured in conventional hydroponic structures, mainly using Nutrient Film Technique (NFT) or Deep Water Culture (DWC), whereas the aquaculture compartment is usually an intensive recirculating aquaculture system (RAS), with fish tanks and a series of filters which makes it possible to transform fish waste into plant nutrients (Love et al. 2015). Although RAS has been successfully integrated with hydroponics in coupled (one-loop) or decoupled (multi-loops) systems (Kloas et al. 2015, Goddek et al. 2016), investment in this technology is expensive and it is not applied to all farm situations (Yeo et al. 2004). Because of this, some variations and alternative types of aquaponics have been studied, for example, the replacement of RAS by biofloc technology (BFT) (Kotzen et al. 2019) (for further details on tilapia and aquaponics, see Chapter 13 of the present book).

In the last few years, there has been an increase in research and publications around the integration of BFT with hydroponics (a variation of aquaponics) (Pinho et al. 2017, Fimbres-Acedo 2019). Some characteristics of BFT make its effluent a promising fertilizer for plants, for example, (i) rich concentration of nutrients, mainly nitrate and phosphates; (ii) diversity of microorganisms that are constantly recycling nutrients; and (iii) the low dependency of filters (Emerenciano et al. 2013, Pinho et al. 2017). Furthermore, the improved zootechnical performance of tilapia reared in bioflocs when compared to RAS (Azim and Little 2008, Luo et al. 2014) suggests BFT+hydroponic potentially more productive and resource-efficient than conventional aquaponics (RAS+hydroponic). To evaluate these hypotheses, some scientific research has been developed where tilapia in BFT have been integrated with plant production. The main results of these studies are summarized in Table 5.

In all articles published with tilapia reared in BFT integrated into a hydroponic system, lettuce was the evaluated plant. This is probably due to its rapid growth and the accessibility of research studies referring to its use in aquaponic and hydroponic systems. Fimbres-Acedo's research is an exception to this; the authors cultivated other plant species and showed that moderately salt tolerant plants can also be grown using BFT effluent. Few authors have explored the productive results of fish or whether integration with plant production could harm the benefits of BFT for fish, except for Pinho (2018) (Table 6). In this study, besides lettuce production, the production of tilapia juveniles was tested and showed that the integration did not affect the benefits of BFT. Another pattern found among these articles is the use of a coupled aquaponic system, i.e. the water from BFT tilapia cultivation was directed to the hydroponic bench and, after that, returned to the fish tank in a constant recirculation of water and nutrients.

The positive results of BFT for vegetable production are not yet unanimous. Barbosa (2017), Lenz et al. (2017), Pinho (2017) and Fimbres-Acedo (2019) reported that BFT+hydroponic integration is technically possible. They also showed that

Table 5. Current information about the use of tilapia biofloc effluents in plant production

Tilapia species (initial-final weight)	Main product (fish, plant or both)	General results	Reference
Oreochromis niloticus	Plant: Lettuce (*Lactuca sativa L.* 'Charles')	Plants cultured with commercial hydroponic solution grew better than those nourished by BFT effluent. Suspended solids in the system were a limiting factor for lettuce growth.	Rahman (2010)
Oreochromis niloticus (70-102 g)	Plant: Three varieties of lettuces (*Lactuca sativa L.*)	The productive performance of lettuce cultured with BFT effluent was better than with RAS effluent, possibly because of the higher nutrient availability provided by the higher microbial activity of the BFT system.	Pinho et al. (2017)
Oreochromis niloticus (166-182 g)	Plant: Two different varieties of lettuces (*Lactuca sativa L*)	The use of mechanical and biological filters in BFT systems benefited plant growth when compared to BFT without filters.	Barbosa (2017)
Oreochromis niloticus (67-98 g)	Plant: Three varieties of lettuces (*Lactuca sativa L.*)	It is possible to use biofloc effluent with low salinity (3 ppm) to produce lettuce in the integrated system.	Lenz et al. (2017)
Oreochromis niloticus (1-30 g)	Both: Tilapia juveniles and lettuce (*Lactuca sativa L.*)	The visual characteristics and growth performance of plants grown with BFT effluent were lower than those grown using RAS effluent. While the zootechnical performance of the tilapia was better in BFT.	Pinho (2018)
Oreochromis niloticus (0.3-60 g and 60-350 g)	Plant: Lettuce, pak-choi, rocket, spinach, and basil	Implementing the biofloc effluents in different trophic levels (chemotrophic, heterotrophic and photoautotrophic). Pak-choi is a good candidate for heterotrophic BFT effluents, rocket and basil for chemotrophic and photoautotrophic treatments.	Fimbres-Acedo (2019)

variations in plant species, water salinity and inclusion of filters can be alternatives to optimize plant yields. However, Rahman (2010) and Pinho (2018) showed problems in this integration. According to their results, lettuces in BFT did not grow better than those grown in hydroponics or conventional (RAS+hydroponics) aquaponics, respectively. Plant production cycles evaluated in both studies were superior to the others (above 40 days or two 21-day plant cycles). This might indicate a negative relationship between time/number of cycles and biofloc environment, which may be related to BFT maturation and flocs volume (suspended solids) in the systems. Although negative results were found, both authors stated that adjustments in the management and engineering of the system may make BFT+hydroponic integration reliable.

Despite the great potential for the BFT effluent utilization and optimization, further work is necessary on some specific barriers. When comparing the ideal parameters for biofloc and, for example, lettuce (main plant produced in the integrated systems) (Table 6), three parameters could be the limiting conditions: the electric conductivity, pH and the salinity. In general, for plants the ideal pH is 5.5 – 6.5, EC (electrical conductivity) 1.0-2.0 dSm^{-1} and salinity < 3.0 ppt (Trejo-Téllez and Gómez-Merino 2012), whereas for biofloc, the observed pH is 6.0-8.0, EC >3.0 dSm^{-1} and the salinity is dependent on the type of culture (Emerenciano et al. 2017). A possible practice could be to harvest the biofloc effluents when they reach the optimal EC value, the use of salinity tolerant plant species (Brown et al. 1999, Shannon and Grieve 1999, Buhmann and Papenbrock 2013, Lenz et al. 2017; Pinheiro et al. 2017), phytoremediation (Ghaly et al. 2005) or implementing a desalination system (Goddek et al. 2019).

Another important point to consider is the management of dissolved solids and suspended solids (Koetzen et al. 2019). This step is of the most importance as the flocs can cover a wide variety of sizes ranging between 50 μm – 1000 μm (Hargreaves 2013).

Ray et al. (2011) suggested that the solids concentration in BFT systems can be managed by an external settling chamber with no additional filtration, but this only settling particles (>100 μm). Lenz et al. (2017) set up two settling tanks, a mechanic filter and biofilter before the hydroponic units, but some solids still passed to the hydroponic benches. An efficient option to solve this problem could be to implement particle membrane filtration (Souza et al. 2019), a flocculation process (Avnimelech 2006), dilution or bioclarification (<30 μm) (Espinal and Matulić 2019) or a mixture of the different processes. Fimbres-Acedo (2015) designed a prototype for the management of biofloc effluents with three steps: (i) sedimentation, (ii) adhesion and (iii) filtration, indicating that some directions are already on track.

The retention of the floc in *ex-situ* aerobic-anaerobic reactors could be a way to recover important nutrients such as P, Mg and Fe (Fimbres-Acedo 2019), which can be used for fertilizer to combat the deficiency of nutrients in hydroponic system. Goddek et al. (2019) described several processes that can be implemented for the management of solid fractions: aerobic-anaerobic bioreactor, distillation or mineralization process. Therefore, for the development of waste management systems it is important to consider the needs which are to be covered within the system and choose the most suitable components before using the biofloc effluents.

Table 6. Comparison of physicochemical parameters between heterotrophic biofloc, hydroponic solution and sensitive plant (Lettuce)

Parameters	BFT (Heterotrophic)	Hydroponic solutions	Sensitive plants (Lettuce)
Oxygen	> 4.0 mg L^{-1}	-	7.0 mg L^{-1***}
pH	6.8-8.0	5.5-6.5	6.0-7.0[*****]
Temperature	28 °C – 30 °C	Depends on the species culture	24 °C – 28 °C
Salinity	Depends on the species culture	0 ppt	Some varieties grow well until 3 ppt *
Alkalinity	> 100 mg L^{-1}	-	0-160 mg L^{-1} (HCO_3) [**]
Conductivity	1.0- > 3.0 $dSm^{-1\infty}$	2.0 dSm^{-1}	1.2-2.2 dSm^{-1***}
Nitrite	< 1 mg L^{-1}		
Nitrate	0.5-20 mg L^{-1} ** Reported values > 100 mg^{l-1}	210 mg L^{-1}	200 mg L^{-1****}
TAN	< 1 mg L^{-1}		
Orthophosphate	0.5-20 mg L^{-1}	31 mg L^{-1}	50 mg L^{-1****}
Settling solids (SS)	5-50 mg L^{-1}	-	-
Total suspended solids (TSS)	< 500 mg L^{-1}	-	229.3-294.4 mg L^{-1***}
References	Emerenciano et al. (2017), Fimbres (2019)	Hoagland and Arnon (1950)	* Lenz et al. (2017), **Roosta (2011), ***Rahman (2010), ****Trejo-Téllez and Gómez-Merino (2012), ***** Singh et al. (2019)

The implementation of BFT for plant nutrition has a lot of potential in both its research and the production sector. Information about this integration is being generated in various parts of the world. Therefore, the described barriers can be easily overcome, and some practical tips and solutions were presented above. Their development and application will help in making the food production more sustainable in the near future.

4.2. Practical tips

- Avoid the excess solids, specifically in the plant production units. Decoupled aquaponic system and well-designed filtration systems are good options for biofloc integrated production aiming to reduce the solid issues.
- Proper fish feed to plant ratio is crucial for the equilibrium of both biofloc and aquaponic systems.
- Mineral supplementation could be an option to decrease total feed input, and thus the initial investments on aquaculture units and production costs (Goddek et al. 2019).

5. Constrains and future perspectives

With the progress and benefits of BFT, there is still possibility for more commercial expansion. During the last two decades, research has intensified significantly but such increase has not been reflected proportionally at the commercial level. For example, in Brazil, it is estimated that less than 1% of tilapia farmers apply BFT in at least one phase of production (Brazilian Fish Farmers Association – PeixeBr, personal communication). In Mexico and Colombia, this number is estimated to be less than 5% of tilapia farmers (A. Miranda-Baeza, personal communication). In terms of comparison with the shrimp industry, in Indonesia it is estimated that at least 20-25% of shrimp production is produced using BFT (Thong 2014). The causes behind such scenario are: (i) lack of practical information; (ii) the higher cost of implementation and production compared to traditional systems (e.g. earthen ponds); and (iii) the difficulties in terms of proper management and implementation of BFT, which requires constant water quality monitoring and more refined technical knowledge (Avnimelech 2015).

On the other hand, BFT so far has proven to be an environmentally friendly technology that optimizes the productivity of cultivated species, including tilapia. In terms of future perspectives, the RD&I is the way to go. Further investigation into better understanding of tolerance levels in terms of water quality when raised in BFT is crucial. Normally, reference levels are derived from conventional systems that use clear water or water exchange and are not necessarily applicable to organisms raised in BFT. The zero or limited water exchange with high levels of solids and interactions with microbiota certainly influences the physiological condition, as well as health and immune response, of the cultivated animals.

Economic analysis performed on commercial scale is key to determine the cost and feasibility of farm's implementation. In this sense, urban farms providing local fresh farmed fish open a huge avenue for biofloc. On the other hand, high energy demand for adequate aeration, water movement and pumping certainly limits the implementation of BFT in some cases. Alternative energy sources (e.g. solar panels) are possibilities to be considered. Other areas of research are the genetics/ breeding of tilapia with better performance and disease resistance (e.g. against pathogenic bacteria).

In terms of wastes, biofloc production systems are recognized for their minimal effluent disposal. However, in some cases due to the complexity of the microbial community, there are high concentrations of solids (biofloc volume) in the water. To prevent it harming the fish growth, solids management by water clarification is required. The use of clarifiers prevents the discharge of solids/effluents which possess a high concentration of nutrients and can be a huge pollutant. After some processing (e.g. mineralization, Goddek et al. 2019), the liquid and solid fractions can be used as fertilizer in decoupled aquaponics or in land-crops, respectively. Some initial research has already been developed in this field (Delaide et al. 2019); however, the results are preliminary and many gaps still exist. Certainly, finding a sustainable destination for this discarded waste is a current challenge in tilapia biofloc production and practical solutions must be investigated.

Exploring the potential of biofloc technology, mainly in the early stages of production, will allow tilapia aquaculture to develop in a more environmentally friendly way and promote reliable production with social and economic benefits.

6. Acknowledgments

The authors thank the Brazilian National Council for Scientific and Technological Development – CNPq (Project number 483450/2013-8), the Scientific and Technological Research Support Foundation of Santa Catarina State - FAPESC (project numbers 2013TR3406 and 2015TR543). M. Emerenciano is CNPq research fellowship. In addition, we also thank the São Paulo Research Foundation (FAPESP 2018/13235-0) for the Ph.D. scholarship provided to Pinho SM and the National Council of Science and Technology of Mexico – CONACYT for the Ph.D. scholarship provided to Fimbres-Acedo YE. Thanks are also due to Timothy Perrin for technical and grammar review of this manuscript.

References cited

Abdel-Tawwab, M., M.H. Ahmad, Y.A.E. Khattab and A.M.E. Shalaby. 2010. Effect of dietary protein level, initial body weight, and their interaction on the growth, feed utilization, and physiological alterations of Nile tilapia, *Oreochromis niloticus* (L.). Aquaculture 298: 267–274.

Al-Hafedh, Y.S. and A. Alam. 2013. Replacement of fishmeal by single cell protein derived from yeast grown on date (*Phoenix dactylifera*) industry waste in the diet of Nile tilapia (*Oreochromis niloticus*) fingerlings. J. Appl. Aquac. 25(4): 346–358.

Alvarenga, É.R., S.C.M. de Sales, T.S. de Brito, C.R. Santos, R.D.S. Corrêa, G.F.O. Alves, et al. 2017. Effects of biofloc technology on reproduction and ovarian recrudescence in Nile tilapia. Aquac. Res. 48: 5965–5972.

Alvarenga, É.R., G.F.O. Alves, A.F.A. Fernandes, G.R. Costa, M.A. Silva, E.A. Teixeira, et al. 2018. Moderate salinities enhance growth performance of Nile tilapia (*Oreochromis niloticus*) fingerlings in the biofloc system. Aquac. Res. 49: 2919–2926.

Amany, A.G., M.A. Elnady, M.A.I. Salem and N.E. Metwally. 2019. Influence of dietary protein level and feed inputs on growth and feeding performance of the Nile tilapia under biofloc conditions. Egyptian Journal of Aquatic Biology & Fisheries 23(3): 483–491.

Anjalee Devi, C. and B. Madhusoodana Kurup. 2015. Biofloc technology: An overview and its application in animal food industry. International Journal Fisheries and Aquaculture Science 5: 1–20.

Arias-Moscoso, J.L., D.A. Cuevas-Acuña, M.E. Rivas-Vega, L.R. Martínez-Córdova, P. Osuna-Amarilas and A. Miranda-Baeza. 2016. Physical and chemical characteristics of lyophilized biofloc produced in whiteleg shrimp cultures with different fishmeal inclusion into the diets. Lat. Am. J. Aquat. Res. 44(4): 769–778.

Arias-Moscoso, J.L., L.G. Espinoza-Barrón, A. Miranda-Baeza, M.E. Rivas-Vega and M. Nieves-Soto. 2018. Effect of commercial probiotics addition in a biofloc shrimp farm during the nursery phase in zero water exchange. Aquacult. Rep. 11: 47–52.

Avnimelech, Y. 1999. Carbon/Nitrogen ratio as a control element in aquaculture systems. Aquaculture 176: 227–235.

Avnimelech, Y. 2006. Bio-filters: The need for a new comprehensive approach. Aquacult. Eng. 34: 172–178.

Avnimelech, Y. and M. Kochba. 2009. Evaluation of nitrogen uptake and excretion by tilapia in biofloc tanks, using 15N tracing. Aquaculture 287: 163–168.

Avnimelech, Y. 2015. Biofloc technology. A Practical Guide Book. World Aquaculture Society. Baton Rouge Third Edition. Louisiana, United States.

Azim, M.E., M.C.J. Verdegem, I. Mantingh, A.A. Van Dam and M.C.M. Beveridge. 2003. Ingestion and utilization of periphyton grown on artificial substrates by Nile tilapia, *Oreochromis niloticus* L. Aquac. Res. 34: 85–92.

Azim, M.E. and D.C. Little. 2008. The biofloc technology (BFT) in indoor tanks: Water quality, biofloc composition, and growth and welfare of Nile tilapia (*Oreochromis niloticus*). Aquaculture 283: 29–35.

Badwy, T.M., E.M. Ibrahim and M.M. Zeinhom. 2008. Partial replacement of fishmeal with dried microalga (*Chlorella* spp. and *Scenedesmus* spp.) in Nile tilapia (*Oreochromis niloticus*) diets. Proc. 8th International Symposium on Tilapia in Aquaculture 801–811.

Baloi, M., R. Arantes, R. Schveitzer, C. Magnotti and L. Vinatea. 2013. Performance of Pacific white shrimp *Litopenaeus vannamei* raised in biofloc systems with varying levels of light exposure. Aquacult. Eng. 52: 39–44.

Barbosa, M.P. 2017. Biofloc technology: Filter elements affect the aquaponic production of lettuce integrated with tilapia. MSc.Thesis. Santa Catarina State University (UDESC), Chapecó, SC, Brazil.

Bob-Manuel, F.G. and J.F. Alfred-Ockiya. 2011. Evaluation of yeast single cell protein (SCP) diets on growth performance, feed conversion and carcass composition of Tilapia *Oreochromis niloticus* (L.) fingerlings. Afr. J. Biotechnol. 10(46): 9473–9478.

Bondari, K. and D.C. Sheppard. 1987. Soldier fly, *Hermetia illucens* L., larvae as feed for channel catfish, *Ictalurus punctatus* (Rafinesque), and blue tilapia, *Oreochromis aureus* (Steindachner). Aquac. Res. 18(3): 209–220.

Brol, J., S.M. Pinho, T. Sgnaulin, K.D.R. Pereira, M.C. Thomas, G.L. de Mello, et al. 2017. Tecnologia de bioflocos (BFT) no desempenho zootécnico de tilápias: efeito da linhagem e densidades de estocagem. Arch. Zootec. 66(254): 229–235.

Browdy, C., A. Ray, J. Leffler and Y. Avnimelech. 2012. Biofloc-based aquaculture systems. Aquaculture Production System. First Edition. Chapter 12: 278–307.

Brown, J.J., E.P. Glenn, K.M. Fitzsimmons and S.E. Smith. 1999. Halophytes for the treatment of saline aquaculture effluent. Aquaculture 175: 255–268.

Buhmann, A. and J. Papenbrock. 2013. Biofiltering of aquaculture effluents by halophytic plants: Basic principles, current uses and future perspectives. Environ. Exp. Botany. 92: 122–133.

Cavalcante, D.H., F.R.S. Lima, V.T. Rebouças and M.V.C. Sá. 2017. Nile tilapia culture under feeding restriction in bioflocs and bioflocs plus periphyton tanks. Acta Scientiarum. Animal Sci. 39(3): 223–228.

Chou, B.S. and S.Y. Shiau. 1996. Optimal dietary lipid level for growth of juvenile hybrid tilapia, *Oreochromis niloticus, Oreochromis aureus*. Aquaculture 143: 185–195.

Correa, A.S., S.M. Pinho, D. Molinari, K.R. Pereira, S.M. Gutiérrez, M.C. Monroy-Dosta, et al. 2019. Rearing of Nile tilapia (*Oreochromis niloticus*) juveniles in a biofloc system employing periods of feed deprivation. J. Appl. Aquaculture DOI: 10.1080/10454438.2019.167931

Crab, R., M. Kochva, W. Verstraete and Y. Avnimelech. 2009. Bio-flocs technology application in over-wintering of tilapia. Aquac. Eng. 40: 105–112.

Crab, R., B. Chielens, M. Wille, P. Bossier and W. Verstraete. 2010a. The effect of different carbon sources on the nutritional value of bioflocs, a feed for *Macrobrachium rosenbergii* postlarvae. Aquac. Res. 41: 559–567.

Crab, R., A. Lambert, T. Defoirdt, P. Bossier and W. Verstraete. 2010b. The application of bioflocs technology to protect brine shrimp (*Artemia franciscana*) from pathogenic *Vibrio harveyi*. J. Appl. Microbiol. 109: 1643–1649.

da Silva, M.A., E.R. Alvarenga, G.F.O. Alves, L.G. Manduca, E.M. Turra, T.S. Brito, et al. 2018. Crude protein levels in diets for two growth stages of Nile tilapia (*Oreochromis niloticus*) in a biofloc system. Aquac. Res. 49: 2693–2703.

Delaide, B., H. Monsees, A. Gross and S. Goddek. 2019. Aerobic and anaerobic treatments for aquaponic sludge reduction and mineralisation. pp. 247–266. *In*: S. Goddek, A. Joyce, B. Kotzen and G.M. Burnell (eds.). Aquaponics Food Production Systems. Springer, USA.

Devic, E., W. Leschen, F. Murray and D.C. Little. 2018. Growth performance, feed utilization and body composition of advanced nursing Nile tilapia (*Oreochromis niloticus*) fed diets containing Black Soldier Fly (*Hermetia illucens*) larvae meal. Aquac. Nutr. 24(1): 416–423.

Doan, H.V., S.H. Hoseinifar, P. Elumalai, S. Tongsiri, C. Chitmanat, S. Jaturasitha, et al. 2018. Effects of orange peels derived pectin on innate immune response, disease resistance and growth performance of Nile tilapia (*Oreochromis niloticus*) cultured under indoor biofloc system. Fish Shellfish Immunol. 80: 56–62.

Doan, H.V., S.H. Hoseinifar, W. Naraballobh, S. Jaturasitha, S. Tongsiri, C. Chitmanat, et al. 2019. Dietary inclusion of orange peels derived pectin and Lactobacillus plantarum for Nile tilapia (*Oreochromis niloticus*) cultured under indoor biofloc systems. Aquaculture 508: 98–105.

Durigon, E.G., A.P.G. Almeida, G.T. Jeronimo, B. Baldisserotto and M.G.C. Emerenciano. 2019. Digestive enzymes and parasitology of Nile tilapia juveniles raises in brackish biofloc water and fed with different digestible protein and digestible energy levels. Aquaculture 506: 35–41.

Durigon, E.G., R. Lazzari, J. Uczay, D.L.A. Lopes, G.T. Jeronimo, T. Sgnaulin, et al. 2020. Biofloc technology (BFT): Adjusting the levels of digestible protein and digestible energy in diets of Nile tilapia juveniles raised in brackish water. Aquac. Fish. 5(1): 42–51.

Ekasari, J., M. Zairin, D. Putri, N. Sari, E. Surawidjaja and P. Bossier. 2013. Biofloc-based reproductive performance of Nile tilapia *Oreochromis niloticus* L. broodstock. Aquac. Res. 46(2): 509–512.

Ekasari, J., M.A. Suprayudi, W. Wiyoto, R.F. Hazanah, G.S. Lenggara, R. Sulistiani, et al. 2016. Biofloc technology application in African catfish fingerling production: The effects on the reproductive performance of broodstock and the quality of eggs and larvae. Aquaculture 464: 349–356.

El-Dahhar, A.A. and R.T. Lovell. 1995. Effect of protein to energy ratio in purified diets on growth performance, feed utilization and body composition of Mozambique tilapia, *Oreochromis mossambicus* (Peters). Aquac. Res. 26: 451–457.

El-Sayed, A-F.M. 2007. Analysis of feeds and fertilizers for sustainable aquaculture development in Egypt. pp. 401–422. *In*: M.R. Hasan, T. Hecht, S.S. De Silva and A.G.J. Tacon (eds.). Study and Analysis of Feeds and Fertilizers for Sustainable Aquaculture Development. FAO Fisheries Technical Paper No. 497. Rome, FAO.

Emerenciano, M.G.C., Y. Avnimelech, R. Gonzalez, A.T.D. Leon, G. Cuzon and G. Gaxiola. 2009. Effect of bio-floc technology (BFT) in ectoparasite control in Nile tilapia *Oreochromis niloticus* culture. CD of Abstracts of World Aquaculture Society Meeting 2009, Veracruz, Veracruz, Mexico.

Emerenciano, M.G.C., G. Cuzon, J. Goguenheim and G. Gaxiola. 2012. Floc contribution on spawning performance of blue shrimp *Litopenaeus stylirostris*. Aquac. Res. 44: 75–85.

Emerenciano, M.G.C., G. Gaxiola and G. Cuzon. 2013. Biofloc technology (BFT): A review for aquaculture application and animal food industry. pp. 301–328. *In*: Matovic, M.D. (ed.). Biomass Now: Cultivation and Utilization. InTech Rijeka Croatia.

Emerenciano, M.G.C., G. Gaxiola and G. Cuzon. 2015. Biofloc technology applied to shrimp broodstock. pp. 215–228. *In*: Avnimelech, Y. (ed.). Biofloc Technology – A Practical Guide Book, 3rd ed. The World Aquaculture Society. Baton Rouge, Louisiana, USA.

Emerenciano, M.G.C., L.R. Martínez-Córdova, M. Martínez-Porchas and A. Miranda-Baeza. 2017. Biofloc technology (BFT): A tool for water quality management in aquaculture. pp. 91–109. *In*: Hlanganani, T. (ed.). Water Quality. InTech, Rijeka.

Espinal, C. and D. Matulić. 2019. Recirculating aquaculture technologies. pp. 35–74. *In*: S. Goddek, A. Joyce, B. Kotzen and G.M. Burnell. (eds.). Aquaponics Food Production Systems. Springer, USA.

Fimbres-Acedo, Y.E. 2015. Caracterización de los nutrientes de interes hidropônico contenidos en la fracción particulada residual de cultivo de tilapia (*Oreochromis* spp.). MSc. Thesis, Centro de Investigaciones Biológicas del Noroeste, La Paz, B.C.S., México.

Fimbres-Acedo, Y.E. 2019. *Oreochromis niloticus* in RAS and BFT integrating aquaculture with hydroponics horticulture in non-recirculating system. Ph.D. Thesis, Centro de Investigaciones Biológicas del Noroeste, La Paz, B.C.S., Mexico.

Furuya, W.M., D. Botaro, R. Maria, G. De Macedo, V. Gomes, L. Carolina, et al. 2005. Aplicação do Conceito de Proteína Ideal para Redução dos Níveis de Proteína em Dietas para Tilápia-do-Nilo (*Oreochromis niloticus*). Rev. Bras. Zootec. 35: 1433–1441.

Furtado, S., L. Poersch and W. Wasielesky. 2015. The effect of different alkalinity levels on *Litopenaeus vannamei* reared with biofloc technology (BFT). Aquacult. Int. 23: 345–358.

Gallardo-Collí, A., C.I. Pérez-Rostro and M.P. Hernández-Vergara. 2019. Reuse of water from biofloc technology for intensive culture of Nile tilapia (*Oreochromis niloticus*): Effects on productive performance, organosomatic indices and body composition. Int. Aquat. Res. 11(1): 43–55.

Gamboa-Delgado, J. and J.M. Márquez-Reyes. 2018. Potential of microbial-derived nutrients for aquaculture development. Rev. Aquacult. 10(1): 224–246.

Gao, W., Y.J. Liu, L.X. Tian, K.S. Mai, G.Y. Liang, H.J. Yang, et al. 2011. Protein-sparing capability of dietary lipid in herbivorous and omnivorous freshwater finfish: A comparative case study on grass carp (*Ctenopharyngodon idella*) and tilapia (*Oreochromis niloticus* × *O. aureus*). Aquac. Nutr. 17: 2–12.

García-Ríos, L., A. Miranda-Baeza, M.G.C. Emerenciano, J.A. Huerta-Rábago and P. Osuna-Amarillas. 2019. Biofloc technology (BFT) applied to tilapia fingerlings production using different carbon sources: Emphasis on commercial applications. Aquaculture 519: 26–31.

Gatlin, D.M., F.T. Barrows, P. Brown, K. Dabrowski, Gaylord, T.G., R.W. Hardy, et al. 2007. Expanding the utilization of sustainable plant products in aquafeeds: A review. Aquac. Res. 38: 551–579.

Ghaly, A.E., M. Kamal and N.S. Mahmoud. 2005. Phytoremediation of aquaculture wastewater for water recycling and production of fish feed. Environ. Intl. 31: 1–13.

Goddek, S., C.A. Espinal, B. Delaide, M.H. Jijakli, Z. Schmautz, S. Wuertz, et al. 2016. Navigating towards decoupled aquaponic systems: A system dynamics design approach. Water 7: 1–29.

Goddek, S., A. Joyce, B. Kotzen and G.M. Burnell. 2019. Aquaponics Food Production Systems. Springer, USA.

Gonçalves, G.S., L.E. Pezzato, M.M. Barros, M. Julia and S. Rosa. 2009. Níveis de proteína digestível e energia digestível em dietas para tilápias-do- nilo formuladas com base no conceito de proteína ideal. Rev. Bras. Zootec. 38: 2289–2298.

Green, B.W., S.D. Rawlesa, K.K. Schraderb, T.G. Gaylordc and M.E. McEntirea. 2019. Effects of dietary protein content on hybrid tilapia (*Oreochromis aureus* × *O. niloticus*) performance, common microbial off-flavor compounds, and water quality dynamics in an outdoor biofloc technology production system. Aquaculture 503: 571–582.

Hargreaves, J.A. 2013. Bioflocs production system for aquaculture. Southern Regional Aquaculture Center (SRAC), Publication No. 4503.

Hoagland, D.R. and D.I. Arnon. 1950. The water-culture method for growing plants without soil. California Agricultural Experiment Station Circular. 347: 1–32.

Hussein, E.E.S., K. Dabrowski, D.M. El-Saidy and B.J. Lee. 2013. Enhancing the growth of Nile tilapia larvae/juveniles by replacing plant (gluten) protein with algae protein. Aquac. Res. 44(6): 937–949.

Inkam, M., N. Whangchai, S. Tongsiri and U. Sompong. 2018. Effects of oil enriched diets on growth, feed conversion ratio and fatty acid content of Nile tilapia (*Oreochromis niloticus*) in biofloc system. Int. J. Agric. Technol. 14(7): 1243–1258.

Jabir, M.D.A.R., S.A. Razak and S. Vikineswary. 2012. Chemical composition and nutrient digestibility of super worm meal in red tilapia juvenile. Pak. Vet. J. 32: 489–493.

Ju, Z.Y., I. Forster, L. Conquest and W. Dominy. 2008. Enhanced growth effects on shrimp, *Litopenaeus vannamei* from inclusion of whole shrimp floc or floc fractions to a formulated diet. Aquac. Nutr. 14: 533–543.

Kim, S., Q. Guo and I. Jang. 2015. Effect of biofloc on the survival and growth of the postlarvae of three Penaeids (*Litopenaeus vannamei, Fenneropenaeus chinensis*, and *Marsupenaeus japonicus*) and their biofloc feeding efficiencies, as related to the morphological structure of the third maxilliped. J. Crustacean Biol. 35(1): 41–50.

Kloas, W., R. Groß, D. Baganz, J. Graupner, H. Monsees, U. Schmidt, et al. 2015. A new concept for aquaponic systems to improve sustainability, increase productivity, and reduce environmental impacts. Aquacult. Environ. Interact. 2: 179–192.

Kotzen, B., M.G.C. Emerenciano, N. Moheimani and G.M. Burnell. 2019. Aquaponics: Alternative types and approaches. *In*: Goddek, S., Joyce, A., Kotzen, B., Burnell G. (eds.). Aquaponics Food Production Systems. Springer, Cham.

Kumar, S., P.S.S. Anand, D. De, A.D. Deo, T.K. Ghoshal, J.K. Sundaray, et al. 2017. Effects of biofloc under different carbon sources and protein levels on water quality, growth performance and immune responses in black tiger shrimp *Penaeus monodon* (Fabricius, 1978). Aquac. Res. 48(3): 1168–1182.

Lenz, G.L., E.G. Durigon, K.R. Lapa and M.G.C. Emerenciano. 2017. Produção de alface (*Lactuca sativa*) em efluentes de um cultivo de tilápias mantidas em sistema BFT em baixa salinidade. B. Inst. Pesca 43: 614–630.

Li, J., G. Liu, C. Li, Y. Deng, M.A. Tadda, L. Lan, et al. 2018. Effects of different solid carbon sources on water quality, biofloc quality and gut microbiota of Nile tilapia (*Oreochromis niloticus*) larvae. Aquaculture 495: 919–931.

López-Elías, J.A., A. Moreno-Arias, A. Miranda-Baeza, L.R. Martínez-Córdova, M.E. Rivas-Vega and E. Márquez-Ríos. 2015. Proximate composition of bioflocs in culture systems containing hybrid red tilapia fed diets with varying levels of vegetable meal inclusion. N. Am. J. Aquacult. 77(1): 102–109.

López-Rodas, V., E. Maneiro and E. Costas. 2006. Adaptation of cyanobacteria and microalgae to extreme environmental changes derived from anthropogenic pollution. Limnetica 25: 403–410.

Love, D.C., J.P. Fry, X. Li, E.S. Hill, L. Genello, K. Semmens, et al. 2015. Commercial aquaponics production and profitability: Findings from an international survey. Aquaculture 435: 67–74.

Lima, E.C.R.D., R.L.D. Souza, P.J.M. Girao, Í.F.M. Braga and E.D.S. Correia. 2018. Culture of Nile tilapia in a biofloc system with different sources of carbon. Rev. Ciênc. Agron. 49(3): 458–466.

Liu, G., Z. Ye, D. Liu, J. Zhao, E. Sivaramasamy, Y. Deng, et al. 2018. Influence of stocking density on growth, digestive enzyme activities, immune responses, antioxidant of *Oreochromis niloticus* fingerlings in biofloc systems. Fish Shellfish Immun. 81: 416–422.

Luo, G., Q. Gao, C. Wang, W. Liu, D. Sun, L. Li, et al. 2014. Growth, digestive activity, welfare, and partial cost-effectiveness of genetically improved farmed tilapia (*Oreochromis niloticus*) cultured in a Recirculating Aquaculture System and an indoor Biofloc System. Aquaculture 422–423: 1–7.

Luo, G., W. Li, H. Tan and X. Chen. 2017. Comparing salinities of 0, 10 and 20 in biofloc genetically improved farmed tilapia (*Oreochromis niloticus*) production systems. Aquac. Fish. 2(5): 220–226.

Mabroke, R.S., O.M. El-Husseiny, A.E-NF.A. Zidan, A.-A. Tahoun and A. Suloma. 2019. Floc meal as potential substitute for soybean meal in tilapia diets under biofloc system conditions. Journal of Oceanology and Limnology 37(1): 313–320.

Mansour, A.T. and M.Á. Esteban. 2017. Effects of carbon sources and plant protein levels in a biofloc system on growth performance, and the immune and antioxidant status of Nile tilapia (*Oreochromis niloticus*). Fish Shellfish Immun. 64: 202–209.

Martínez-Córdova, L.R., M.G.C. Emerenciano, A. Miranda-Baeza and M. Martínez-Porchas. 2015. Microbial-based systems for aquaculture of fish and shrimp: An updated review. Rev. Aquacult. 7: 131–148.

Martínez-Córdova, L.R., M. Martínez-Porchas, M.G.C. Emerenciano, A. Miranda-Baeza and T. Gollas-Galván. 2016. From microbes to fish the next revolution in food production. Crit. Rev. Biotechnol. 37: 287–295.

Martins, G.B. 2016. Concentração de dureza da água durante berçário de tilápia do Nilo *Oreochromis niloticus* (L.) em sistema de bioflocos. *In*: Controle de pH e dureza total em sistema de bioflocos, avaliando a qualidade de água e o desempenho de tilápia do Nilo *Oreochromis niloticus* (L.). Ph.D dissertation, Rio Grande Federal University (FURG), RS, Brazil.

Martins, G.B., C.E. Rosa, F.M. Tarouco and R.B. Robaldo. 2019. Growth, water quality and oxidative stress of Nile tilapia *Oreochromis niloticus* (L.) in biofloc technology system at different pH. Aquac. Res. 50: 1030–1039.

Mirzakhani, N., E. Ebrahimi, S.A.H. Jalali and J. Ekasari. 2019. Growth performance, intestinal morphology and nonspecific immunity response of Nile tilapia (*Oreochromis niloticus*) fry cultured in biofloc systems with different carbon sources and input C:N ratios. Aquaculture 512: 734235.

Miranda-Baeza, A., M.D.L.A. Mariscal-López, J.A. López-Elías, M.E. Rivas-Vega, M.G.C. Emerenciano, A. Sánchez-Romero, et al. 2017. Effect of inoculation of the cyanobacteria *Oscillatoria* sp. on tilapia biofloc culture. Aquac. Res. 48(9): 4725–4734.

Monroy-Dosta, M.D.C., D. Lara-Andrade, J. Castro-Mejía, G. Castro-Mejía and M.G.C. Emerenciano. 2013. Composición y abundancia de comunidades microbianas asociadas al biofloc enun cultivo de tilapia. Revista debiología marina y oceanografía 48(3): 511–520.

Muin, H., N.M. Taufek, M.S. Kamarudin and S.A. Razak. 2017. Growth performance, feed utilization and body composition of Nile tilapia, *Oreochromis niloticus* (Linnaeus, 1758) fed with different levels of black soldier fly, *Hermetia illucens* (Linnaeus, 1758) maggot meal diet. Iran. J. Fish. Sci. 16(2): 567–577.

National Research Council (NRC). 2011. Nutrient Requirements of Fish and Shrimp. National Academy Press, Washington DC.

Ng, W.-K. and N. Romano. 2013. A review of the nutrition and feeding management of farmed tilapia throughout the culture cycle. Rev. Aquacult. 5: 220–254.

Nhi, N.H.Y., C.T. Da, T. Lundh, T.T. Lan and A. Kiessling. 2018. Comparative evaluation of Brewer's yeast as a replacement for fishmeal in diets for tilapia (*Oreochromis niloticus*), reared in clear water or biofloc environments. Aquaculture 495: 654–660.

Norambuena, F., A. Estevez, F.J. Sanchez-Vazquez, I. Carazo and N. Duncan. 2012. Self-selection of diets with different contents of arachidonic acid by Senegalese sole (*Solea senegalensis*) broodstock. Aquaculture 364–365: 198–205.

Ogunji, J.O. 2004. Alternative protein sources in diets for farmed tilapia. Nutrition Abstracts and Reviews. Series B, Livestock Feeds and Feeding (Vol. 74, No. 9). CAB International.

Pacheco-Veja, J.M., M.A. Cadena-Roa, J.A. Leyva-Flores, O.I. Zavala-Leal, E. Pérez-Bravo and J.M.J. Ruiz-Velazco. 2018. Effect of isolated bacteria and microalgae on the biofloc characteristics in the Pacific white shrimp culture. Aquacult. Rep. 11: 24–30.

Palm, H.W., U. Knaus, S. Appelbaum, S. Goddek, S.M. Strauch, T. Vermeulen, et al. 2018. Towards commercial aquaponics: A review of systems, designs, scales and nomenclature. Aquacult. Inter. 3: 813–842.

Pasco, J.J.M., J.W. Carvalho Filho, C.M. de Espirito Santo and L. Vinatea. 2018. Production of Nile tilapia *Oreochromis niloticus* grown in BFT using two aeration systems. Aquac. Res. 49(1): 222–231.

Pérez-Fuentes, J.A., M.P. Hernández-Vergara, C.I. Pérez-Rostro and I. Fogel. 2016. C:N ratios affect nitrogen removal and production of Nile tilapia *Oreochromis niloticus* raised in a biofloc system under high density cultivation. Aquaculture 452: 247–251.

Piñeros-Roldan, A.J., M.C. Gutierrez-Espinosa, M.L. Viana and M.G.C. Emerenciano. 2020. Aireación en la tecnología biofloc (BFT): Princípios básicos, aplicaciones y perspectivas. Revista Politécnica, 16(31): 29–40.

Pinheiro, I., R. Arantes, C.M. do Espírito Santo, F.N. Vieira, K.R. Lapa, L.V.G. Fett, et al. 2017. Production of the halophyte *Sarcocornia ambigua* and Pacific white shrimp in an aquaponic system with biofloc technology. Ecol Eng. 100: 261–267.

Pinho, S.M., D. Molinari, G.L. Mello, K.M. Fitzsimmons and M.G.C. Emerenciano. 2017. Effluent from a biofloc technology (BFT) tilapia culture on the aquaponics production of different lettuce varieties. Ecol. Eng. 103: 146–153.

Pinho, S.M. 2018. Tilapia nursery in aquaponics system applying the biofloc technology. M.Sc. Thesis. Sao Paulo State University (UNESP), Jaboticabal, SP, Brazil.

Rahman, S.S.A. 2010. Effluent waste characterization of intensive tilapia culture units and its application in an integrated Lettuce aquaponic production facility. M.Sc. Thesis. Auburn University, Auburn, Alabama, USA.

Ray, A., G. Seaborn, J. Leffler, S. Wilde, A. Lawson and C. Browdy. 2010. Characterization of microbial communities in minimal-exchange, intensive aquaculture systems and the effects of suspended solids management. Aquaculture 310: 130–138.

Ray, A.J., K.S. Dillon and J.M. Lotz. 2011. Water quality dynamics and shrimp (*Litopenaeus vannamei*) production in intensive, mesohaline culture systems with two levels of biofloc management. Aquacult. Eng. 45: 127–136.

Roosta, H.R. 2011. Interaction between water alkalinity and nutrient solution pH on the vegetative growth, chlorophyll fluorescence and leaf magnesium, iron, manganese, and zinc concentrations in lettuce. J. Plant Nutr. 34: 717–731.

Samocha, T.M., S. Patnaik, M. Speed, A.M. Ali, J.M. Burger, R.V. Almeida, et al. 2007. Use of molasses as carbon source in limited discharge nursery and grow-out systems for *Litopenaeus vannamei*. Aquacult. Eng. 36: 184–191.

Schveitzer, R., R. Arantes, P.F.S. Costódio, C. Santo, L.V. Arana, W.Q. Seiffert, et al. 2013. Effect of different biofloc levels on microbial activity, water quality and performance of *Litopenaeus vannamei* in a tank system operated with no water exchange. Aquacult. Eng. 56: 59–70.

Shannon, M.C. and C.M. Grieve. 1999. Tolerance of vegetable crops to salinity. Sci. Hort. 78: 5–38.

Silva, U.L., D.R. Falcon, M.N.D.C. Pessôa and E.D.S. Correia. 2017. Carbon sources and C:N ratios on water quality for Nile tilapia farming in biofloc system. Revista Caatinga, 30(4): 1017–1027.

Simon, C.J., D. Blyth, N.A. Fatan and S. Suri. 2019. Microbial biomass (Novacq™) stimulates feeding and improves the growth performance on extruded low to zero-fishmeal diets in tilapia (GIFT strain). Aquaculture 501: 319–324.

Singh, H., B. Dunn and M. Payton. 2019. Hydroponic pH modifiers affect plant growth and nutrient content in leafy greens. J. Hortic. Res. 27(1): 31–36.

Sivonen, K. and G. Jones. 1999. Cyanobacterial toxins. *In*: Chorus I. and Bartram (eds.). Toxic Cyanobacteria in Water: A Guide to Their Public Health Consequences, Monitoring and Management. WHO, St Edmundsbury, Great Britain.

Sousa, A.A., S.M. Pinho, A.N. Rombenso, G.L. Mello and M.G.C. Emerenciano. 2019. Pizzeria by-product: A complementary feed source for Nile tilapia (*Oreochromis niloticus*) raised in biofloc technology? Aquaculture 501: 359–367.

Sousa, A.A., D.L.A. Lopes, M.G.C. Emerenciano, L. Nora, C.F. Souza, M.D. Baldissera, et al. 2020. Phosphatidylcholine in diets of juvenile Nile tilapia in a biofloc technology system: Effects on performance, energy metabolism and the antioxidant system. Aquaculture 515: 734574.

Souza, R.L.D., E.C.R.D. Lima, F.P.D. Melo, M.G.P. Ferreira and E.D.S. Correia. 2019. The culture of Nile tilapia at different salinities using a biofloc system. Rev. Ciênc. Agron. 50(2): 267–275.

Sparks, R.T., B.S. Shepherd, B. Ron, N.H. Richman III, L.G. Riley, G.K. Iwama, et al. 2003. Effects of environmental salinity and 17α-methyltestosterone on growth and oxygen consumption in the tilapia, *Oreochromis mossambicus*. Comp. Biochem. Physiol. B 136: 657–665.

Stokstad, E. (2010). Down on the shrimp farm. Science 328: 1504–1505.

Sweilum, M.A., M.M. Abdella and S.A. Salah El-Din. 2005. Effect of dietary protein-energy levels and fish initial sizes on growth rate, development and production of Nile tilapia, *Oreochromis niloticus* L. Aquac. Res. 36: 1414–1421.

Thong, P. 2014. Biofloc technology in shrimp farming: Success and failure. Aquaculture Asia Pacific Magazine 10(4): 13–16.

Trejo-Téllez, L. and F.C. Gómez-Merino. 2012. Nutrient solutions for hydroponics systems. pp. 1–22. *In*: T. Asato (ed.). Hydroponics – A Standard Methodology for Plant Biological Researches. USA.

Tubin, J.S.B. 2017. Insect meal in the feed tilapia in biofloc technology and recirculating aquaculture system (in Portuguese). M.Sc. Thesis. Santa Catarina State University (UDESC), Chapecó, SC, Brazil.

Tubin, J.S.B., D. Paiano, G.S.O. Hashimoto, W.E. Furtado, M.L. Martins, E. Durigon, et al. 2020. *Tenebrio molitor* meal in diets for Nile tilapia juveniles reared in biofloc system. Aquaculture 519: 734763.

Verma, A.K., A.B. Rani, G. Rathore, N. Saharan and A.H. Gora. 2016. Growth, non-specific immunity and disease resistance of *Labeo rohita* against *Aeromonas hydrophila* in biofloc systems using different carbon sources. Aquaculture 457: 61–67.

Verschuere, L., G. Rombaut, P. Sorgeloos and W. Verstraete. 2000. Probiotic bacteria as biological control agents in aquaculture. Microbiol. Mol. Biol. Rev. 64: 655–671.

Yeo, S.E., F.P. Binkowski and J.E. Morris. 2004. Aquaculture effluents and waste by-products characteristics, potential recovery, and beneficial reuse recommended. *In*: NCRAC Technical Bulletins North Central Regional Aquaculture Center (Vol. 6). http://lib. dr.iastate.edu/ncrac_techbulletins/6

Yep, B. and Y. Zheng. 2019. Aquaponic trends and challenges – A review. J. Clean. Prod. 228: 1586–1599.

Walker, D.U., M.C. Morales-Suazo and M.G.C. Emerenciano. 2020. Biofloc technology (BFT): Principles focused on alternative species and the case study of Chilean river shrimp *Cryphiopscae mentarius*. Rev Aquacult. doi:10.1111/raq.12408.

Ward, B.A. 2019. Mixotroph ecology: More than the sum of its parts. PNAS 116 (13): 5846–5848.

Webster, C.D., S.D. Rawles, J.F. Koch, K.R. Thompson, Y. Kobayashi, A.L. Gannam, et al. 2016. Bio-Ag reutilization of distiller's dried grains with solubles (DDGS) as a substrate for black soldier fly larvae, *Hermetia illucens*, along with poultry by-product meal and soybean meal, as total replacement of fish meal in diets for Nile tilapia, *Oreochromis niloticus*. Aquac. Nutr. 22(5): 976–988.

Zhang, Z.Q., B. Lin, S.Q. Xia, X.J. Wang and A.M. Yang. 2007. Production and application of a bioflocculant by multiple-microorganism consortia using brewery wastewater as carbon source. J. Environ. Sci. 19: 660–666.

15

Tilapia Processing – Relevant Aspects Pointing to Promising Horizons

Maria Emília de Sousa Gomes*, Amanda Maria Teixeira Lago and
Francielly Corrêa Albergaria

Department of Food Science, Federal University of Lavras, University Campus,
Fish Processing Pilot Plant, Post Office Box 3037, Lavras, Minas Gerais, 37200-900, Brazil

1. Tilapia as food: Chemical composition and nutritional quality

Tilapia has become one of the most important species for fish farming due to its rapid growth rate, easy adaptability to various breeding conditions and good meat sensory characteristics, with a mild flavored fillet with high nutritional value and without Y-shaped intramuscular bones (Eltholth et al. 2015, AGMRC 2018). Therefore, it has high potential for industrialization. Tilapia musculature can contain 60 to 85% moisture, water being the largest constituent, having an inversely proportional relationship with the amount of fat. Moisture is related to meat stability, quality, and composition and may affect storage, packaging and processing (Tacon and Metian 2013).

The smallest fraction is the carbohydrates, which are present at around 0.3 to 1.0%. Glucose, the main monosaccharide, acts as an energy substrate and is stored as liver and muscle glycogen. Blood glucose is an indicator for stress responses because the largest reserves are mobilized to meet energy needs, which increase mainly under stress conditions. Thus, glucose levels in fish at their death are also related to the final meat quality because during capture, if energy wears out, it can lead to complete glycogen depletion, resulting in a short pre-rigor phase and a rapid establishment of the *rigor mortis* phase, leading to accelerated deterioration (Ogawa and Maia 1999, Gonçalves 2011).

The greatest interest in tilapia is also due to the nutritional quality. The high nutritional value attributed has been proven and justified by the complete and balanced profile of essential amino acids, namely: arginine, histidine, isoleucine, leucine, lysine, methionine, phenylalanine, threonine, tryptophan, and valine. In addition,

*Corresponding author: maria.emilia@ufla.br

tilapia meat has some similarity, in protein ratio (17 - 20%), to other meats such as beef (22%), pork (23%), and poultry (20%); however, it has a higher contribution of the myofibrillar fraction, whose availability is greater than that of connective tissue proteins. The low amount of connective tissue, around 3%, contributes to better nutritional quality, since such tissue is a protein difficult to digest. Thus, the mean digestibility is 96% for tilapia, 90% for poultry and 87% for cattle (Contreras-Guzmán 2002, Gonçalves 2011, Tacon and Metian 2013).

Another important factor is the low to medium fat content, between 1 and 4%. A large part of this fat is composed of fatty acids with 14 to 22 carbon atoms with a high degree of unsaturation, containing 4 to 6 double bonds. This composition, which is highly dependent on feed during fish farming, has significant nutritional advantages. It represents the largest source of polyunsaturated fatty acids beneficial to the human organism, with an emphasis on the omega-3 series (eicosapentaenoic acid – EPA, docosahexaenoic acid – DHA and alpha-linolenic acid – LNA) and on the omega-6 series (Avisi et al. 2017). From a nutritional point of view, polyunsaturated fatty acids present in tilapia meat prevent cardiovascular disease, ensure the best digestion and ready assimilation by the tissues of the organisms (ATA 2020). On the other hand, the presence of these lipids is one of the influential factors in the shelf life of the products and their acceptance by the consumer. This is due to the complex process of lipid oxidation that can be caused by different pathways such as enzymatic action, photo-oxidation or self-oxidation, whereas the latter occurs most commonly in foods rich in unsaturated fatty acids. Lipid oxidation, besides being responsible for the development of oxidative rancidity, compromises the nutritional value, integrity, and safety of food through the formation of potentially toxic compounds (Amaral et al. 2018).

The ash contents, or fixed mineral residues, in tilapia fillets vary from 0.1 to 3.3%. Such difference in mineral content is directly related to the state at which the animal is analyzed, i.e. fillet or whole, with or without the bones and with or without skin (Contreras-Guzmán 1994). Tilapia meat is considered a valuable source of calcium and phosphorus, and has reasonable amounts of sodium, potassium, manganese, cobalt, zinc, iron, and iodine. Freshwater fish such as tilapia eventually contain lower levels of sodium and potassium when compared to saltwater varieties (Tacon and Metian 2013, ATA 2020). The chemical composition of tilapia fillets is shown in Table 1.

Table 1. Chemical composition of the tilapia fillets

Source	Chemical composition (%)			
	Moisture	Protein	Fat	Ash
Zapata et al. (2017)	74.30	18.02	2.37	2.35
Costa (2017)	75.60	19.30	4.00	1.15
Almeida (2018)	75.88	20.18	1.71	1.28
Lago et al. (2019)	76.32	16.79	2.42	1.09
Schneider (2019)	78.96	17.05	2.09	1.01

As for vitamins, tilapia is a good source, especially of the water-soluble B complex, such as thiamine and riboflavin. Considered a lean fish, it has less fat-soluble vitamins such as A and D. It stands out that, in conservation processes, losses may occur due to leaching, temperature, light, oxygen, and presence of enzymes (ATA 2020). Based on this context, therefore, from a nutritional point of view tilapia is considered an indispensable nutrient source, a food that reduces the risk factors of chronic diseases and that serves as the basis for functional food preparation. It is noteworthy that the composition of tilapia may vary depending on seasonal changes, physiological status, age, sex, and size of the fish, as well as on the composition of the fish diet, pre-slaughter and slaughter management (Ng and Romano 2013).

2. Tilapia meat quality: Pre-slaughter and slaughter management and muscle conversion

Tilapia meat that reaches the consumer comes mainly from fish farming, or fishing of this species in water sources. Fishing does not guarantee quality because there is no evidence of the safety of this food, which may have been caught in polluted waters and handled in precarious and inadequate conditions. Thus, in order to provide quality tilapia, it is important to discuss this topic about the good manufacturing practices in pre-slaughter and slaughter, and post-slaughter muscle conversions.

Pre-slaughter procedures directly affect tilapia life. These procedures include fasting, harvesting, transport, and the stunning procedure. Fasting before harvesting consists of not providing food for a period of time, for the purpose of emptying the gastrointestinal tract to reduce ammonia excretion and oxygen consumption, especially during transport to the industrial plant. In addition, it reduces the risk of bacterial contamination, which promotes rapid deterioration of tilapia, also decreasing its shelf life. The fasting time for tilapia is around 24 hours and may be shorter if the off-flavor intensity is low (Castro et al. 2017, Ferreira et al. 2018).

Fish harvesting involves catching tilapia when it reaches commercial size. It may be total or partial, manual (with trawl using ropes or netting panels, nets) or mechanical (tractor coupled nets, munck assisted suspension nets, suction pumps, among others) and varies by fish farming system. When performed incorrectly, harvesting promotes acute stress, which enhances the use of muscle energy reserves, which interferes with pre-rigor and onset of rigor mortis. In addition, increased stress weakens the fish immune system, leaving them susceptible to opportunistic pathogens present in the growth water, increasing mortality during and after this practice. It should therefore be carried out in the early hours of the morning, carefully and quickly, ensuring the shortest possible out-of-water time, seeking to ensure animal welfare and tilapia quality (Chandroo et al. 2014, Ferreira et al. 2018).

Regarding transport after harvesting, also considered a stressing factor, it is important to emphasize that when live fish are transported to industry, their quality is preserved. Therefore, the quality of the transport water is extremely important and its temperature should be monitored before, during and after. High temperatures increase the oxygen consumption by the fish. A widely used practice to keep the water temperature below 20°C is the addition of ice to the transport water. In addition, a number of actions can contribute to the success of the transport: tilapia should be transported in light colored boxes, and the transport time and the density of fish to be transported should be planned in advance; for a maximum density of 150 kg fish/m³, 8 g of sodium chloride should be used to aid osmoregulation, decrease energy expenditure and increase production of mucus covering the outside of the fish and serve as a barrier against opportunistic pathogens (Kubitza 2016, Ferreira et al. 2018).

It is also worth mentioning the use of diluted anesthetics in water to minimize transport stress, such as benzocaine (cheaper and more widely used) and clove oil. For the latter, the dose considered sufficient to anesthetize tilapia is 5 ml/L. Immediately after transport, a rest period of 24 to 48 hours (which will depend on the transport time and conditions) is recommended in order to guarantee the meat quality. If not, *rigor mortis* will be accelerated and the pH will be close to neutrality, also reducing the useful life. It should be noted that if transport has been very long, the fish should be fed during this rest so that there is no significant weight loss (Oliveira et al. 2009).

Tilapia is a filtering fish and as such uses phytoplankton and zooplankton present in water as a food source. Thus, if they come from intensive cultivation in nurseries, on arrival at the refrigerator they should undergo purification before marketing. This practice is carried out to reduce off-flavor (undesirable odors and/or flavors acquired during cultivation, by absorption of substances dissolved in water or ingested during feeding, such as geosmine and methyl isoborneol). The fish remain fasting in a tank with continuous flow of clean aerated water for 12 to 24 hours, depending on the initial intensity of the off-flavor, which is determined by sensory evaluation (Souza et al. 2012).

Once cleared, they will be forwarded for desensitization. In order for slaughter to be considered humanitarian, this procedure is essential. Besides reducing suffering for a long period, fear and pain are reduced, which are also considered stressors that can accelerate *rigor mortis* and influence the life of the product. The most commercially used and most studied method is thermonarcose, which consists of immersing fish in water and ice at a 1:1 ratio. This method reduces the animal's body temperature, oxygen consumption and metabolism. Several other methods have been evaluated, such as concussion, electronarcosis, narcosis in CO_2, CO and N, each with its pros and cons, but all more expensive than thermonarcose. The important thing is that the time from stunning to death be as short as possible, in order to prevent the fish from regaining consciousness during the process. Some researchers have pointed out that the most efficient method to ensure animal welfare at the time of slaughter is to insert a sharp knife into one of the fish's operculae until it reaches the medulla and then break it (Pedrazzani et al. 2009, Viegas et al. 2012, Freire and Gonçalves 2013).

The actual killing consists of the bleeding of the fish, which can be done either by cutting the gills, perforating the heart or by removing the head. After bleeding, the fish should be placed in tanks with water and ice for approximately three minutes and then pre-washed in chlorinated water. Evisceration and scaling should be carried out, in that order, to prevent the contents of the viscera from contaminating the carcasses (Freire and Gonçalves 2013). Tilapia have gallbladder and additional care must be taken to prevent it from breaking, which may give the meat a bitter taste.

From slaughter, the transformation of muscle into meat begins, which involves a series of molecular and biochemical events, which will reflect the efficiency of all procedures performed until then. Fish are the products of animal origin with the highest perishability due to the following characteristics: pH close to neutrality; high water activity in the tissues; high nutrient content easily used by microorganisms; high unsaturated lipids content; rapid destructive action of enzymes present in the tissues; and high metabolic activity of microbiota. When fish die, the muscle does not immediately convert to meat because the adenosine triphosphate (ATP) present continues to provide energy for a certain period of time (Poli et al. 2005, Oetterer et al. 2014a). Post-mortem changes can be divided into three phases: pre-rigor, *rigor mortis* and post-rigor.

The pre-rigor is the stage between the death of the animal and the beginning of muscle contraction. In this phase, the muscle remains flexible (responds to electrical stimuli); residual oxygen is consumed; anaerobic glycolysis begins; lactic acid accumulation occurs; ATP degradation occurs and a drop in the creatine phosphate level occurs. The duration of this phase depends on the ATP and glycogen reserves at the time of death. The good nutritional condition of the fish and the cooling immediately after slaughter contribute to a longer time in this phase. *Rigor mortis* is the result of complex biochemical reactions in muscle, characterized by muscle stiffening (extreme and irreversible contraction of its fibers, i.e. formation of actomyosin) and depletion of energy sources. At that stage, the muscle pH reaches a minimum, around 6.2. The degree of exhaustion and the condition of death greatly influence its duration. Immediate cooling contributes to its prolongation. The post-rigor is the result of the action of endogenous proteases. In this phase, the recovery of muscle extensibility, the increase in protein solubility and the increase in muscle pH occur. Thus, the pH changes from neutral to alkaline, the uncontrolled action of enzymes (proteolytic and lipolytic) occurs as well as the action of microorganisms and the oxidation of unsaturated fatty acids. This whole process culminates in total deterioration (putrefaction) of fish (Viegas et al. 2012, Oetterer et al. 2014a, Castro et al. 2017).

In conclusion, if good practices in pre-slaughter management and slaughter are applied and the fish is immediately cooled after slaughter, the speed of these changes will be minimized and the fish will stay fresh longer, resulting in quality tilapia meat.

3. Tilapia quality: Evaluation methods and quality control

The way the fish is handled determines the intensity of the enzymatic, oxidative and microbiological alterations. The speed with which each of these changes develops depends on how the basic principles of conservation, hygiene and maintenance of the cold chain, as well as the species caught and capture methods, have been applied. The concept of fish quality is more closely linked to the freshness of the raw material or the useful life of the product. The fish freshness determines the quality of derived products and significantly limits it. Due to the complexity of the fish decomposition process, it is impossible to use only one method to evaluate its quality; therefore, the combined use of some methods is the most feasible because each one has its specific advantages and disadvantages (Ordóñez et al. 2005, Soares and Gonçalves 2012).

Therefore, to evaluate the quality, in terms of freshness and hygiene or safety of tilapia, it is necessary to carry out a series of evaluations involving both physical-chemical and microbiological and sensory analyses. In this way, it is possible to obtain a more precise result regarding the quality of the raw material or product. There are two main methods, subjective (sensory) and objective (non-sensory), available to evaluate the freshness and quality of fish. Non-sensory methods can be mentioned: chemicals (total volatile nitrogen bases – TVNB, non-protein nitrogen – NPN; hypoxanthine – Hx, K value, biogenic amines, free amino acids, ammonia, hydrogen sulphide – H_2S, among others); physicochemical (pH and electrical properties); physical (muscle fiber tension, muscle hardness, viscosity of juice extracted from meat, among others); histological (degree or rate of myofibril destruction); and microbiological (total viable count and deteriorating organisms) (Gonçalves 2011, Ocaño-Higuera et al. 2011, Oetterer et al. 2014a, Gokoglu and Yerlikaya 2015).

The closest concept to quality evolution is called biochemical freshness, the phase between capture and the end of *rigor mortis*, and microbiological freshness, the phase corresponding to *post-rigor* quality changes. The compounds that characterize the state of freshness in the first phase are of muscle enzymes of autolytic origin and impossible to avoid; however, there are no undesirable changes, only a small decrease in quality. The compounds of the second phase are products of microbial activity that, although not possible to eliminate, can be delayed in order to enable the fish marketing or processing. Due to the subjectivity of the sensory methods and the delay and high cost for microbiological tests, chemical methods that quantify products derived from endogenous and bacterial enzymatic activity have been applied in the evaluation of fish freshness (Gonçalves 2011, Galvão et al. 2014).

Another factor related to the deterioration of fish are the odors; generally, many are related to the products of amino acid degradation. The bacterial decomposition of amino acids, which contain sulfur, cysteine and methionine, results in the formation of hydrogen sulfide (H_2S), methyl mercaptan (CH_3SH) and dimethyl sulfide ($CH_3)_{2S}$, respectively. Among these, the relevant parameter to assess quality and infer about possible failures in the cold chain is the formation of hydrogen sulphide gas. These volatile compounds of unpleasant odor are sensorially detectable at ppb (part per billion) levels, so that small quantities have a considerable effect on quality. The main components of this fraction are TVNB, such as ammonia, putrescine, cadaverine, spermidine, creatine, nucleotides and purine bases (Shahidi and Botta 2012, Thorn and Greenman 2012).

In addition to the flavor/odor factor that is very relevant to consumers, there are also other fish meat quality and freshness indicators. The NPN is the first fraction to be affected by the action of microorganisms, which use it as an energy source. This fraction comprises low molecular weight substances of various origins, such as free amino acids and peptides, and can represent 0.5 to 10% of the edible part of fish. The enzyme activity may cause a change in the concentration of these compounds or give rise to different ones. The detection of progressive changes to these substances in fish muscle during storage is the first requirement to consider such substances as potential freshness indices (Contreras-Guzmán 2002, Savay-da-Silva 2009).

Among the biogenic amines found in food, histamine, formed mainly by bacterial activity, requires special attention for causing intoxication, at levels above

100 ppm (parts per million). The quantitative determination of biogenic amines such as histamine, putrescine, cadaverine, spermidine and spermine is a quality indicator from a health point of view. Thus, a quality indicator has been formulated which has good correlation with sensory evaluation. The choice of this index is based on the knowledge that concentrations of histamine, putrescine and cadaverine increase as decomposition progresses, while spermidine and spermine decrease. The value of K is also a good indicator of fish freshness as it is based on the determination of compounds resulting from the degradation of adenosine triphosphate (ATP):

$$K = \frac{HxR + Hx}{\underbrace{(ATP + ADP + AMP + IMP)}_{Nucleotides} + HxR + Hx} \times 100$$

where ADP – adenosine diphosphate, AMP – adenosine monophosphate, IMP– inosine monophosphate, HxR – inosine, and Hx – hypoxanthine. The value of K<5% indicates fresh fish (freshly caught); from 5%≤K≤20% indicates that the fish is still fresh and can be consumed; 20%≤ K≤60% indicates that the fish should be cooked before consumption or processing; and from 60%≤K≤80% indicates signs of putrefaction. The principle of this method is related to the fact that, with the death of the fish and all the phosphorus reserve being consumed, ATP cannot be regenerated, following a degenerative route regulated by enzymes from muscle tissue (Gonçalves 2011, Ocaño-Higuera et al. 2011, Gokoglu and Yerlikaya 2015, Feddern et al. 2019).

The lipid content of fish has, as a characteristic, the presence of unsaturated fatty acids which, although being a positive nutritional characteristic, is worrying because they are more susceptible to the peroxidation process. These oxidative changes can be detected by the chemical test for thiobarbituric acid reactive substances (TBARS) which quantifies the malondialdehyde compound, one of the main products formed during the oxidative process in fish. With high TBARS values, there can be the formation of toxic and carcinogenic compounds such as ketones, aldehydes, alcohols, acids and hydrocarbons, and the formation of unpleasant flavors (rancidity) that compromise the acceptance of the product. It is worth noting that fish is a food of high nutritional value; however, it can pose a high risk to consumer health if contaminated with inorganic compounds such as heavy metals (arsenic, cadmium, chromium, copper, mercury and lead). The determination of inorganic contaminants can be performed by different analytical chemical techniques (Secci and Parisi 2016, Reitznerová et al. 2017).

Regarding the physicochemical method, the determination of pH represents important data in the evaluation of the quality of several foods such as fish, since the concentration of hydrogen ions is almost always altered when the hydrolytic, oxidative or fermentative decomposition of its muscle is processed. The higher the pH, the higher the bacterial activity, which is probably related to the accumulation of substances like ammonia. The chemical composition is therefore an important aspect of fish quality and influences both the maintenance of quality and its technological characteristics. In addition, chemical changes from deterioration can sometimes be translated into physical changes, which are easily visually measurable or measurable by not very sophisticated equipment (Nollet and Toldrá 2010, Oetterer et al. 2014a).

Examples of these physical changes are those that can be observed in the state of stiffness or flaccidity of the muscle, determined by the index of *rigor mortis* (IR). For IR assessment, the whole fish should be placed on a table or smooth surface in a horizontal plane so that half of the body (caudal region) is suspended; thus, the assessment of the degree of *rigor mortis* occurs by measuring bending. At the beginning of the *rigor mortis*, the tail part lifts and the bending gradually reduces until it reaches maximum *rigor*. Other physical methods used to help check the freshness and quality of fish flesh are the determination of texture (consistency) and color to determine the structural change in muscle and color, respectively. Variations in these parameters can be monitored sensorially; however, objective instrumental techniques have been developed (Contreras-Guzmán 1994, Ogawa and Maia 1999, Gonçalves 2011).

In relation to texture, the changes found can be explained as a result of the installation of *rigor mortis,* which is characterized by muscle contraction that provides muscle stiffening, or by the breakdown of muscle structure, which causes the loss of consistency or softening of the product, caused by proteolytic enzymes. During storage there is a decrease in the meat texture, resulting from the advance of the degradation state, a direct effect of the activation of proteolytic activity. The autolytic and microbiological changes that occur during degradation cause changes in the color of the fish. Fish meat may be significantly lighter in color due to acidification of the environment caused by denaturing of muscle proteins, or a marked darkening when the proteins interact with the compounds from lipid oxidation (Oetterer et al. 2014a, Castro et al. 2017).

The deterioration of the fish occurs in phases, in which each event occurs in a logical and increasing sequence. Autolytic, proteolytic, oxidative and microbiological changes trigger sensory changes and, on the other hand, the metabolites formed by these changes favor microbial development. Some microorganisms are related to diseases through the consumption of fish, so they need to be checked (Gonçalves 2011). Normally, these are the microorganisms: *Salmonella, Staphylococcus aureus,* coliforms at 45°C and mesophilic aerobic bacteria (FAO 2012). These criteria may be supplemented when establishing monitoring and tracing programs for pathogenic and sanitary quality micro-organisms in products. Microbiological tests have limitations in the control of fish quality, regarding the response time. As the deterioration processes have not occurred until the microorganisms have multiplied to levels capable of producing unpleasant odors, the freshness of the fish correlates with the sensory analyses together with the bacterial count (Françoise and Jean-Jacques 2011, Bolivar et al. 2017).

Sensory perception is the oldest and most reliable method for evaluating the freshness of fish, being widely used in the industry routine, due to the speed in judging batches of raw material and finished product, as well as the ease of non-destructive and low cost execution. The health inspection of the fish is based mainly on sensory observations, giving preference to sight, touch, smell and taste for the verification of presentation, appearance, consistency, resistance, odor and taste. This practice allows the release, to the retail trade or the food industry, only of fish in good sanitary conditions (FAO 1999, Cheng et al. 2014).

In general terms, the following sensory characteristics can be established for fresh fish: specific smells (light and pleasant), reminiscent of sea plants or, sometimes, clay; clean body surface, with relative metallic shine; transparent, shiny and prominent eyes, completely occupying the orbits; rosy or red gills, moist and shiny, with a natural odor, proper and smooth; round belly, leaving no lasting impression when finger pressed; shiny scales, well adhered to the skin, and fins showing some resistance to movements caused; they should present firm flesh, elastic consistency, the color of the species; complete viscera, perfectly differentiated; and closed anus (Martinsdóttir et al. 2001, FAO 2012). When consumed, the taste must be specific to the species (mild taste, in the case of tilapia) until the bitter, ammoniacal or rancid taste appears, as well as viscous secretions, which is formed during deterioration (Oetterer et al. 2014b).

Among the sensory analysis methods with fish, the quality index method (QIM) stands out for being species-specific, which makes it more reliable and, consequently, it has been widely applied in recent decades. When producing the QIM protocol for a certain species, it is necessary to associate it with analyses, mainly microbiological, physical-chemical and chemical, the results of which will confer the accuracy of the product quality for the establishment of its commercial useful life. Therefore, the information obtained sensorially should be interpreted together with the data derived from other analyses. Meeting these standards will prevent the marketing of low quality products that are harmful to consumer health (Martinsdóttir et al. 2001, Freitas and Amaral 2011, Bernardi et al. 2013).

4. Tilapia processing: Traditional and emerging technologies

Tilapia is a very versatile fish, which can be found in various forms in national and international commerce. The processing to produce fillets of this species can be divided into the following steps: pre-treatment, which consists of gutting the fish, removing the head, tail, fins and washing (to reduce contamination); filleting, which is the removal of the fillet itself; removal of trimmings, which consists of removing skin and thorns; packaging and storage, which can be as chilled or frozen.

Because it is extremely perishable, tilapia, as well as other types of fish, needs the maintenance of the cold chain throughout processing. The most basic form of conservation by cold is the use of ice, which prevents dryness, reduces weight loss and maintains humidity. Its use is laborious because as the ice melts in contact with the fish, it has to be constantly replaced. With the use of ice, the temperature of the fish is kept slightly above the freezing point. In practice, the tilapia:ice ratio is 2:1. The latter must be laid out in layers on the fish, with ice always at the bottom of the container and in the top layer. As such, it is possible to control enzymatic processes, delay microbial deterioration and minimize pathogen growth, besides causing mild sensorial and nutritional impacts. In refrigeration, another cold storage technology, the temperature of tilapia is also kept close to the freezing point, but storage takes place in cold chambers regulated at temperatures close to 0°C. In households, as well as retail locations, temperatures are kept below 5°C, ensuring the freshness of the fish (Cribb 2018).

There is a great demand in the world market today for "fresh" fish (kept fresh). Therefore, as an alternative to tilapia, minimal processing, which is an emerging and mild technology, where the fish is kept under refrigeration, results in a product of convenience and safety. The product obtained is not sterile, as there is only a moderate decrease in the microbial flora present during processing, requiring a rigorous cold process to prevent the growth of pathogenic microorganisms (Kumar et al. 2018). To produce minimally processed tilapia, in addition to controlled refrigeration, some supporting techniques are used, such as acidification and irradiation, which are barriers to microorganism growth. For tilapia, these studies are already very advanced. Acetic acid and lactic acid are used as acidulants and their presence is effective for bacterial inhibition, provided that good hygienic conditions are maintained (Oetterer et al. 2014b).

As for irradiation, also already successfully applied to whole tilapia and fillets, it has been found that the dose of 5 kGy, half of the maximum allowed by the legislation, is already sufficient to increase the tilapia shelf life. Both acidification and irradiation require special packaging conditions to ensure their safety. Some of them are commonly found in the Brazilian market, such as the Styrofoam tray, composed of polystyrene foam, covered with polyvinyl chloride plastic film. This packaging combines quality, price, logistics and delivery time, guaranteeing the economic viability of the acquisition. However, there are various methods and materials for packaging, being an area of great growth and innovation at the technological level (Gonçalves 2011, Oetterer et al. 2014c).

These innovations aim to maintain food quality for a longer period of time, while at the same time seeking to produce lower environmental impact through new applications of materials obtained mainly from renewable sources (biopolymers). Packaging in modified atmosphere and vacuum packaging also contributes to extending product shelf life. This last process consists of exposing tilapia to the absence of air, controlling the development of microorganisms, enzymatic action and oxidation, the main deterioration mechanisms (Elgadir et al. 2017).

Besides minimally processed tilapia, which is strictly dependent on refrigerated storage and keeps the tilapia fresh for longer, several other technologies can be applied, individually or together, to obtain a safe and quality product. In freezing, for example, the temperature of the fish is below the freezing point and thus there is the formation of ice crystals. This formation retards deterioration, ensures non-seasonal supply, preserves for long periods and contributes to the retention of flavor, color and nutritional value. There are various freezing techniques, but generally tilapia is frozen at $-25°C$ and kept stored at $-18 °C$. The shelf life of the frozen product varies according to the freezing temperature and also the fish type. In general, the lower the freezing temperature, the longer the life of the product. This guarantees a lifetime of months or even years, depending on the type of product. For freezing, it is very common to perform pre-treatments with phosphates and cryoprotectants, which ensure the retention of natural fluids and color. After freezing, for fillets and derivatives, it is also common to use ice glazing and differentiated packaging in order to protect against excessive water loss and also to prevent the occurrence of oxidation and freezer burn. This glazing technique consists of applying an ice coating to the surface of the frozen product. For transport of frozen tilapia, the temperature of the transport chamber must be kept at –18 °C. In retail locations, this temperature must be maintained (Maciel et al. 2013, Oetterer et al. 2014b).

Canning and "retort pouch" (sterilizable flexible bags), and heat conservation technologies, are also applied to tilapia. These are very efficient because, in addition to producing a high sensory quality food, it inactivates microorganisms and enzymes. In addition, canned tilapia or retort pouch packaging does not require refrigeration. The packaging used is hermetically sealed and the temperature applied is sufficient to eliminate all pathogenic microorganisms. Its material is resistant to corrosion, laceration and storage. The main goal is to prevent the growth of *Clostridium botulinum* (FDA 2020).

Some tilapia processing technologies, such as salting, drying and smoking, promote the reduction of water activity (FAO 2018). Salting consists of the osmotic dehydration of fish. Its basic principle is therefore the removal of water from tissues by osmosis and partial replacement by salt through diffusion. In this way, there is enzyme inactivation – a decrease in the speed of chemical reactions, a decrease of, or impediment to, fish deterioration, and an increase in palatability and accentuation of the characteristic flavor. Thus, there is a reduction in protein solubility and WRC (water retention capacity) in tissues. For tilapia, wet or brine salting is most commonly used. The process can last from 24 to 60 hours and is considered complementary to pressing (for 24 to 48 hours), and drying operations. The latter can be done by the natural method, more rustic and demanding under very specific climatic conditions such as temperature of 30 to 50°C and relative humidity of 65 to 70%, or by the artificial method, carried out in greenhouses, fully controlled, with temperatures between 30 and 40°C and relative humidity of 45 to 55% (Gonçalves 2011, Bhat et al. 2012).

Smoking is an ancient method of preservation where salt, smoke and drying are combined. In this method, a protective layer is formed on the product caused by surface dehydration, protein coagulation, the layer of resin formed and the action of the smoke constituents. Smoked products are characterized by water loss, typical coloring, special taste and aroma and surface dryness. The smoking operation comprises three distinct phases: brining (which slows down the autolysis phenomena), drying (where the formation of film occurs which will prevent the excessive loss of intrinsic substances) and smoking (conservation and absorption of all the smoke constituents whether they are preservatives, antioxidants and/ or flavorings). When smoking hot, temperatures above 40°C are used, reaching 80-85°C. Because it involves pasteurization and cooking, the product can be consumed immediately, without prior cooking. However, the useful life is restricted. This conventional process can be replaced by liquid smoke, which has different compositions and, therefore, provides products with different, and more uniform, sensory characteristics (Gonçalves 2011, Oetterer et al. 2014c).

There are also other emerging technologies, which are under study for application in tilapia processing, such as high pressure processing. In some studies, these technologies have proven effective in inactivating deterioration and pathogenic microorganisms, while maintaining nutritional and sensory properties (Kumar et al. 2018, Zhao et al. 2019).

5. Use of waste from tilapia processing

Over the past decades, environmental issues have become some of the main responsibilities of government bodies and society due to changes in ideologies, ethical values and regulations. In the course of the fish production chain, a considerable amount of solid waste is generated, the indiscriminate handling of which constitutes a serious global problem because it provides environmental, social and economic vulnerability. The proper destination of this material is of paramount importance, since it is considered a source of animal protein of excellent nutritional quality (FAO 2020). Thus, viable technologies are presented that enable sustainable management of the tilapia agribusiness, besides proposing the development of co-products of interest and high value for the food industries (Olsen et al. 2014, WordFish 2017, Santos 2018).

Overall, much of the fish production is processed through filleting, canning and curing. A large proportion of tilapia cultivation is marketed in the form of fillets, and the filet yield in this species varies from 30 to 37%. From these data, it can be seen that the processing of tilapia produces a large amount of waste which, when improperly employed, aggravates environmental problems (Gonçalves 2011, Vidotti 2011). Inadequate management of agroindustrial waste presents itself as one of the most critical problems, since the generation rate is much higher than the degradation rate. It is worth noting that the waste from tilapia filleting is discarded by the industries due to ignorance of its true potential. Therefore, there is a need for research that focuses on the development and/or introduction of new technologies that provide better use of these raw materials until the final product, and also provide the development of new products from liquid and solid waste (Oetterer et al. 2014d, Lago et al. 2019).

A great innovation in fish waste recovery technology was the appearance of equipment, an electric deboner, capable of separating muscle material still attached to the bones, after filleting. The minced fish, obtained in this mechanical process, is free of viscera, scales, skin and bones. Thus, much research has been carried out to recycle this material, with the objective of increasing its use, as a low cost and high nutritional potential raw material, in the elaboration of food for human consumption (Borgogno et al. 2017, Secci et al. 2017, Cribb et al. 2018). Over the past few years, at the Fish Processing Pilot Plant of the Federal University of Lavras (Lavras, Minas Gerais, Brazil), products have been developed from solid waste mainly from tilapia filleting. Some products can be seen in Figure 1.

In recent studies, different formulations of fish sausage (emulsified product) (Lago et al. 2017, 2018, 2019) and fish mortadella (Abud 2019, Vidal 2016, Vieira 2019, Zanutto 2017), fried fish balls (Zumas 2017) and breaded fish (Abreu et al. 2015, Fukushima et al. 2014) have been elaborated, all with increasing inclusion of minced fish form tilapia. The authors found good sensory acceptance by consumers, increased nutritional value and excellent product stability. Several studies have even been developed in order to improve the texture of the fish mortadella. Research has evaluated the inclusion of flour of minced fish from tilapia in chocolate cake (Figueiredo 2017), in loaf bread (Fukushima et al. 2011), and in dry pasta noodles (Reis 2013). The authors concluded that the studied foods

Fig. 1. Products prepared from waste from tilapia processing: fish sausage (emulsified product) (A); breaded fish (B); cooked ham (restructured fish) (C); fish mortadella (D); fried fish balls (E); fish sausage (embedded product) (F) made with minced fish; chocolate flavor cake (G); dry pasta noodles (H) enriched with minced fish flour (I); fish bouillon cubes from tilapia residues (J); fish hamburgers made with minced fish (K); and oil from different tilapia residues extracted by acid silage (L).

obtained satisfactory acceptance scores; therefore, the waste flour can be used in the enrichment of products for human consumption.

In the seasoning sector, broths were developed in the form of fish bouillon cubes based on solid residues, ground and pulped from tilapia (Fabrício et al. 2013). Based on the analyses carried out, the fish bouillon cubes prepared had excellent nutritional characteristics and stability. The extraction of oil from different tilapia wastes (head, viscera, carcass and whole fish still with scales) was performed through acid silage, even obtaining the flour as a remaining product (Oliveira 2015). The oils and flours obtained in the process were found to be of high quality, presenting applicability in human foodstuffs. An exception would be products obtained from tilapia viscera, which must be destined for animal feed.

Another project refers to the feasibility of extracting oil from different tilapia wastes for the production of biofuel (Inoue et al. 2016). Preliminary studies have shown that the oil extracted from the fish viscera showed a higher yield and consequently a better potential for obtaining this biofuel. Future research and studies already under development aim at expanding the possibilities of using discards. This would include the preparation of several industrialized meat products (restructured fish cooked ham, salami, sausage-embedded product, and hamburgers) with the addition of minced fish from tilapia, obtaining of gelatin extracted from the tilapia leather aiming at the development of an edible coating, process of obtaining fried tilapia skin, and extraction of concentrate and protein isolate from the tilapia scraps.

Although the approaches are different for the types of exploitation, there is a concern to significantly reduce waste. Based on this context, the applicability of processing technologies has become an attractive alternative for the use of discards generated along the production chain of tilapia, due to the commercial appreciation of the co-products obtained and the reduction, to practically zero, of the residual volume. It is perceived that these tests fulfill the commitment of environmental protection by satisfying the triad of sustainability, that is, they are economically viable, socially just and environmentally correct.

6. Marketing and sales: Current status, safety, quality, convenience, practicality, availability and sustainability as marketing strategies and their reflections on tilapia sales

Thinking about collaborating with the expansion of the fish market in Brazil, events such as "Sell your Fish" and "Fish Week" were created with the intention of expanding the fish market, as well as promoting actions aimed at the valorization of practicality in the preparation of fish and the emphasis on nutritional qualities and incentive to the marketing of fish and seafood in retail and *food service*. The "Eat more fish" campaign is also developed in the country to stimulate the increase of cultivated fish consumption. Booklets are released with tips on consumption and how to choose and store the fish properly (PeixeBR 2018, 2020, Hammitt 2020).

The consumer is increasingly demanding, looking for practical products that preserve the *in natura* characteristics, that are healthy, with higher added value, sustainable and provided by companies that value animal welfare and sustainability in the production chain (Oetterer et al. 2014d). In order to meet these needs, as well as stimulate fish consumption, the industry has invested in new technologies. The main innovations that have gained a large market niche are linked to convenience products that deal with ease of preparation and that present novelties regarding packaging (Han et al. 2018). The products with health appeal are considered as "natural", those without additives and preservatives and those that provide a source of omega-3. This is in addition to ethical products, which aim at environmentally friendly processing, as well as tracked products aimed at ensuring food quality and safety (Costa et al. 2020).

In this sense, companies have been preparing ready-to-eat products, such as complete meals with tilapia fillet, in which the preparation is only to place it in the microwave. The right selection of packaging materials and technologies maintains quality, freshness, color, and struture of the product during distribution and storage, as well aretains liquids and gases, maintaining the juiciness of the product when processed. Thus, the development of Skin Pack type packaging has been explored, in which the product is sealed in a stable bottom film or a pre-molded tray with a special skin, which combines a good presentation and a considerable increase in the shelf life of the fish, mainly in fractionated in natura products or ready-to-eat foods. (Camilo 2019).

In addition, food fortification is a strategy used in several countries to meet the nutrient deficiencies presented by the population. In this segment, omega-3 fatty acids are showing strong growth as one of the most popular functional ingredients, yet still low in consumption. Thus, industries have used the incorporation of omega-3 in product formulations, making them high cost and difficult for consumers to access. An alternative that has been applied is to offer tilapia feed rich in these fatty acids, in order to incorporate them into muscle tissues and thus obtain a raw material with more expressive content of polyunsaturated fatty acids of the omega-3 series, for the preparation of various products (Ayisi et al. 2017, Duarte 2017, Stoneham et al. 2018).

The application of less aggressive anesthetic agents is an appeal towards animal welfare by the consumer market, which increasingly demands a product with a seal of proof of this nature. Thus, following the line of natural products, plant extracts are fundamental in the advancement of research, since some essential oils that perform this function are already well established, such as eugenol and menthol compounds, coming mainly from clove oil (*Eugenia caryphollata*) and mint (*Mentha* sp.), respectively. Menthol should not be considered the anesthetic of choice for *tilapia* because it causes hyperglycemia, which causes major stress in this species. Thus, the research has published other essential oil options, such as Lúcia-lima (*Aloysia tryphila*), Brazilian citronella (*Lippia alba*), melaleuca (*Melaleuca alternifolia*), alfavaca (*Ocimum gratissimum*) and jambú (*Spilanthes acmella*), which present satisfactory results (Correia et al. 2018, Silva et al. 2019).

Finally, in view of the generation of residues from the fishing industries and as these residues, in general, have a high concentration of organic material, their release into water bodies may provide a decrease in the concentration of dissolved oxygen in this medium. In this sense, the fishing sector has been engaged in the zero waste emission proposals (ZERI – *Zero Emissions Research and Initiatives*), providing alternatives for the management of the waste that may be generated, becoming a differential factor for companies, ensuring the diversification of the product line, sustainable growth and socio-environmental responsibility (Pires et al. 2016, Costa et al. 2017).

Based on the context presented, the tilapia is recognized for the excellence of its meat due to its sensory and nutritional properties, so it has future expectations. Therefore, many technical aspects need to be followed as well as improved, aiming at animal health and, consequently, the quality and safety of the final product. There is also a concern to significantly reduce the waste from tilapia processing, highlighting the importance of different types of use, which generate new products with high added value. Such scenario is fundamental for the tilapia industry and the productive chain, since it provides all the necessary security to achieve better productivity, competitiveness, sustainability, and quality of the final product, which are directly related to the marketing and sales strategies.

Acknowledgments

The authors wish to thank the financial support provided by the Coordenação de Aperfeiçoamento de Pessoal de Nível Superior (CAPES, Brasília, DF, Brazil)-Finance Code 001, the Conselho Nacional de Desenvolvimento Científico e Tecnológico (CNPq, Brasília, DF, Brazil), and the Fundação de Amparo à Pesquisa do Estado de Minas Gerais (FAPEMIG, Belo Horizonte, MG, Brazil) for the research.

References cited

Abreu, I.L., R.M.E. Oliveira, T.C. Mesqueita, I.E. Nogueira, A.M.T. Lago and M.E.S.G. Pimenta. 2015. Utilização da carne mecanicamente separada de pescado proveniente de resíduos da filetagem de tilápia (*Oreochromis niloticus*) na fabricação de empanados. Rev. Hig. Alim. 29: 182–187.

Abud, E.J.M. 2019. Avaliação da textura de embutidos cárneos cozidos tipo mortadela de tilápia contendo carragena e goma guar. B.N. Monography, Universidade Federal de Lavras, Lavras, Brazil.

AGMRC (Agricultural Marketing Resource Center). 2018. Aquaculture Fin Fish Species: Important Aquaculture Fish Species. AGMRC, Ames, USA.

Almeida, P.R.M.B. 2018. Desempenho produtivo e qualidade da carne de tilápias do Nilo (*Oreochomis niloticus*) submetidas a dietas suplementadas com *Arthrospira platensis* (Spirulina) na fase de terminação. Ph.D. Thesis, Universidade Federal de Goiás, Goiânia, Brazil.

Amaral, A.B., M.V. Silva and S.C.S. Lannes. 2018. Lipid oxidation in meat: Mechanisms and protective factors – A review. Food Sci. Technol. 38: 1–15.

ATA (Americas Tilapia Alliance). 2020. Nutritional Info. ATA, Martinsville, USA.

Ayisi, C.L., J. Zhao and E.J. Rupia. 2017. Growth performance, feed utilization, body and fatty acid composition of Nile tilapia (*Oreochromis niloticus*) fed diets containing elevated levels of palm oil. Aquac. Fish. 2: 67–77.

Bath, R., A.K. Alias and G. Paliyath. Osmotic dehydration: Theory, methodologies, and applications in fish, seafood, and meat products. pp. 161–189. *In*: R. Bath, A.K. Alias and G. Paliyath (eds.). Progress in Food Preservation. John Wiley & Sons, Hoboken, USA.

Bernardi, D.C., E.T. Mársico and M.Q. Freitas. 2013. Quality Index Method (QIM) to assess the freshness and shelf life of fish. Braz. Arch. Biol. Technol. 56: 587–598.

Bolivar, A., J.C.C.P. Costa, G.D. Posada-Izquierdo, F. Pérez-Rodríguez, I. Bascón, G. Zurera, et al. 2017. Characterization of foodborne pathogens and spoilage bacteria in Mediterranean fish species and seafood products. pp. 21–39. *In*: O.V. Singh (ed.). Foodborne Pathogens and Antibiotic Resistance. John Wiley & Sons, Hoboken, USA.

Borgogno, M., Y. Husein, G. Secci, S. Masi and G. Parisi. 2017. Technological and nutritional advantages of mechanical separation process applied to three European aquacultured species. LWT-Food Sci. Technol. 84: 298–305.

Camilo, A.N. 2019. Embalagens de pescado seguem evoluindo no Brasil. Seafood Brasil, São Paulo, Brazil.

Castro, P.L., V. Lewandowski, M.L.R. Souza, M.F. Coradini, A.A.C. Alexandre, C. Sary, et al. 2017. Effect of different periods of pre-slaughter stress on the quality of the Nile tilapia meat. Food Sci. Technol. 37: 52–58.

Chandroo, K., I.J. Duncan and R. Moccia. 2014. Can fish suffer: Perspectives on sentience, pain, fear and stress. Appl. Anim. Behav. Sci. 86: 225–250.

Cheng, J., D.-W. Sun, Z. Han and X.-A. Zeng. 2014. Texture and structure measurements and analyses for evaluation of fish and fillet freshness quality: A review. Compr. Rev. Food Sci. Food Saf. 13: 52–61.

Contreras-Guzmán, E.S. 1994. Bioquímica de pescados e derivados. FUNEP, Jaboticabal, Brazil.

Contreras-Guzmán, E.S. 2002. Bioquímica de pescados e invertebrados. CECTA, Universidad de Santiago de Chile, Santiago.

Correia, A.M., A.S. Pedrazzani, R.C. Mendonça, A. Massucatto, R.A. Ozório and M.Y. Tsuzuki. 2018. Basil, tea tree and clove essential oils as analgesics and anaesthetics in *Amphiprion clarkii* (Bennett, 1830). Braz. J. Biol. 78: 436–442.

Costa, A.C.P.B., F.S. Macêdo and G. Honczar. 2020. Fatores que influenciam o consumo de alimentos, pp. 23–38. *In*: L.C. Moraes (ed.). Brasil Food Trends 2020. ITAL/FIESP, São Paulo, Brazil.

Costa, F.E.S., G.F. Bitencourt, S.G.S. Moraes and P.P. Mendonça. 2017. Aproveitamento de resíduos de pescado, pp. 348–365. *In*: U.R. Vianna, J.O. Carvalho and J.R. Carvalho (eds.). Tópicos especiais em Ciência Animal VI. UNICOPY, Alegre, Brazil.

Cribb, A.Y., J.T. Seixas Filho and S.C.R.P. Mello. 2018. Manual técnico de manipulação e conservação do pescado. Embrapa, Brasília, Brazil.

Duarte, F. 2017. Caracterização da carne de tilápia do Nilo (*Oreochromis niloticus*) submetida a dietas suplementadas com óleo de peixe. M.Sc. Dissertation, Universidade Federal de Goiás, Goiânia, Brazil.

Elgadir, M.A., A.A. Al-Hassan, M.Z.I. Sarker and M.J.H. Akanda. 2017. Shelf life extension of various types of fish meat using selected modified atmosphere packaging (MAP) methods, review. Int. J. Food Nut. Sci. 6: 89–97.

Fabricio, L.F.F., M.E.S.G. Pimenta, T.A. Reis, T.C. Mesquita, K.L. Fukushima, R.M.E. Oliveira, et al.. 2013. Elaboração de caldo de peixe em cubos compactados utilizando pirambeba (*Serrasalmus brandtii*) e tilápia (*Oreochromis niloticus*). Semina: Ciênc. Agrár. 34: 241–252.

FAO (Food and Agriculture Organization). 1999. Guidelines for the sensory evaluation of fish and shellfish in laboratories (CAC/GL 31-1999). FAO/WHO, Rome, Italy.

FAO (Food and Agriculture Organization). 2012. Code of practice for fish and fishery products. FAO/WHO, Rome, Italy.

FAO (Food and Agriculture Organization). 2018. Standard for smoked fish, smoke-flavoured fish and smoke-dried fish. FAO/WHO, Rome, Italy.

FAO (Food and Agriculture Organization). 2020. The State of World Fisheries and Aquaculture (SOFIA) – Sustainability in action. FAO/WHO, Rome, Italy.

FDA (Food and Drug Administration). 2020. Fish and Fishery Products Hazards and Controls Guidance. FDA, Gainesville, USA.

Feddern, V., H. Mazzuco, F.N. Fonseca and G.J.M.M. Lima. 2019. A review on biogenic amines in food and feed: Toxicological aspects, impact on health and control measures. Anim. Prod. Sci. 59: 608–618.

Ferreira, N.A., R.V. Araújo and E.C. Campos. 2018. Boas práticas no pré-abate e abate de pescado. Pubvet 12: 1–14.

Figueiredo, A.F. 2017. Bolo sabor chocolate enriquecido com farinha de carne mecanicamente separada de tilápia do Nilo (*Orechromis niloticus*). B.N. Monography, Universidade Federal de Lavras, Lavras, Brazil.

Françoise, L. and J.-J. Joffraud. 2011. Microbial degradation of seafood, vol. 2, pp. 47–72. *In*: D. Montet and R.C. Ray (eds.). Aquaculture Microbiology and Biotechnology. CRC Press/Taylor & Francis Group, Enfield, USA.

Freire, C.E.C. and A.A. Gonçalves. 2013. Diferentes métodos de abate do pescado produzido em aquicultura, qualidade da carne e bem estar do animal. Holos 6: 33–41.

Freitas, D.G.C. and G.V. Amaral. 2011. Método do Índice de Qualidade (MIQ) para a Avaliação Sensorial da Qualidade de Pescado. Embrapa Agroindústria de Alimentos, Rio de Janeiro, Brazil (Documentos 112, Embrapa).

Fukushima, K.L., L.M. Torres, R.S. Leal, M.E.S.G. Pimenta, B.V. Laurenti and T.O. Oliveira. 2011. Aplicação da farinha do resíduo da filetagem de tilápia na fabricação de pães de forma – Análise sensorial. Rev. Hig. Alim. 25: 1331–1332.

Fukushima, K.L., R.M.E. Oliveira, M.E.S.G. Pimenta, R.B.S. Oliveira, T.A. Reis and A.M.T. Lago. 2014. Características químicas, microbiológicas e sensoriais de empanados formulados à base de polpa de tilápia do Nilo (*Oreochromis niloticus*). Rev. Hig. Alim. 28: 181–186.

Galvão, J.A., M. Oetterer and A. Matthiensen. 2014. Sustentabilidade na produção do pescado: Qualidade da água. pp. 1–30. *In*: J.A. Galvão and M. Oetterer (eds.). Qualidade e processamento de pescado. Elsevier, São Paulo, Brazil.

Gokoglu, N. and P. Yerlikaya. 2015. Seafood Chilling, Refrigeration and Freezing: Science and Technology. Wiley-Blackwell, Hoboken, USA.

Gonçalves, A.A. 2011. Tecnologia do pescado: Ciência, tecnologia, inovação e legislação. Atheneu, São Paulo, Brazil.

Hammitt, E. 2020. Globefish: Fish Health Campaigns. FAO/WHO, Rome, Italy.

Han, J.-W., L. Ruiz-Garcia, J.-P. Qian and X.-T. Yang. 2018. Food packaging: A comprehensive review and future trends. Compr. Rev. Food Sci. Food Saf. 17: 860–877.

Inoue, M.H., A.M.T. Lago and M.E.S.G. Pimenta.2016. Viabilidade técnico-econômica de biodisel elaborado com óleo proveniente de resíduos da filetagem de tilápia. Cong. Rede Bras. Tecnol. Biodiesel, Brasil 6: 110.

Kubitza, F. 2016. Common salt a useful tool in aquaculture – Part 1. Global Aquaculture, Portsmouth, USA.

Kumar, K.A., K.R. Sreelakshmi, K. Elavarasan and C. O. Mohan. 2018. Training Manual on seafood value addition. ICAR, Kerala, India.

Lago, A.M.T., A.C.C. Vidal, M.C.E.V. Schiassi, T.A. Reis, C.J. Pimenta and M.E.S.G. Pimenta. 2017. Influence of the addition of minced fish on the preparation of fish sausage: Effects on sensory properties. J. Food. Sci. 82: 492–499.

Lago, A.M.T., M.E.S.G. Pimenta, I.E. Aoki, A.F. Figueiredo, M.C.E.V. Schiassi and C.J. Pimenta. 2018. Fish sausages prepared with inclusion of Nile tilapia minced: Correlation between nutritional, chemical, and physical properties. J. Food. Process. Preserv. 42: e13716.

Lago, A.M.T., J.T. Teixeira, B.J.G. Olimpio, M.C.E.V. Schiassi, C.J. Pimenta and M.E.S. Gomes. 2019. Shelf life determination of frozen fish sausage produced with fillet and minced fish derived from the Nile tilapia processing. J. Food. Process. Preserv. 43: e13984.

Maciel, E.D.S., J.S. Vasconcelos, L.K. Savay-Da-Silva, J.A. Galvão, J.G. Sonati., J.C. Christofoletti, et al. 2013. Label designing for minimally processed tilapia aiming the traceability of the productive chain. Bol. Cent. Pesqui. Process. Aliment. 30: 157–168.

Martinsdóttir, E., K. Sveinsdóttir, J. Luten, R. Schelvis-Smit and G. Hyldig. 2001. Reference manual for the fish sector: Sensory evaluation of fish freshness. QIM Eurofish, Ijmuiden, Netherlands.

Ng, W.-K. and N. Romano. 2013. A review of the nutrition and feeding management of farmed tilapia throughout the culture cycle. Rev. Aquacul. 5: 220–254.

Nollet, L.M.L. and F. Tóldra. 2010. Handbook of seafood and seafood products analysis. CRC Press/Taylor & Francis Group, Boca Raton, USA.

Ocaño-Higuera, V.M., A.N. Maeda-Martínez, E. Marquez-Ríos, D.F. Canizales-Rodríguez, F.J. Castillo-Yáñez, E. Ruíz-Bustos, et al. 2011. Freshness assessment of raw fish stored in ice by biochemical, chemical and physical methods. Food Chem. 125: 49–54.

Oetterer, M., J.A. Galvão and L.K. Savay-da-Silva. 2014a. Qualidade do pescado: Sistema para padronização. pp. 31–71. *In*: J.A. Galvão and M. Oetterer (eds.). Qualidade e processamento de pescado. Elsevier, Rio de Janeiro, Brazil.

Oetterer, M., J.A. Galvão and L.K. Savay-da-Silva. 2014b. Tilápia: Controle de qualidade, beneficiamento e industrialização. Tilápia minimamente processada. pp. 183–206. *In*: J.A. Galvão and M. Oetterer (eds.). Qualidade e processamento de pescado. Elsevier, Rio de Janeiro, Brazil.

Oetterer, M., J.A. Galvão and L.K. Savay-da-Silva. 2014c. Tilápia: Controle de qualidade, beneficiamento e industrialização. Radiações Ionizantes e Defumação. pp. 211–237. *In*: J.A. Galvão and M. Oetterer (eds.). Qualidade e processamento de pescado. Elsevier, Rio de Janeiro, Brazil.

Oetterer, M., J.A. Galvão and L.F.A. Sucasas. 2014d. Sustentabilidade na cadeia produtiva do pescado: Aproveitamento de resíduos. pp. 97–118. *In*: J.A. Galvão and M. Oetterer (eds.). Qualidade e processamento de pescado. Elsevier, Rio de Janeiro, Brazil.

Ogawa, M. and E.L. Maia. 1999. Manual de pesca: Ciência e tecnologia do pescado. Varela, São Paulo, Brazil.

Oliveira, J.R., J.L. Carmo, K.K.C. Oliveira and M.C.F. Soares. 2009. Cloreto de sódio, benzocaína e óleo de cravo-da-índia na água de transporte de tilápia-do-nilo. Rev. Bras. Zootec. 38: 1163–1169.

Oliveira, R.M.E. 2015. Caracterização de óleos e farinhas, obtidos da silagem ácida de resíduos da filetagem de tilápia (*Oreochromis niloticus*). Ph.D. Thesis, Universidade Federal de Lavras, Lavras, Brazil.

Olsen, R.L., J. Toppe and I. Karunasagar. 2014. Challenges and realistic opportunities in the use of by-products from processing of fish and shellfish. Trends Food Sci. Tech. 36: 144–151.

Ordóñez, J.A., M.I.C. Rodriguez, L.F. Álvarez, M.L.G. Sanz, G.D.G.F. Minguillón, L.H. Peralesand, et al. 2005. Tecnologia de alimentos: Alimentos de origem de animal. Artmed, Porto Alegre, Brazil.

Pedrazzani, A.S., P.C.F. Carneiro, P.G. Kirschnik and C.F.M. Molento. 2009. Impacto negativo de secção de medula e termonarcose no bem-estar e na qualidade da carne da tilápia-do-Nilo. Rev. Bras. Saúde Prod. Anim. 10: 188–197.

Peixe BR. Associação Brasileira da Piscicultura. 2018. Peixe BR lança campanha para aumento do consumo de peixes de cultivo no país. Peixe BR, São Paulo, Brazil.

Peixe BR. Associação Brasileira da Piscicultura. 2020. Coma mais peixe. Peixe BR, São Paulo, Brazil.

Pires, D.R., A.C.N. Morais, J.F. Costa, L.C.D.S.A. Góes and G.M. Oliveira. 2014. Aproveitamento do resíduo comestível do pescado: Aplicação e viabilidade. Rev. Verde Agroecologia Desenvolv. Sustent. 9: 34–46.

Poli, B.M., G. Parisi, F. Scappiniand and G. Zampacavallo. 2005. Fish welfare and quality as affected by preslaughter and slaughter management. Aquac. Int. 13: 29–49.

Reis, T.A. 2013. Caracterização de macarrão massa seca enriquecido com farinha de polpa de pescado. M.Sc. Dissertation, Universidade Federal de Lavras, Lavras, Brazil.

Reitznerová, A., M. Šuleková, J. Nagy, S. Marcinčák, B. Semjon, M. Čertík, et al. 2017. Lipid peroxidation process in meat and meat products: A comparison study of malondialdehyde determination between modified 2-thiobarbituric acid spectrophotometric method and reverse-phase high-performance liquid chromatography. Molecules 22: 1988.

Santos, E. 2018. Brazil starts the biggest research project ever elaborated to develop aquaculture. Embrapa Fishery and Aquaculture, Brasília, Brazil.

Savay-da-Silva, L.K. 2009. Desenvolvimento do produto de conveniência: Tilápia (*Oreochromis niloticus*) refrigerada minimamente processada embalada a vácuo – padronização para a rastreabilidade. LG, Piracicaba, Brazil.

Schneider, T.L. 2019. Influência da depuração na qualidade do filé de tilápia do Nilo (*Oreochomis niloticus*), sob os aspectos físico-químicos, microbiológicos e sensoriais. M.Sc. Dissertation, Universidade Federal do Paraná, Palotina, Brazil.

Secci, G. and G. Parisi. 2016. From farm to fork: Lipid oxidation in fish products: A review. Ital. J. Anim. Sci. 15: 124–136.

Secci, G., M. Borgogno, S. Mancini, G. Paci and G. Parisi. 2017. Mechanical separation process for the value enhancement of Atlantic horse mackerel (Trachurus trachurus), a discard fish. Innov. Food Sci. Emerg. Technol. 39: 13–18.

Shahidi, F. and J.R. Botta. 2012. Seafoods: Chemistry, Processing Technology and Quality. Springer, New York, USA.

Silva, E., G. Deschamps, A. Jatobá, D.D. Schleder, A.L. Preto, F.G. Carvalho, et al. 2019. Uso de óleos essenciais como alternativa para anestésicos sintéticos na aquicultura. Aquac. Bras. 11: 17–21.

Soares, K.M.P. and A.A. Gonçalves. 2012. Seafood quality and safety. Rev. Inst. Adolfo Lutz 71: 1–10.

Souza, S.M.G., V.D. Mathies and R.F. Fioravanzo. 2012. Off-flavour por geosmina e 2-metilisoborneol na aquicultura. 2012. Semina 33: 835–846.

Stoneham, T.R., D.D. Kuhn, D.P. Taylor, A.P. Neilson, S.A. Smith, D.M. Gatlin, et al. 2018. Production of omega-3 enriched tilapia through the dietary use of algae meal or fish oil: Improved nutrient value of fillet and offal. Plos One 13: e0194241.

Tacon, A.G.J. and M. Metian. 2013. Fish matters: Importance of aquatic foods in human nutrition and global food supply. Rev. Fish. Sci. 21: 22–38.

Thorn, R.M.S. and J. Greenman. 2012. Microbial volatile compounds in health and disease conditions. J. Breath Res. 6: 024001.

Vidal, A.C.C. 2016. Embutido cárneo cozido tipo mortadela elaborado com carne mecanicamente separada de tilápia: Características físicas e sensoriais. B.E. Monography, Universidade Federal de Lavras, Lavras, Brazil.

Vidotti, R.M. 2011. Processamento e aproveitamento integral de tilápias. pp. 205–245. *In:* L.M.S. Ayroza (ed.). Piscicultura: Manual Técnico. CATI, Campinas, Brazil.

Viegas, E.M.M., F.A. Pimenta, T.C. Previero, L.U. Gonçalves, J.P. Durães, M.A.R. Ribeiro, et al. 2012. Métodos de abate e qualidade da carne de peixe. Arch. Zootec. 61: 41–50.

Vieira, N.B. 2019. Avaliação química e físico-química de embutidos cárneos cozidos tipo mortadela elaborados com filé, carne mecanicamente separada do Nilo (*Oreochromis niloticus*) e caseína. B.E. Monography, Universidade Federal de Lavras, Lavras, Brazil.

WordFish. 2017. FISH: CGIAR Research Program on fish agrifood systems. CGIAR, Penang, Malaysia.

Zanutto, L.D. 2017. Avaliação da textura de embutidos cárneos cozidos tipo mortadela elaborados com filé, carne mecanicamente separada de tilápia do Nilo (*Orechromis niloticus*) e carragena. 2017. B.E. Monography, Universidade Federal de Lavras, Lavras, Brazil.

Zapata, K.P., L.O. Brito, P.C.M. Lima, L.A. Vinatea, A.O. Galvez and J.M.V. Cárdenas. 2017. Cultivo de alevinos de tilápia em sistema de bioflocos sob diferentes relações carbono/ nitrogênio. Bol. Inst. Pesca 43: 399–407.

Zhao, Y.-M., M. Alba, D.-W. Sun and B. Tiwari. 2019. Principles and recent applications of novel non-thermal processing technologies for the fish industry – A review. Crit. Rev. Food Sci. Nutr. 59: 728–742.

Zumas, A.A.R. 2017. Utilização de carne mecanicamente separada de tilápiasoreochromisniloticus na produção de quibes. B.E. Monography, Universidade Federal de Lavras, Lavras, Brazil.

Index

Milton Keynes UK
Ingram Content Group UK Ltd.
UKHW022037141024
449569UK00014B/641